TP 1089

Deep Foundation Improvements: Design, Construction, and Testing

Melvin I. Esrig and Robert C. Bachus, editors

ASTM
1916 Race Street
Philadelphia, PA 19103

Library of Congress Cataloging-in-Publication Data

Deep Foundation Improvements: Design, Construction, and Testing
Melvin I. Esrig and Robert C. Bachus, editors.
(STP ; 1089)

"ASTM publication code (PCN) 04-010890-38. "
Includes bibliographical references and index.
ISBN 0-8031-1392-7
1. Foundations. 2. Columns—Foundations. 3. Soil mechanics.
4. Soil stabilization. I. Esrig, Melvin I. II. Bachus, Robert C.
(Robert Charles). 1951- III. Series: ASTM special technical
publication ; 1089.
TA775,D45 1991
624. 1'5—dc20   91-12795
                     CIP

Photocopy Rights

Authorization to photocopy items for internal or personal use, or the internal or person:
use of specific clients, is granted by the AMERICAN SOCIETY FOR TESTING AND MATERIAL
for users registered with the Copyright Clearance Center (CCC) Transactional Reportin
Service, provided that the base fee of $2.50 per copy, plus $0.50 per page is paid directl
to CCC, 27 Congress St., Salem, MA 01970; (508) 744-3350. For those organizations tha
have been granted a photocopy license by CCC, a separate system of payment has bee
arranged. The fee code for users of the Transactional Reporting Service is 0–8031–1416–
/91 $2.50 + .50.

Peer Review Policy

Each paper published in this volume was evaluated by three peer reviewers. The autho:
addressed all of the reviewer's comments to the satisfaction of both the technical editor(:
and the ASTM Committee on Publications.

The quality of the papers in this publication reflects not only the obvious efforts of th
authors and the technical editor(s), but also the work of these peer reviewers. The ASTV
Committee on Publications acknowledges with appreciation their dedication and contributio
to time and effort on behalf of ASTM.

Printed in Baltimore, MD
June 1991

Foreword

This publication, *Deep Foundation Improvements: Design, Construction, and Testing*, contains papers presented at the symposium on Design, Construction, and Testing of Deep Foundation Improvement: Stone Columns and Other Related Techniques, held in Las Vegas, NV on 25 January 1990. The symposium was sponsored by ASTM Committee D-18 on Soil and Rock. Dr. Melvin I. Esrig of Woodward-Clyde Consultants in Wayne, NJ and Dr. Robert C. Bachus of Geosyntec Consultants in Norcross, GA presided as symposium chairmen and are editors of the resulting publication.

Contents

OVERVIEW

The Symposium on Design, Construction and Testing of Deep Foundation Improvements was held on January 25, 1990 in Las Vegas, Nevada. The focus of the symposium was to be the improvement of deep foundation conditions through the construction of stone columns and/or similar inclusions. In fact, a broader perspective was provided by the authors, enhancing the value of the Symposium and the papers included herein.

The twenty-two papers presented at the Symposium and included in this publication are grouped into two main categories: those related to deep foundation improvements through the formation of "composite ground" and those related to improvement through "compaction and densification." Composite ground is formed by the introduction of columns of sand or stone through a "weak" deposit of sand, silt or clay or by the introduction of cement grout or chemical grout into a soil mass through an injection or mixing process. Compaction and densification techniques are used for fill or natural materials and do not involve the introduction of other materials. Nineteen of the papers address some aspect of composite ground. The remaining three are on compaction and densification techniques, specifically, deep vibratory compaction of loose sand deposits. For this publication the papers are subdivided as follows: (i) five papers on construction and analysis of composite ground; (ii) two papers on construction guidelines and specifications; (iii) eight papers on case history presentations; (iv) four papers on introducing grout to the soil; and (v) three papers on compaction and densification techniques.

The papers by Barksdale and Takefumi, Enoki et al., Aboshi et al., and Terashi et al., provide information on the design, construction, testing and performance of ground stabilized by sand piles or sand compaction piles. These four papers, combined with the paper by Priebe in which design criteria for stone columns are provided, present important information on analysis of the stability of composite ground. It is quite clear that a consensus has not been reached on the best approach for analyzing composite ground and, therefore, the presentation in this publication of diverse approaches is believed by the editors to be of considerable value to practitioners.

Stark and Yacyshyn suggest guidelines for performance specifications for the construction and testing of stone columns in cohesive soils. These were developed from a critical review of specifications used in U.S. and British practice. Additional information on the performance of equipment during installation of stone columns as well as on load testing techniques is provided by Slocombe and Moseley. This latter contribution is based on British practice. Taken together, these papers are likely to be of help to engineers seeking to specify the installation of stone columns.

The eight papers that follow are case histories of the use of stone columns in several engineering applications in the United States and the United Kingdom. Papers by Allen et al., Hayden and Welch, and Goughnour, Sung and Ramsey describe how stone columns were used for natural ground and for slope stabilization. The papers by Davie, et al. and Snethen and Homan report on the stabilization of waste materials in areas once associated with coal mining. Of more-than-passing interest is the report by Greenwood of the full-scale loading response of ground improved by stone columns. Greenwood's observations are significant for those who design stone column installations and specify load tests or field instrumentation. Additional comments on the types of commonly referenced load tests are provided by Watts and Charles. Hussin and Baez present the procedures and results of short term load tests conducted on several recent U.S. stone column projects. In all, a wide range of valuable experience is provided in these papers.

Four papers follow in which the focus is on the introduction of grout into soil for purposes of ground modification and stabilization. Seismic testing methods to evaluate ground improvement from compaction grouting are described and evaluated by Byle, et al. Gambin describes a new, more scientific approach to compaction grouting which has been developed in France and suggests that the approach be applied in future practice. Munfakh provides a case study that led to the use of chemical stabilization to protect an old tunnel from the effects of constructing two new tunnels within 2m of its invert. Babasaki et al. describe the use of deep cement mixing in Japan to provide the foundation for a three-story building on loose silty sand and to avoid liquefaction in an earthquake.

The last three papers deal with the compaction of loose granular soils by vibratory means. Castelli reports on the relative effectiveness of several methods of deep compaction. One of them is the vibro wing system, also reported on by Massarsch, in which the vibration frequency can be varied during

construction. The benefits for compaction using a Y-
shaped probe are described by Neely and Leroy. These
three papers provide valuable information to the
profession on the new tools available for compaction of
deep soil deposits and on the results that can be
expected. Furthermore, they describe the effectiveness
of the methods and the variability of results.
Massarsch cautions that the high energies associated
with deep compaction may potentially densify sands
beneath existing buildings near construction sites
causing settlements.

The editors are grateful to the authors for their
efforts in preparing the excellent papers included in
this publication. The papers in this volume provide a
valuable contribution to the literature and effectively
demonstrate the advances that have been achieved since
publication of proceedings from a similar Syumposium
sponsored by the Institution of Civil Engineers in
1976. The authors have helped provide to the
profession significant information on which to base
current decisions on deep foundation improvement and on
which to build for future developments.

Robert C. Bachus
Geosyntec Consultants
Norcross, GA.

Melvin I. Esrig
Woodward-Clyde Consultants
Wayne, New Jersey

Richard D. Barksdale and Tahara Takefumi

DESIGN, CONSTRUCTION AND TESTING OF SAND COMPACTION PILES

REFERENCE: Barksdale, R. D., and Takefumi, T., "Design, Construction, and Testing of Sand Compaction Piles," _Deep Foundation Improvements: Design, Construction, and Testing_, _ASTM STP 1089_, Melvin I. Esrig and Robert C. Bachus, Eds., American Society for Testing and Materials, Philadelphia, 1991.

ABSTRACT: Sand compaction piles can be used to improve marginal sites for stability, liquefaction, and settlement applications. They have been employed extensively in Japan for many years to improve land reclaimed from the sea. The advantages and disadvantages of using sand compaction piles are compared with other vibro-compaction techniques such as stone columns. Methods are described for construction of sand compaction piles on land and over water. Design theories are summarized for the utilization of sand compaction piles at sites underlain by both cohesionless and cohesive soils. Practices and design criteria are presented.

KEYWORDS: Sand Compaction Piles, Mammoth Compaction Piles, Vibro-Composer, Sand Densification, Sand Vibration, Construction, Design

INTRODUCTION

This paper describes the state-of-the-art of sand compaction pile design, construction, and utilization in Japan. Sand compaction piles are used extensively primarily to prevent stability failure, decrease the time of consolidation and prevent liquefaction. They are also used to reduce settlement although this appears usually to be a secondary objective since preloading is generally utilized. Sand compaction piles are also employed to a much lessor extent in Taiwan, Korea and China. This method, however, has not been used in the United States where stone columns have gained popularity. Stone columns are constructed using an open-graded stone while sand compaction piles are constructed using sand. In Japan over 60,960,000 m ($2x10^8$ ft.) of sand compaction piles and sand drains were constructed during about the past 25 years by one construction company. Over 90 percent of the sand

Dr. Barksdale is a professor at the Georgia Institute of Technology, School of Civil Engineering, Atlanta, GA 30332. Tahara Takefumi is an engineer with the Okinawa government.

compaction piles are used in Japan to support stockpiles of heavy
materials, tanks, embankments for roads, railways, and harbor
structures. Only 4 percent of sand compaction piles are used to
support buildings and warehouses. One of the most important uses of
this technique is in improving land reclaimed from the sea.

The primary advantages of sand compaction piles compared to
conventionally constructed stone columns are: (1) Construction of the
sand column is extremely fast; (2) The column is constructed of sand
which is usually considerably cheaper than stone; (3) The hole is fully
supported by a casing during construction that eliminates the
possibility of hole collapse. Also, the possibility of intrusion
and/or erosion of the soil surrounding the completed sand column is
significantly reduced compared with stone columns. Whether movement of
soil occurs due to erosion depends upon the gradation of both the sand
compaction pile and the surrounding soil and also the existing flow
conditions into the column. Use of a bottom feed type of stone column
[1] which can construct sand columns removes most of the advantages of
sand compaction piles except rapid construction.

The primary disadvantages of sand compaction piles are: (1)
Because of the use of sand, the column has a lower angle of internal
friction and stiffness than stone columns; hence, in general, a larger
percentage replacement of weak soil is required using sand compaction
piles to provide the same level of improvement as stone columns; (2)
Driving the casing through a clay layer causes "smear" along the
boundary of the column that reduces lateral permeability and hence its
effectiveness as a drain. The level of smear caused by the
construction of a sand column is greater than for constructing a stone
column by the vibro-replacement (wet) method. Even then an important
decrease in time for primary consolidation occurs when sand compaction
piles are used; (3) Following the design criteria of Seed and Booker
[2], sand compaction piles do not have a sufficiently high permeability
to function as vertical drains during earthquakes whereas properly
designed stone columns can. Sand compaction piles do provide shear
resistance of the column during an earthquake. Also, in sands the
vibration applied during construction densities surrounding loose sands
which also contribute to earthquake resistance.

SAND COMPACTION PILE CONSTRUCTION

Sand compaction piles are usually constructed by driving a pipe
having a special end restriction through a loose to firm sand or a very
soft to firm silt or clay stratum using a vibrator located at the top
of the casing. During or immediately after driving, the pipe is filled
with sand. The native sand is then densified by repeatedly raising and
lowering the vibrating pipe as it is withdrawn from the ground.
Mammoth compaction piles are similar to sand compaction piles except
they are usually larger and constructed over water using larger
equipment (Table 1). Strong sand piles are installed using the same
procedure as for sand compaction piles but are further densified using
a horizontal vibrator at the bottom of the casing. Sand compaction
piles and mammoth compaction piles are the most commonly used sand pile
techniques for improving poor sites. Due to equipment limitations,
mammoth compaction piles are constructed to a maximum depth of 35 m
(115 ft.) on land and 65 m (210 ft.) offshore. Because of the 65 m

TABLE 1 -- Design Conditions for Sand and Mammoth Compaction Piles.

Factor	Sand Compaction Piles	Mammoth Compaction Method
Site	Land, Offshore	Land, Offshore
SPT N-Value	5-10	0-10
Replacement Ratio, a_s	≤ 0.3 - 0.5	≥ 0.5 - 0.8
Sand Column Diameter	700 - 800 mm φ (on land) 1200 - 2000 mm φ (offshore)	700 - 800 mm φ (on land) 1200 - 2000 mm φ (offshore)
Steel Casing Diameter	600 - 1000 mm φ	800 - 1300 mm φ

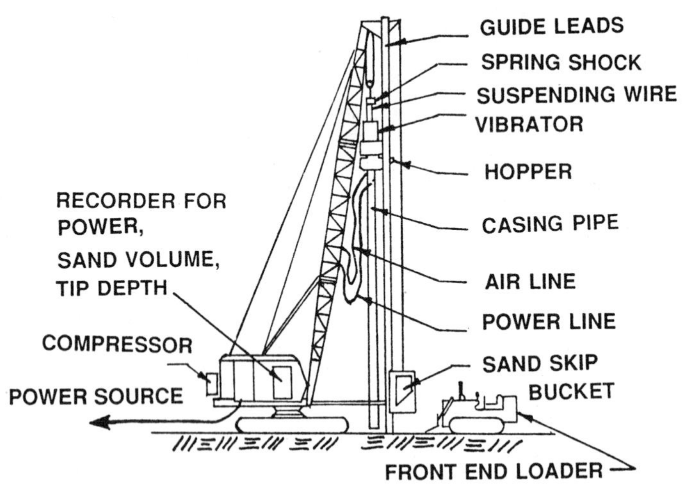

Figure 1. Typical Sand Compaction Pile Equipment (After Tanimoto, Reference 7).

(210 ft.) high structure required, vessels employed to construct
mammoth compaction piles are susceptible to damage from typhoons and
are also hard to navigate.

Sand Compaction Piles: For constructing sand compaction piles, a
40 kN to 53 kN (4.5 - 6-ton) hydraulic or electric vibrator is attached
to the top of a 400 to 600 mm (16 to 24 in.) diameter steel pipe (Fig.
1). The pipe casing is slightly longer than the desired length of the
sand compaction pile so as to protrude out of the ground after reaching
the design depth. The pipe fully supports the surrounding soil at all
times during construction. Water jets having pressures up to 85 kg/cm^2
(1200 psi) are sometimes used on the outside of the pipe when layers
are encountered that have a SPT N-value greater than about 15 to 20.

The casing with attached vibratory hammer is suspended from a
crawler crane and is guided by leads. A 310 to 355 kN (35 - 40 ton)
crawler crane is used for constructing 20 m (65 ft.) long sand
compaction piles. A coil spring shock absorber is fastened to the top
of the vibrating hammer to dampen the shock as the casing is pulled
from the ground by the crane. During driving, the cable from the crane
to the casing is kept slack so that the pipe-vibrator assembly is free
floating. Proper pipe alignment is maintained by a guide attached to
the vibrator that moves up and down the crane leads.

A low frequency, high amplitude vertical vibrator is used having a
frequency of 500 to 600 cpm and amplitude during idling of 15 to 18 mm
(0.6 to 0.7 in.). The amplitude is defined as one-half the total tip
movement. The commonly used vibrators are driven by 90 to 120 kW (120
to 160 hp) motors and have unbalanced forces varying from 400 to 600 kN
(90,000 to 135,000 lb.). Large units having greater than 90 kW (120
hp) vibrators are usually used for only the construction of mammoth
compaction piles over water.

Methods used to prevent soil from entering the lower end of the
casing during driving include a finger end, restricted open end and
bucket end plate [3]. To minimize construction time, the steel casing
is usually filled with sand as it is being driven down so that
extraction of the casing can begin immediately upon reaching the
required depth. After filling the skip with a front end loader, the
sand is mechanically lifted and dumped into the hopper located at the
top of the pipe as the pipe is being vibrated down. Using this
efficient process, a 15 m (50 ft.) sand compaction pile is constructed
in about 20 minutes. Approximately 45 sand compaction piles per day
can be constructed using 6 m (20 ft.) long piles.

Upon filling the casing with sand, 3 to 5 kg/cm^2 (40-70 psi) of
air pressure is applied to the top of the sand column. To develop the
required air pressure on top of the sand column, a pressure of about 7
kg/cm^2 (100 psi) is developed at the air compressor. A special valve
is employed to keep the pressurized sand hopper sealed from the
atmosphere. A sand compaction pile is constructed using a stroking
motion of the casing as it is withdrawn (Fig. 2). The casing is first
pulled up 2 to 3 m (6 to 10 ft.) using the crane and then vibrated back
down 1 to 2 m (3 to 7 ft.). This up and down stroking motion is
repeated until the casing has been completely withdrawn from the
ground.

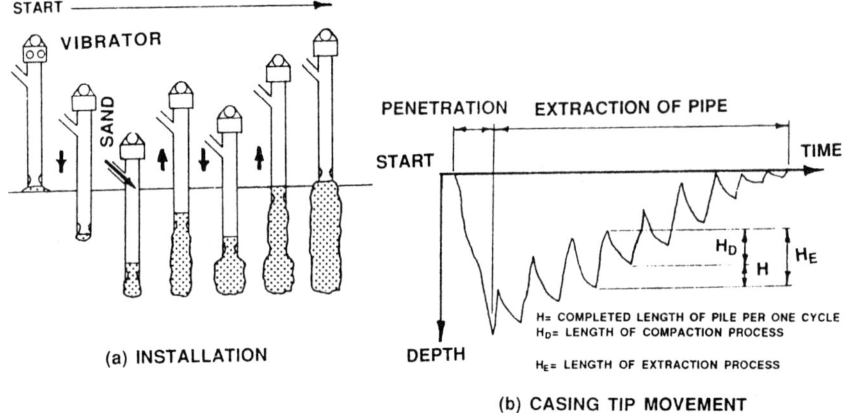

Figure 2.   Construction Sequence for Sand Compaction Piles (after Tanimoto, Reference 7).

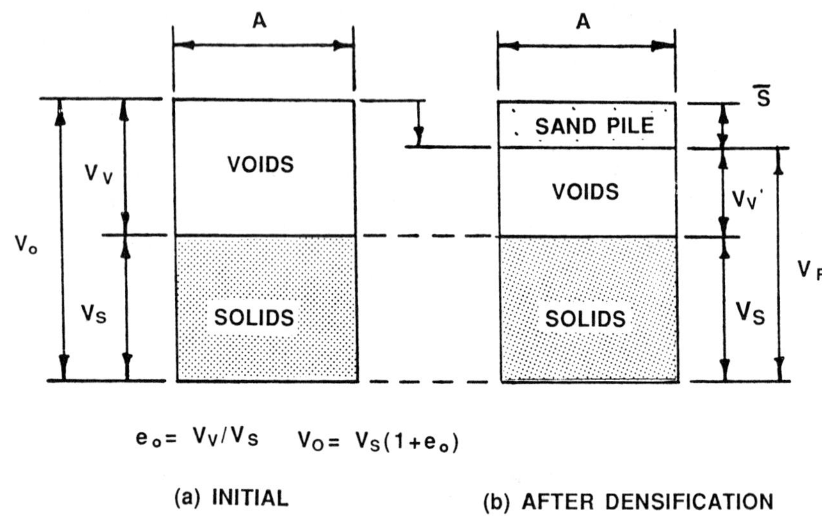

Figure 3.   Volume Block Diagram of In-Situ Sand Before and After Sand Pile Construction.

<u>Strong Sand Piles</u>:   The strong sand pile method, developed in 1973, is used in soft silts and clays in the same way as sand compaction piles. A horizontal vibrator is placed in the bottom of the casing in addition to the vertical vibrator located at the top. Because of horizontal vibration, a higher degree of densification of the sand pile can be obtained compared to sand compaction piles [4]. Adding water to the sand in this method helps to achieve a higher density during vibration allowing the sand to detach from the casing smoothly. After filling the casing with sand by a front-end loader (on land) or a belt conveyor from a barge (over water), the inlet is closed, and pressurized air is applied to the top of the sand pile. The vibrator at the bottom of the casing is then actuated causing a horizontal and circular vibration. As the casing is gradually extracted, the lids at the bottom of the casing open outward and compacted sand comes out forming the pile. The horizontal vibrator operates continuously during extraction. At selected intervals extraction of the casing may be stopped to achieve additional densification of the sand.

<u>Mammoth Compaction Piles</u>: The stabilization of very soft ocean sediments with mammoth compaction piles is accomplished using materials and construction techniques generally similar to sand compaction piles constructed on land. From two to four mammoth compaction piles are constructed simultaneously from a large barge. Each casing is driven to the required tip elevation using a large vertical vibrator. The vibrator is mounted at the top of the casing and typically driven by a 120 to 300 kW (160 to 400 hp) motor. In very soft sediments the casings are sometimes pulled down by a cable rather than using the vertical vibrator.

<u>General Considerations</u>: Offshore projects having sand compaction piles are frequently constructed in very soft cohesive soil. Sand compaction piles cause displacement of soft cohesive soils resulting in surface heave. Proper treatment of this heaved soil must be implemented to minimize sea water pollution which is caused by erosion of the heaved soil which is in a remolded state. The height of heaving can be estimated by the simple equations proposed by Sogabe [5]. A reduction of shear strength of soil occurs due to the disturbance and heaving from sand compaction pile construction. The disturbed soil usually takes about 30 to 90 days after construction to approach its initial shear strength. The diameter and density of sand compaction piles are varied by changing the casing diameter, extraction rate, compressed air pressure, time of vibration and the size of the vibrator. During construction the diameter of the sand pile, volume of sand supplied and discharged, casing tip elevation, and the power (amps) used by the vibrator are constantly recorded automatically.

DESIGN THEORY FOR SAND COMPACTION PILES

The volume of loose sand or soft clay replaced by sand is one of the most important factors in improving weak ground using sand compaction piles, strong sand piles or mammoth compaction piles (and also stone columns). To quantify the amount of soil replacement, define the replacement ratio a_s as the total area of soil tributary to pile divided by the area of the sand compaction pile. Sand compaction pile design can be based on either standard penetration resistance or relative density since the two quantities can be related to each other.

Sites Underlain by Sand

For sites underlain by sands the required sand compaction pile spacing and diameter are estimated using the theoretical approach described by Aboshi, et al. [6], Tanimoto [7], Ichimoto [8], and Barksdale [3]. This approach is based on the fact that the strength and settlement properties of a cohesionless soil are primarily determined by relative density. The importance of stress history, grain size, gradation, angularity, and other characteristics should, of course, not be forgotten. For each cohesionless soil a single void ratio is associated with each value of relative density. If the required increase in relative density of a loose sand is determined from stability, settlement, or liquefaction considerations, the required reduction in void ratio of the native sand can be estimated using basic relationships described in this section. The required replacement ratio a_s is determined by assuming the total volume tributary to the pile remains constant, neglecting any increase in relative density of the native sand due to vibration from pile construction and assuming the loose sand displaces only laterally during the construction of sand compaction pile. The loose native sand is thus densified due only to lateral displacement equal to the volume of sand pile added. Using the relationships shown in Fig. 3

$$\bar{s} = V_o - V_f \ (e_o - e_f) \tag{1}$$

$$\bar{s} = (V_v/e_o) \ (e_o - e_f) \tag{2}$$

$$A\,L = V_v + V_s = V_v + (V_v/e_o) \quad (\text{where } L = 1) \tag{3}$$

$$a_s = \bar{s}/A = (e_o - e_f)/(1 + e_o) \tag{4}$$

where:

e_o = initial void ratio of loose sand before improvement
e_f = final void ratio of loose sand after improvement
V_o = total volume of soil per unit length of depth
V_f = final volume of soil excluding volume of s
V_v = volume of voids
V_s = volume of solids
$\bar{s}$ = volume of sand compaction pile per unit length
D = sand pile diameter

Equation (4) can be changed to a more useful form for design by considering a unit length of sand pile construction. A column spacing s for a square sand compaction pile grid can be expressed by using the following equation:

$$s = \sqrt{\bar{V_o}} = \sqrt{\frac{1+e_o}{e_o-e_f}} \cdot \bar{s} = \sqrt{\frac{1+e_o}{e_o-e_f}} \ \frac{\pi D^2}{4} \tag{5}$$

For an equilateral pile spacing multiply equation (5) by 1.08.

Due to waste and densification of the added sand during construction, the volume of loose sand brought to the site for sand pile construction must typically be greater than 20 to 30 percent of the volume of compacted sand pile. The sand compaction pile diameter used

in practice is typically between 600 and 800 mm (24 and 32 in.). The
constructed sand compaction pile diameter is usually taken as 1.5 to
1.6 times the inside diameter of the pipe.

<u>Empirical Approach</u>:   An alternative approach used in Japan for
designing sand compaction piles is based on previously observed
increases in standard penetration resistance (SPT N-value) of sand
compaction piles during construction (Fig. 4). The design procedure
requires the SPT N-value to be corrected for effective overburden
pressure [3,7]. This empirical approach is employed together with the
theoretical method given by equation (5).

<u>Casing Extraction</u>:   Sand compaction piles are densified using a
stroking motion as the casing is extracted from the ground. The
required distance to redrive the pipe can be estimated using the
following empirical expression

$$H_D = H_E \left(1 - n' \frac{A_p}{A_s}\right) \qquad (6)$$

where:

$$H_D = \text{distance casing must be redriven downward}$$
$$H_E = \text{distance casing is extracted before redriving}$$
$$n' = \text{empirical factor that can be taken as about } 0.8$$
$$A_p = \text{inside area of the casing}$$
$$A_s = \text{area of the completed sand compaction pile}$$

Because of its approximate nature, the above expression should be
used together with standard construction practice for specifying values
of H_E and H_D. Usually in practice the casing is extracted 2 to 3 m
(6.5 - 10 ft.) and redriven about one-half that height. The largest
constructor of sand compaction piles typically extracts the casing 2 m
(6.5 ft.).

Sites Underlain by Cohesive Soils

<u>Stress Concentration</u>:   Both field and laboratory studies show that
upon loading an important concentration of stress occurs in the
relatively stiff sand compaction pile constructed in a soft cohesive
soil. The stress concentration develops as consolidation takes place
since the total settlement in the sand and clay is approximately the
same [3,6]. This concentration of stress n forms the basis of design
for both stability and settlement of a sand compaction pile reinforced
cohesive soil and is defined, similarily to stone columns, as

$$n = \sigma_s / \sigma_c \qquad (7)$$

$$\sigma_c = \sigma / [1 + (n-1) a_s] = \mu_c \sigma$$

$$\sigma_s = n\sigma / [1 + (n-1) a_s] = \mu_s \sigma$$

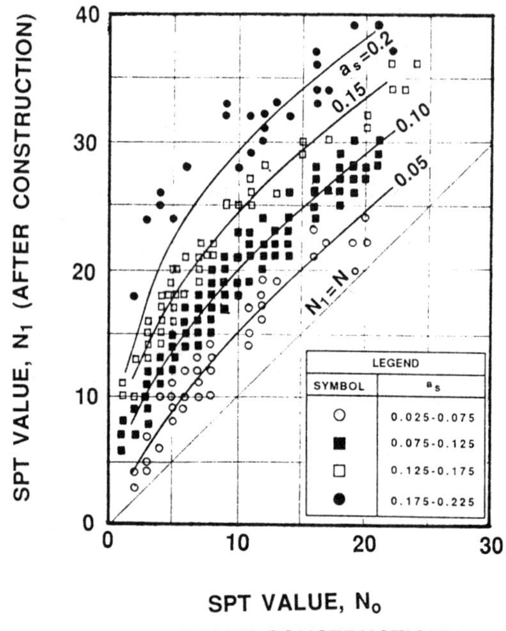

Figure 4.   SPT N-Value at Center of Sand Compaction Piles [4].

TABLE 2 -- Proposed ϕ_S and n Values for Selected Replacement Ratios
(After Ichimoto and Suematsu [9].

a_S	ϕ_S	n
0 - 0.3	30	3
0.3 - 0.7	30	2
0.7 -	35	1

where:

$$n \ = \ \text{stress concentration factor}$$
$$\sigma \ = \ \text{average stress over tributary area}$$
$$\sigma_s \ = \ \text{vertical stress in the sand compaction pile}$$
$$\sigma_c \ = \ \text{vertical stress in the clay}$$

Knowing the stress concentration factor n, the stress in the clay and sand pile is calculated using expressions given above. Once the stress in the clay has been determined, the consolidation settlement is estimated using conventional theory as described by Barksdale [3]. Immediately after rapid loading, the value of n equals 1 and reaches a maximum value as consolidation occurs and load is shifted to the more rigid sand compaction pile. Ichimoto and Suematsu [9] propose using the n values given in Table 2.

Shear Strength: Aboshi [6] has described the basic procedure used to calculate shear strength of sand compaction piles. This general approach has been used, for example, by Murayama [10] and Matsuo [11]. Sogabe [5] has proposed different shear strength equations based on the type of soil and area replacement ratio (Table 3).

For a normally consolidated cohesive soil exhibiting a linear increase in shear strength with effective stress, the increase in undrained shear strength c with time due to consolidation can be expressed for sand compaction pile improved ground as

$$c \ = \ K \ (\Delta \sigma \mu_c) \ U$$

where $\Delta \sigma$ is the average increase in vertical stress in the unit cell on the shear surface due to the applied loading, μ_c is the stress concentration factor in the clay, U is the degree of consolidation of the clay with time, and K is the constant of proportionality defining the linear increase in shear strength with effective stress. A value of K = 0.25 and U = 0.8 has been found to give good results for $a_s \geq$ 0.3. Upon loading before any consolidation takes place, the stress concentration ratio n equals 1. As the soil consolidates n increases until it reaches a maximum value. Yoshikuni [12] has found the maximum value of n to occur at a degree of consolidation equal to about 80 percent. Mogami and Nakayama [13] found n = 3 to 5 based on laboratory and in-situ tests (refer also to Table 2).

GENERAL CRITERIA AND PRACTICES

The construction of sand compaction piles is typically performed on soils with shear strengths from 0.1 to 0.15 kg/cm^2 (200 to 300 psf). Fills, embankments and tanks are routinely placed on these very soft cohesive soils; strengths as low as 0.05 kg/cm^2 (100 psf) have been used. Settlements of land reclaimed from the sea are frequently 2 to 3 m (6.6 to 10 ft.). Typically used sand pile geometrics are given in Table 1.

Strength Loss: Construction of sand compaction piles in cohesive soils causes a significant loss of strength due to remolding of the soil surrounding the pile. Field measurements indicate that typically from 2 to 20 weeks are required to regain initial strength with perhaps about 4 weeks being more usual [6,8]. For mammoth compaction piles the effect of remolding is even more severe than for sand compaction piles, but strength gain still occurs [8]. Ichimoto [8] recommends waiting in cohesive soils at least a month before loading sand compaction piles.

<u>Stability Performance</u>: Sand compaction piles, mammoth compaction piles, and strong sand piles have a record of good performance in Japan with respect to stability with only a few exceptions. One construction company routinely uses $\phi_s = 30^\circ$ for design although it is admitted that ϕ could be as great as 35°. A very large steel company uses 30° for dumped sand and 40° for vibrated sand having an N-value of 20 to 25. Aboshi, et al. [6] have described a well-known trial embankment constructed on 11 m (36 ft.) of organic silt having a shear strength of 0.1 to 0.15 kg/cm^2 (200 to 300 psf). This was underlain by 29 m (95 ft.) of peat with a shear strength varying from 0.07 to 0.13 kg/cm^2 (150 to 250 psf). A test section without sand compaction piles failed at an embankment height of 6 m (21 ft.). Another section of the embankment was constructed using a square grid of sand compaction piles at a 2 m (6.6 ft.) spacing resulting in a small area-replacement ratio of only 0.1. Nevertheless, failure did not occur under an embankment height of 14.6 m (48 ft.). In 15 test embankments supported on soils having shear strengths varying from 0.07 to 0.15 kg/cm^2 (150 to 300 psf) with a_s = 0.16 to 0.20, calculated stability safety factors varied from 0.99 to 1.59 with seven having safety factors less than 1.10. None of the embankments failed.

<u>Sand Gradation</u>: Sand is usually used for site improvement work although gravel and crushed stone have been employed on a very limited basis. Typical sand gradation specifications for Mammoth compaction piles require a well-graded fine to medium sand with D_{10} between about 0.2 to 0.8 mm and D_{60} between about 0.7 and 4 mm. An almost linear relation exists between increasing effective grain size D_{60} and SPT N-value after densification [3]. Saito [14] has presented extensive field data showing the detrimental effect that fines content of a sand has on the vibro-rod method of construction (Fig. 5). The vibro-rod method is used to densify sands following a procedure somewhat similar to the Terra Probe [15]. The vibro-rod consists of a closed pipe having a number of outward protrusions that is driven by a vertical vibrator located at the top of the pipe.

<u>Residual Lateral Stress</u>: Construction of sand compaction piles typically results in the SPT N-value increasing by a factor of 2 to 5. Based on extensive field measurements, Saito [14] has concluded that the large measured increase in SPT N-value is not accompanied by a corresponding large increase in relative density. The misleadingly large standard penetration resistances measured after site improvement is attributed to the significant increase of lateral pressure during densification of the sand. The important effect of residual lateral stress on SPT N-value has been previously reported [16-18] and deserves more study.

QUALITY CONTROL

For sand compaction piles constructed in cohesive soils the penetration resistance is usually measured down the center of the pile

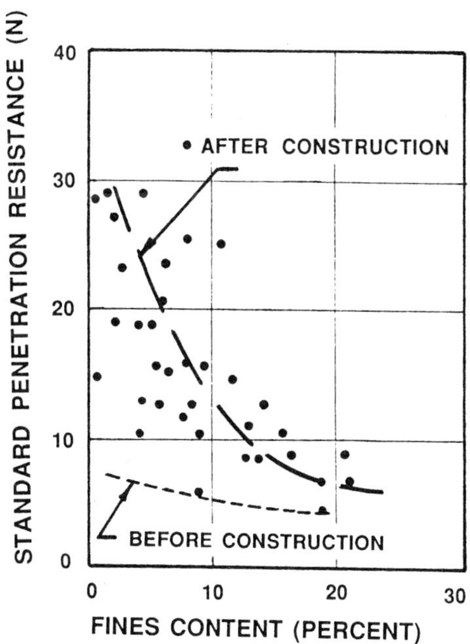

Figure 5.   Influence of Fines Content on Standard Penetration
Resistance -- Vibro-Rod Method (After Saito, Reference 14).

TABLE 3 -- Equations for Calculation of Shear Strength in Stability
Analysis (After Sogabe, [5]).

$\tau = (1-a_s)(C_o + kz + \mu_c\sigma_z\Delta c/ p\ U + \gamma_s Z + \sigma_z\mu_s)\ a_s\ \tan\phi\ \cos^2\alpha$
(composite foundation of sand and clay, $a_s < 0.3$)

$\tau = (\gamma_m + \sigma_z)\ \tan\phi_s\ \cos^2\alpha$ (uniform sand layer, $a_s \geq 0.7$)

$\tau = (\gamma_m + \sigma_z)\ \mu_s\ a_s\ \tan\phi_s\ \cos^2\alpha$ (sand layer, $a_s \geq 0.7$)

$\tau = (1-a_s)\ (c_o + kz) + (\gamma_m z + \sigma_z)\ \mu_s\ a_s\ \tan\phi\ \cos^2\alpha$ (clayey sand, $a_s \geq 0.3$)

$\tau = (\gamma_s z + \sigma_z\mu_s)\ \tan\phi_s\ \cos^2\alpha$ for sand columns, $\tau = c_o + kz$ for clay
(separate foundation of and columns and clay, $a_s \geq 0$)

where:

σ_z = average pressure on improved surface at depth Z
$C/\Delta p$ = strength increase ratio
U = average degree of consolidation
γ_s = unit weight of sand piles
γ_m = average unit weight of composite soil = $\gamma_s\ a_s + \gamma_c\ (1-a_s)$
γ_c = unit weight of the original clay

every 2 m (6.6 ft.); tests between columns are usually not performed. Specifications typically require a SPT N-value between 10 and 20 (apparently the SPT N-value uncorrected for overburden pressure is still frequently used). At sites underlain by sands, the penetration resistance between compaction piles is usually required to be greater than 10 to 15; SPT N-values are measured between the columns at a vertical interval of 2 m (6.6 ft.).

One company uses, as a rule-of-thumb, one test pile for every linear 1000 m (3280 ft.) of pile constructed. A harbor authority requires testing 1 to 1.5 columns out of 100. SPT values near the edge of a sand compaction pile are usually higher than those at the center. SPT values near the surface are larger than those near the bottom. Extra sand is sometimes used in column construction to increase the SPT N-value. Grain size distribution is checked before construction and once per 1000 m^3 ($3.5x10^4$ ft^3) of completed column. Usually load and density tests are not performed.

For construction of two LNG tanks at separate sites underlain by sand, specifications required N to be greater than 15 at the center of the sand compaction pile grid. At a waste disposal station site being reclaimed from the sea, three borings were put down in every third mammoth compaction pile; originally one boring was placed in each pile which proved inadequate. A SPT N-value greater than 15 was required in the center of each pile. At a land reclamation site for apartment houses, 3 or 4 test borings were to be made for a total of 1,584 piles. Specifications required N to be greater than 10. For a large number of mammoth compaction pile jobs, an average N value of 18.9 was observed with the standard deviation being 5.4.

SUMMARY

Extensive use is made in Japan of sand compaction piles to prevent stability failures in soft clays having shear strengths typically varying from 0.1 to 0.15 kg/cm^2 (200 to 300 psf). When settlement is of concern, preloading is often used after the sand compaction piles have been constructed. Sand compaction piles are constructed employing a fully supported hole. Construction of sand compaction piles, which is fast and efficient, utilizes low-cost, often locally available sand.

Since jetting is not normally used, construction of sand compaction piles does not result in a large quantity of excess muddy water. Erosion of fines into the sand compaction pile should also not be a problem for normally used gradations.

ACKNOWLEDGEMENTS

This work was supported by the Federal Highway Adminstration and the U.S. Army Corps of Engineers, Waterways Experiment Station. Special appreciation is extended to Mr. Mizutani of Kensetsu Kikai Chosa Co., Ltd., Osaka, Japan for his assistance.

REFERENCES

[1] Dobson, T., "Case Histories of the Vibro Systems to Minimize
     Risk of Liquefaction", _Soil Improvement - A Ten Year
     Update_, ASCE, April, 1987.
[2] Seed, H.B., and Booker, J.R., _Stabilization of Potentially
     Liquefiable Sand Deposits_, University of California,
     Berkeley, Report EERC 76-10, Earthquake Engineering
     Research Center, 1976.
[3] Barksdale, R.D., _State-of-the-Art For Design and Construction
     of Sand Compaction Piles_, U.S. Army Engineer Waterways
     Experiment Station, Technical Report REMR-GT-4, Nov, 1987.
[4] Personal Communication with Mr. K. Hayashi and Mr. Misutani,
     Kensetsu Kikai Chosa Co., Osaka, Japan, March, 1981.
[5] Sogabe, T., "Problems in the Design and the Construction of
     Sand Compaction Method", The 36th Annual Meeting of
     Japanese Society of Civil Engineers at Hiroshima
     University, August, 1981, (in Japanese).
[6] Aboshi, H., et al., "The Compozer - A Method to Improve
     Characteristics of Soft Clay by Inclusion of Large Diameter
     Sand Column", _Proceedings_, International Conference on Soil
     Reinforcement, Reinforced Earth and Other Techniques, Vol. I,
     Paris, 1979, p. 211-216.
[7] Tanimoto, K., _Introduction to the Sand Compaction Pile Method
     as Applied to Stabilization of Soft Foundation Grounds_",
     Division of Applied Geomechanics, GSIRO, Technical Report
     No. 16, Australia, 1973.
[8] Ichimoto, A., "Construction and Design of Sand Compaction
     Piles", _Soil Improvement_, General Civil Laboratory, Vol. 5,
     June, 1980 (in Japanese).
[9] Ichimoto, E., and Suematsu, N., Sand Compaction Pile Method,
     Fudo Construction Co., Technical Report, Japan, 1981.
[10] Murayama, S., "An Analysis on Vibro-Compozer Method on Cohesive
     Soils", _Mechanization of Construction_, Japanese Association
     of Construction Machines, No. 150, 1962 (in Japanese).
[11] Matsuo, M., "Stability Analysis of the Clay Foundation Improved
     by Sand Columns", _Soils and Foundations_, Vol. 15, No. 12,
     1976 (in Japanese).
[12] Yoshikuni, H., Multi-Dimensional Consolidation Theory and Its
     Axial Symmetrical Problems, PhD Thesis, 1973.
[13] Mogami, T., and Nakayama, J., "Composition Foundation Model
     in Laboratory Test (Part 1,2)", _Soils and Foundations_, Vol.
     16, No. 8, 1968 (in Japanese).
[14] Saito, A., "Characteristics of Penetration Resistance of a
     Reclaimed Sandy Deposit and Their Change Through Vibratory
     Compaction", _Soils and Foundations_, Vol. 17, No. 4,
     December, 1977, p. 31-43.
[15] Brown, R.E., and Glenn, A.J., "Vibroflotation and Terra-Probe
     Comparison", _Geotechnical Journal_, ASCE, Vol. 102, No. GT10,
     October, 1976.
[16] Rodenhauser, J., "The Effect of Mean Normal Stress on the
     Blow Count of the SPT in Dense Chattahoochee Sand", Duke
     University, Project Report, 1974.
[17] Zolkof, E., and Wiseman, G., "Engineering Properties of Dune
     and Beach Sands and the Influence of Stress History",
     Proceedings, 6th International Conference on Soil Mechanics
     and Foundation Engineering, 1964.

[18] Schmertmann, J.H., "The Measurement of Insitu Shear Strength-
State-of-the-Art Presentation to Session 3", *Proceedings*,
ASCE Specialty Conference on Insitu Measurements of Soil
Properties, Raleigh, Vol. II, 1975.

Meiketsu Enoki, Norio Yagi, Ryuich Yatabe and Eizaburo Ichimoto

SHEARING CHARACTERISTIC OF COMPOSITE GROUND AND ITS APPLICATION TO STABILITY ANALYSIS

REFERENCE: Enoki, M., Yagi, N., Yatabe, R., and Ichimoto, E., "Shearing Characteristic of Composite Ground and Its Application to Stability Analysis," _Deep Foundation Improvements: Design, Construction, and Testing_, _ASTM STP 1089_, Melvin I. Esrig and Robert C. Bachus, Eds., American Society for Testing and Materials, Philadelphia, 1991.

ABSTRACT: Shearing characteristics of composite ground, that is clayey soft ground improved by stone or sand columns, are studied. Triaxial compression tests and simple shear tests on composite ground elements are carried out. The result suggests that sand and clay fail individually when the element fails, and that the element has anisotropy of shear strength parameters. Stabilities of embankments on several composite grounds are analyzed by the proposed method in which anisotropy of shear strength parameters is introduced. The result shows that the proposed method gives lower safety factor than the conventional method.

KEYWORDS: composite ground, sand columns, shear strength, anisotropy, stability analysis

INTRODUCTION

To design composite ground, which is clayey soft ground including columns of sand or soil-cement mixture, etc., two methods are known. One is the conventional method, in which stress concentration on a column is assumed, and shear strength is obtained by proportional sum of ones of clay and sand along a common slip surface. This method is used as a routine method for practical problems. Another is finite

Dr. Enoki is an associate professor of Ocean Engineering, Ehime University, Bunkyo-cho 3, Matsuyama, Ehime, Japan; Dr. Yagi is a professor, ditto; Dr. Yatabe is a research associate, ditto; Mr. Ichimoto is a managing director of Fudo Construction Co. Ltd., Taito 1, Taito-ku, Tokyo, Japan.

element method (elastic, plastic or elasto-plastic), but this is not a routine method and only used for an additional check. The finite element method predicts individual failure of the sand and clay elements; a common slip surface is not predicted. Different failure mechanisms are assumed in these methods. This paper treats only sand columns to simplify the discussion.

Theoretical study on the shearing behavior of composite ground elements under triaxial stress conditions, is carried out to find the actual mechanism. A composite ground element is defined to be composed of a column and ambient clay, and is thin enough to neglect its weight. Triaxial tests and simple shear tests on composite ground elements are also carried out.

From the mechanism supported by these studies, the authors propose to treat composite ground as anisotropic c-ϕ material. The proposed method is then applied to stability analysis of an embankment on composite ground.

BACKGROUND

Equal Strain Assumption

Ground improvement using columns of sand, stone, soil-cement mixture or other materials, resembles a pile foundation in appearance. Enoki [1] suggests that (1) with pile foundations, the essentially incompressible pile moves downward relative to the surrounding soil, thereby generating skin friction, while (2) with composite ground, the vertical strain of the column does not differ significantly from the vertical strain of the surrounding soft soil. Accordingly, vertical stress concentrates on, but is not entirely borne by, the columns in composite ground (Fig. 1). This vertical strain compatibility is characteristic of ground improvement.

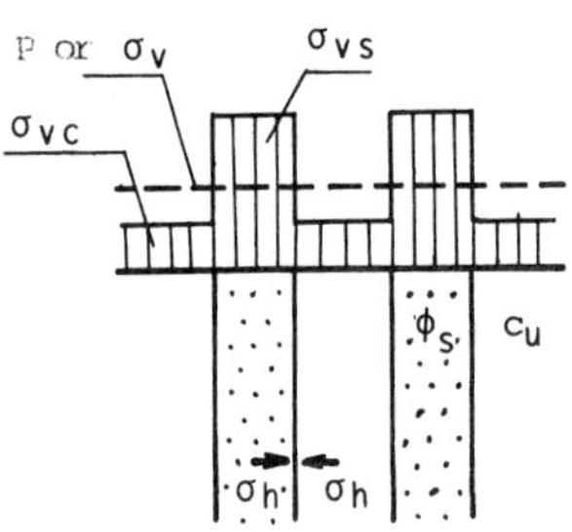

FIG. 1 -- Stress distribution in composite ground.

Outline of the Conventional Design Method

In Japan stabilities of structures on improved ground have been analyzed conventionally as follows [2]. In the method, stress concentration on a column is assumed as $d\sigma_{vs}=m \cdot d\sigma_{vc}$, where m is called the stress distribution ratio, as shown in Fig. 1. $d\sigma_{vs}$ and $d\sigma_{vc}$ are increments of stress on the column and on the ambient soil from initial state, respectively. When a circular slip surface is assumed as shown in Fig. 2, shear strength s on the slip surface is obtained as follows.

$$s=(1-a_s)\,c_u+a_s\,(\gamma_s\,z+p\mu)\tan\phi_s\cos^2\alpha \qquad (1)$$

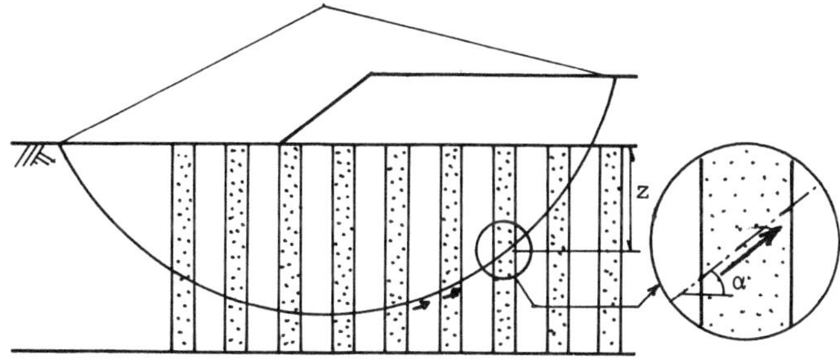

FIG. 2 -- Slipping mass along circular surface.

where, a_s is area ratio of sand column, c_u is undrained shear strength of ambient clay, ϕ_s is friction angle of sand, γ_s is unit weight of sand, p is loading intensity, $\mu = m/\{(m-1)a_s+1\}$ and α is inclination of slip plane. Total shear resistance is obtained by integration of shear strength s along the slip plane. Stability of the slipping mass is analyzed by Fellenius method [3]. The value of m is assumed to be 3 for a_s=15-40%, 2 for a_s=40-70% and 1 for a_s>70%, which were obtained experimentally. Strength parameter of sand ϕ_s =30° is usually used, corresponding to residual state of sand, because the sand may be in residual state when clay fails.

Finite element analysis is also used to estimate stability of composite ground. After FEM analysis, the stability of the whole structure is judged by the number and position of failing elements, but this method cannot be a routine method, because a proper basis to judge the stability of the structure cannot be found.

INVESTIGATION

Two Failure Mechanisms of Composite Ground

Two failure mechanisms are possible for a composite ground element. The mechanisms are demonstrated for clayey ground improved by installation of sand columns or c-material improved by ϕ-material.

One mechanism assumes a common slip plane along which column and ambient soil slip, as shown in Fig. 3. This takes into consideration the same assumption as in the conventional design method shown above. Eq. 1 is modified into Eq. 3 for a specimen of composite ground element under triaxial compression, considering the effect of confining pressure σ_h.

$$\tau = \frac{\sigma_v - \sigma_h}{2}\sin 2\alpha \qquad (2)$$

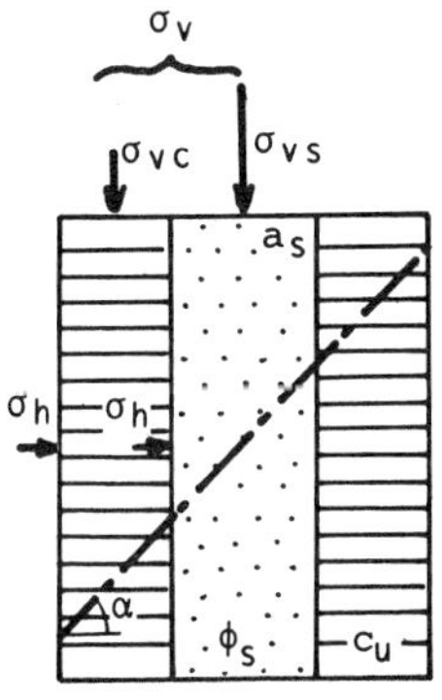

FIG. 3 -- Schematic diagram of composite ground element.

$$s = c_u(1-a_s) + \left(\frac{\sigma_{vs}+\sigma_h}{2} + \frac{\sigma_{vs}-\sigma_h}{2}\cos 2\alpha\right)a_s\tan\phi_s \qquad (3)$$

Total vertical stress σ_v is obtained by a weighted sum of σ_{vs} and σ_{vc} with respect to their areas, as shown in Fig. 1.

$$\sigma_v = a_s\sigma_{vs} + (1-a_s)\sigma_{vc} \qquad (4)$$

Since the specimen is now slipping along the slip plane, the safety factor must be 1.0 and it must be minimized on the plane. Then, the following equations are obtained.

$$F_s = s/\tau = 1 \qquad (5)$$

$$dF_s/d\alpha = 0 \qquad (6)$$

An assumed value for m is necessary, of course. From Eqs. 2 to 6, it is possible to obtain σ_{vsmax} for a given σ_h and equivalent c_{eq} and ϕ_{eq} of the element for various σ_h.
Values of c_{eq}/c_u and ϕ_{eq} are shown in Fig. 4 for various a_s for $\phi_s = 30°$, $c_u = 30$ kPa, $m=3$ and 5, and $\sigma_h = 100$ to 150 kPa. Suffix (a) means active or axial compression failure and (p) means passive or axial extension failure in the figure.

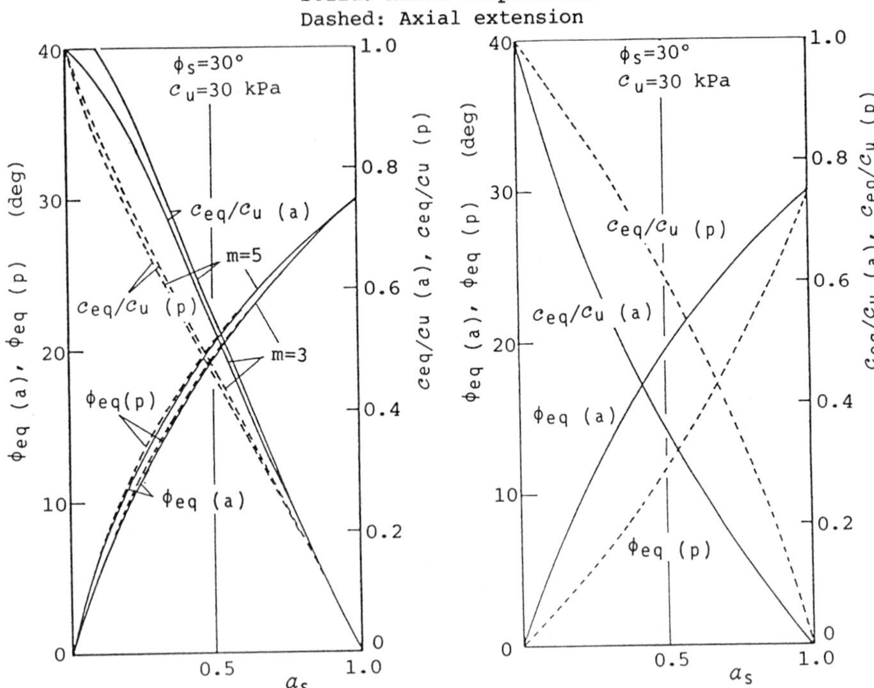

FIG. 4 -- Equivalent ϕ_{eq} and c_{eq}/c_u of composite ground element on the first mechanism.

FIG. 5 -- Equivalent ϕ_{eq} and c_{eq}/c_u of composite ground element on the second mechanism.

The second mechanism assumes individual failures of sand and clay. This takes into consideration the same situation as in finite element analysis of composite ground, because sand and clay will fail individually in the analysis. Equations for this failure mechanism are summarized in Table 1.

TABLE 1 -- Equations for the proposed failure mechanism of composite ground.

	Axial compression	Axial extension
Failure condition sand:	$\sigma_{vs}=\dfrac{1+\sin\phi_s}{1-\sin\phi_s}\sigma_h$ (7)	$\sigma_{vs}=\dfrac{1-\sin\phi_s}{1+\sin\phi_s}\sigma_h$ (14)
clay:	$\sigma_{vc}=\sigma_h+2c_u$ (8)	$\sigma_{vc}=\sigma_h-2c_u$ (15)
Stress distribution	$\sigma_v=a_s\sigma_{vs}+(1-a_s)\sigma_{vc}$ (4)	
Equivalent ϕ_{eq}:	$\phi_{eq}=\dfrac{\pi}{2}-2\mathrm{Tan}^{-1}\sqrt{Ka}$ (10)	$\phi_{eq}=2\mathrm{Tan}^{-1}\sqrt{Kp}-\dfrac{\pi}{2}$ (16)
c_{eq}:	$c_{eq}=(1-a_s)c_u\sqrt{Ka}$ (11)	$c_{eq}=(1-a_s)c_u\sqrt{Kp}$ (17)
	$Ka=\dfrac{1-\sin\phi_s}{1+(2a_s-1)\sin\phi_s}$ (12)	$Kp=\dfrac{1+\sin\phi_s}{1-(2a_s-1)\sin\phi_s}$ (18)
Stress distribution ratio m': ($m'=\sigma_{vs}/\sigma_{vc}$)	$m'=\dfrac{1+\sin\phi_s}{1-\sin\phi_s}\cdot\dfrac{1}{1+2c_u/\sigma_h}$ (13)	$m'=\dfrac{1-\sin\phi_s}{1+\sin\phi_s}\cdot\dfrac{1}{1-2c_u/\sigma_h}$ (19)

Eq. 7 and 14 are failure condition of sand for axial compression and for axial extension, respectively.
Eq. 8 and 15 are failure condition of clay for axial compression and for axial extension, respectively.
Eq. 4 is total vertical stress σ_v obtained by a weighted sum of σ_{vs} and σ_{vc} with respect to their areas, as shown in Fig. 1.
Eliminating σ_{vs} and σ_{vc} from Eqs. 4, 7 and 8, the following equation is obtained for axial compression of the element.

$$\sigma_v=\left\{a_s\frac{1+\sin\phi_s}{1-\sin\phi_s}+(1-a_s)\right\}\sigma_h+2c_u(1-a_s) \qquad (9)$$

Comparing Eq. 9 with failure equation of general c-ϕ material, the composite ground has equivalent c_{eq} and ϕ_{eq} when it is axially compressed. These parameters are expressed by Eqs. 10 and 11. In the same manner equivalent c_{eq} and ϕ_{eq} are obtained for axial extension. These parameters are expressed by Eqs. 16 and 17. In this mechanism

stress distribution ratio m' is determined by strength parameters of
sand and clay and confining stress σ_h of composite ground element, as
written by Eqs. 13 and 19. Note that this stress distribution ratio m'
is defined by whole stresses, and differs from m defined by stress
increments which appears in Eq. 1. ϕ_{eq} and c_{eq}/c_u obtained by this
mechanism are shown in Fig. 5 for various as on the identical
condition with Fig. 4. (Fig. 5 lies next to Fig. 4 for comparison.)

In the latter mechanism, anisotropy of shear strength parameters,
ϕ_{eq} and c_{eq}, is clearly shown for a middle range of a_s. To satisfy the
latter mechanism, if clay and sand are both in active failure or in
passive failure, a strict condition of equal vertical strain is not
necessarily required. As failure strains of sand and clay differ from
one another in general, strength parameters to be used in Eqs. 7 to 19
should correspond to the critical or residual states.

TRIAXIAL TESTS AND SIMPLE SHEAR TESTS ON COMPOSITE GROUND ELEMENTS

Triaxial compression tests were carried out on composite ground
elements. The apparatus was modified from triaxial test cell, as shown
in Fig. 6. Specimen had 10 cm diameter and 10 cm height, including

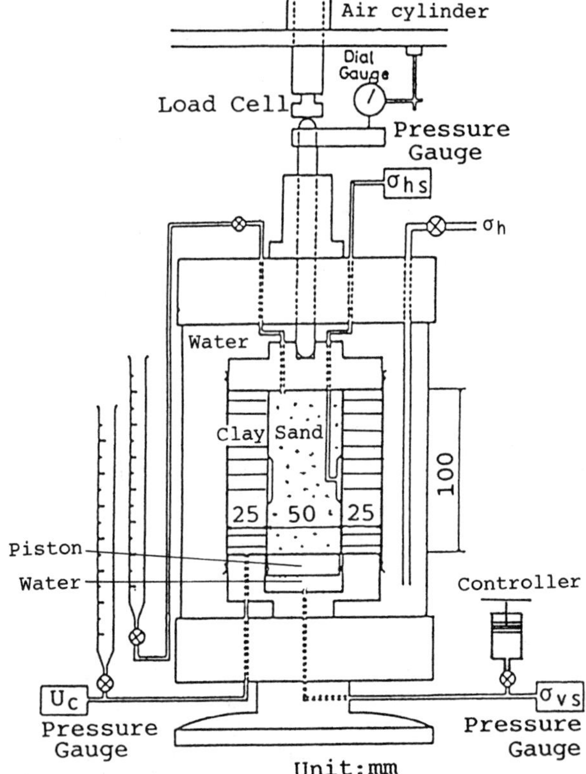

FIG. 6 -- Triaxial test apparatus for composite ground element.

sand column of 5 cm diameter. Sand column portion and clay cylinder portion were covered with caps, pedestals and rubber membranes, separately, so that vertical stresses, volume changes or pore pressures could be obtained, separately. The existence of the rubber membrane between sand column portion and clay cylinder portion might not affect the friction, because the same displacement were given by the rigid cap, and no relative displacement was allowed.
Testing procedure was as follows; (1)Hollow cylinder of clay was prepared in another chamber under preconsolidation pressure at 50 kPa. (2)The clay cylinder portion was seated on the pedestal, and the inside and outside were covered with rubber membranes. (3)A column of saturated sand of void ratio 0.68 or 0.80 was installed inside the hollow. (4)After cap was put on, the sample was consolidated isotropically, at 100 or 200 kPa. During the consolidation, the piston within the pedestal was controlled to keep σ_{vs} equal to the consolidation pressure. (5)The sample is compressed or sheared. During the shear, drainage was allowed for the sand column portion of the sample and prevented for the clay cylinder portion of the sample, and pore water pressure generated in the clay cylinder portion of the sample and volume change in the sand column portion were measured.
Toyoura sand and Fujinomori clay were used to make the sample.

Fig. 7 shows a typical result of the triaxial test. After examinations of deviator stress of clay or stress ratio of sand, it is obvious that clay and sand fail individually.
Confining pressure of sand column is a little higher than cell pressure due to three-dimensional effect (circumferential stress in ambient clay).

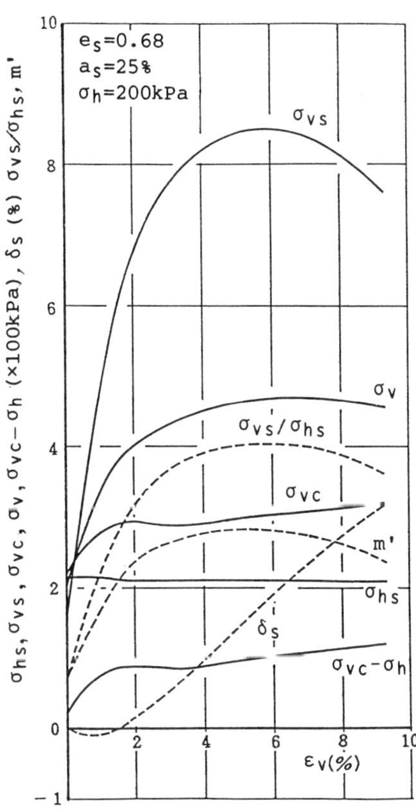

FIG. 7 -- Typical results of triaxial test on composite ground element.

Stress states of composite ground element at failure are shown in Fig. 8. In the figure, σ_v for total element is used. Test results are indicated by solid circles. Calculated stress states on the former mechanism in which the observed stress distribution ratios are used, are indicated by dotted circles. Calculated stress states on the latter mechanism in which the observed confining pressures for sand column are used, are indicated by thick circles. The failure lines calculated on the latter mechanism are added, too. The latter mechanism agrees better with the test results than the former.

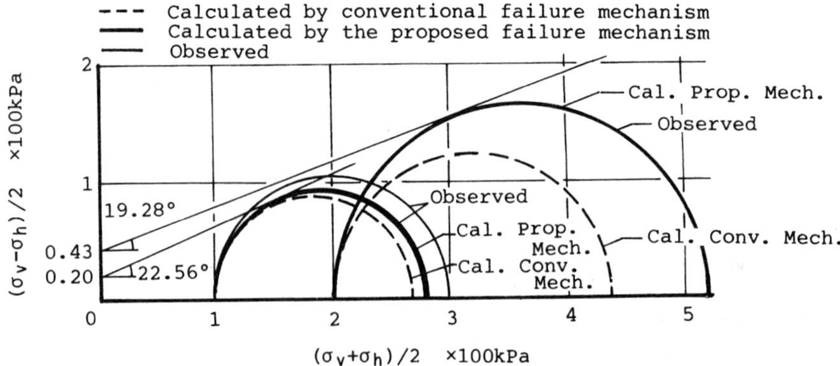

FIG. 8 -- Stress state of composite ground element at failure
in triaxial test.

Simple shear tests were also carried out on composite ground
elements to study variation of strength parameters with direction.
The apparatus shown in Fig. 9 is developed recently by the authors.
The strength parameters obtained by the tests using the new equipment
have been shown to agree with ones obtained by triaxial tests [4]. The
required specimen dimensions were 5 cm diameter and 2.5 cm height,
including dry sand column of 2.5 cm diameter covered with rubber
membrane. Shearing rate of 3 mm/min resulted in undrained shearing of
clay. The inclination angle θ of major principal stress direction to
the vertical was controlled by following equation on the assumption
that mean stress is not affected by application of shear stress τ.

$$\sigma_v=\sigma_h+2\tau/\tan2\theta \tag{20}$$

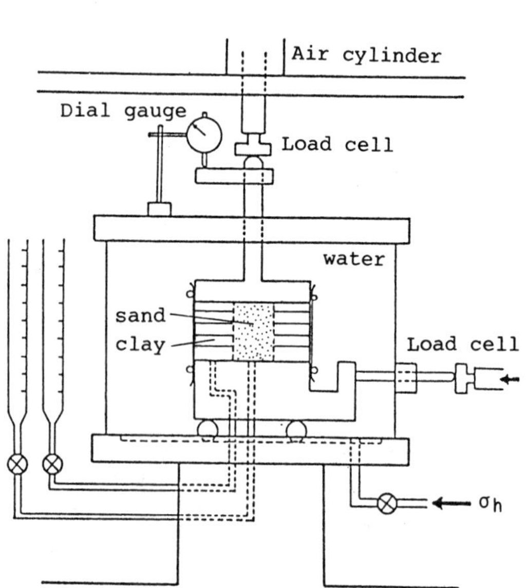

FIG. 9 -- Simple shear apparatus for composite ground element.

Typical stress states of composite ground elements at failure are shown in Fig. 10. Although c component cannot be obtained with reasonable   accuracy, ϕ component clearly varies with direction,   as shown in Fig. 11. As the strength parameters of a composite ground element only for axial compression or axial extension can   be theoretically determined using Eqs. 10, 11, 16 and 17, the strength parameters for intermediate direction are predicted using elliptical functions of directional distribution in Eqs. 21 and 22.

$$\phi_\theta = \tan\phi_a \tan\phi_p / \sqrt{\tan^2\phi_a \sin^2\theta + \tan^2\phi_p \cos^2\theta} \quad (21)$$

$$c_\theta = c_a c_p / \sqrt{c_a{}^2 \sin^2\theta + c_p{}^2 \cos^2\theta} \quad (22)$$

$\phi_a : \phi_{eq(a)}$
$\phi_p : \phi_{eq(p)}$
$c_a : c_{eq(a)}$
$c_p : c_{eq(p)}$

Solid arcs in Fig. 11 are these elliptical functions assuming $\phi_s = 30°$, 35° or 40°.
Although Eqs. 7 to 19 are derived assuming two-dimensional deformation of the element, actual deformation is three-dimensional   and interfacial normal stress between sand and clay is larger than confining pressure, σ_h, in axial compression and smaller in axial extension. Therefore, actual anisotropy might be more emphasized than Fig. 11.

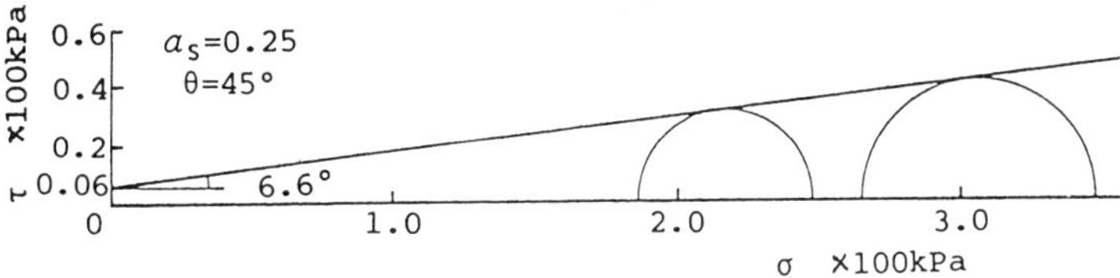

FIC. 10   Stress state of composite ground element at failure in simple shear test.

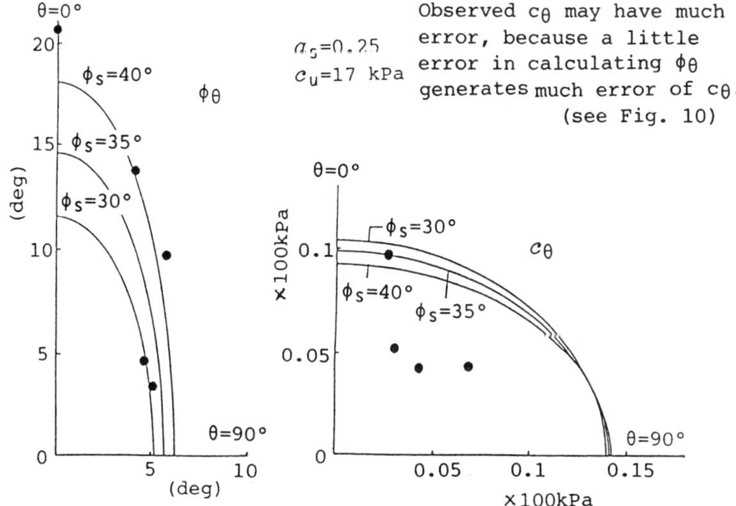

FIG. 11 --Anisotropy of strength parameters obtained by simple shear test.

From these observations, it is clear that composite ground element fails in the latter mechanism.

COMPARISON OF THE PROPOSED METHOD WITH THE CONVENTIONAL METHOD IN STABILITY ANALYSIS

The authors have proposed to treat composite ground as an anisotropic c-ϕ material in the above section. In this section stability of an embankment on a composite ground is analyzed, using the proposed method and the conventional method to take the shear strength of the composite ground. In general, errors in stability analysis of an embankment on composite ground may be caused by (1)misjudgment of shear strength of composite ground and by (2)improper analysis method of stability.

Fellenius method is the most common in practical design. The authors suggested that the method has a tendency to overestimate the bearing capacity of the clayey ground of which strength increases with depth, and to underestimate the bearing capacity of the homogeneous sandy ground [5], using Generalized Limit Equilibrium Method proposed by the authors [6]. The clayey ground improved by sand columns can be considered to be a kind of mixture of the clayey ground of which strength increases with depth and the homogeneous sandy ground. Then, Fellenius method may possibly underestimate or overestimate the bearing capacity of the improved ground. However, as it is significant for practical use to clarify the influence of method to take shear strength of composite ground on the bearing capacity, stability of an embankment on composite ground is studied, here, using Fellenius Method.

When a method of stability analysis is used assuming slip surface, it is easy to introduce anisotropy of ground into the analysis. Inclination angle α of slip surface to the horizontal has following relation with inclination angle θ of major principal stress to the vertical and friction angle ϕ of the material, assuming slip surface is a failure surface of a small element including the slip surface.

$$\alpha - \pi/4 + \phi_\theta/2 + \theta = 0 \qquad (23)$$

From Eqs. 21, 22 and 23, θ, ϕ_θ and c_θ for a given α can be obtained after some calculation.

Here, comparisons of the conventional method which uses Eq. 1 and the proposed method, both using Fellenius method, are carried out. The object is short term stability of embankment on composite ground of 10 m thickness. The ground is originally soft clay having shear strength, c_u, of 10 kPa, and is improved by sand columns having friction angle, ϕ_s, of 30°. Loading intensity of embankment is chosen among 100, 150 and 200 kPa for obtained safety factor to be in a range from about 1.0 to 1.5. Only basement failure is considered.

The first comparison concerns embankment on composite ground which spread out right and left infinitely, as shown in Fig. 12. In

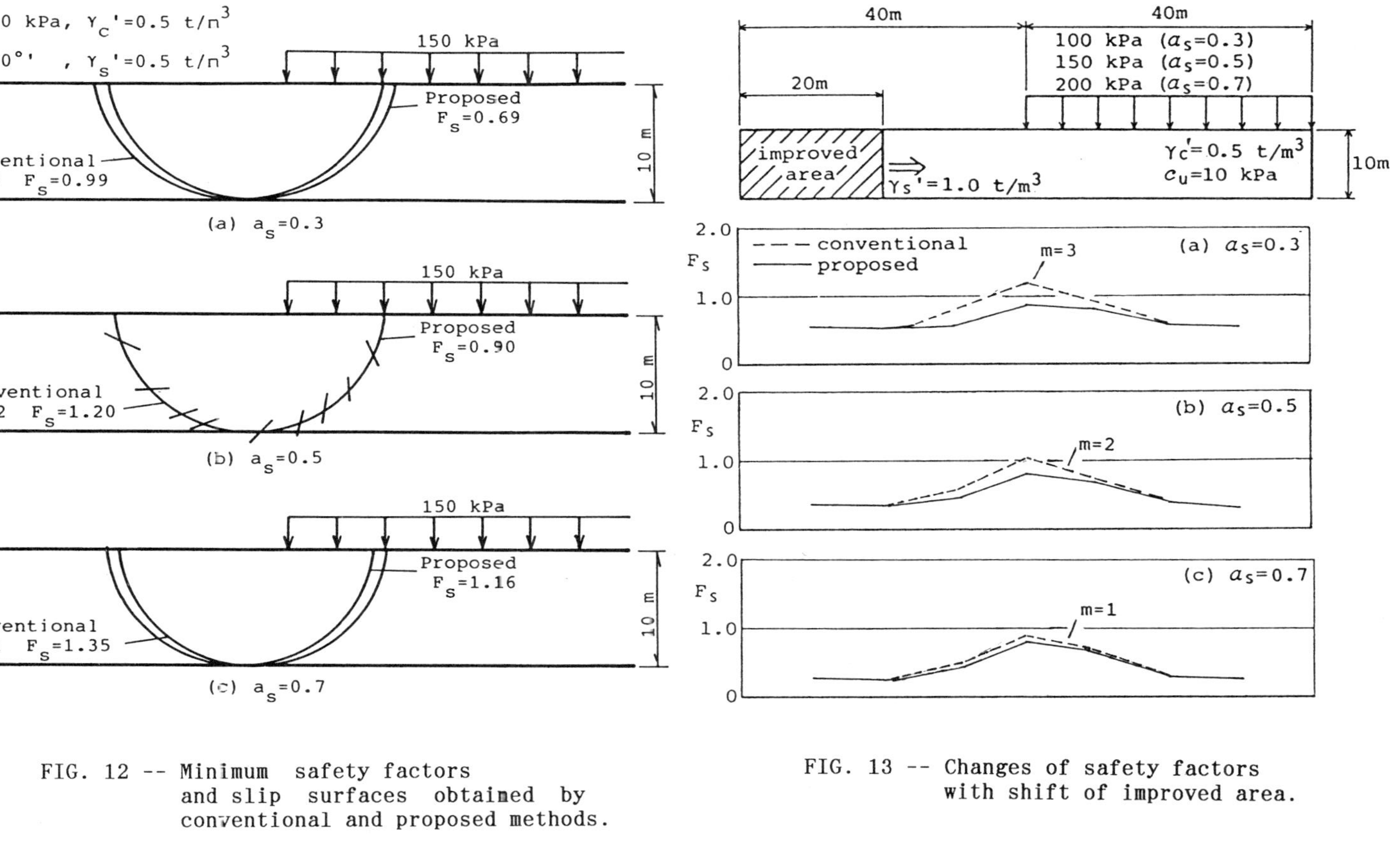

FIG. 12 -- Minimum safety factors and slip surfaces obtained by conventional and proposed methods.

FIG. 13 -- Changes of safety factors with shift of improved area.

the figure, minimum safety factors and corresponding slip surfaces are demonstrated for area ratios 0.3, 0.5 and 0.7. Stress distribution ratio m necessary for the conventional method, is chosen among 1, 2 and 3. In every comparison safety factor obtained by the conventional method is larger than one obtained by the proposed method. The proposed method to take shear strength of the improved ground resembles the lower bound formulation, while the conventional method resembles the upper bound formulation. This difference of the characteristics may cause the difference of the safety factors. Slip surfaces which give minimum safety factors are different a little from one another. In Fig. 12 (b), directions of major principal stress are drawn, too. The direction seems to be reasonable.

The second comparison concerns the embankment on composite ground which is limited in area. Safety factors of embankments are obtained when composite ground of 20 m width is shifting toward right, and are plotted at the center of the composite ground, as shown in Fig. 13. In the cases of a_s=0.3 and 0.5, proposed method gives lower safety factors than conventional method, but in the case of a_s=0.7 two methods give nearly same safety factors. Position of the composite ground to give the highest safety factor obtained by the proposed method is similar to that by the conventional method.

Which method is valid, the proposed method or the conventional method to take the shear strength of the improved ground, cannot yet be decided by these simple comparisons. However, the characteristic of the improved ground is clarified in part by the proposed method in which the improved ground is treated as an anisotropic c-ϕ ground, as follows:
  1. Since ϕ-material displays its strength under high confining pressure, the improved area should be located at the place where confining stress or mean stress σ_m increases by loading and where shear deformation may be caused, as shown in Fig. 13.
  2. Improved area should also be located at the place where active failure is expected, because apparent friction angle of composite ground is higher in active state than in passive state, as shown in Fig. 5.

Broms [7] threw doubt upon the effect of sand columns from the reason that the confining pressure provided by soft clay may be very low. However, when clay is in undrained condition and its strength is negligible, loading pressure on clay will induce total lateral stress of the same magnitude on sand-clay interface, and effective lateral stress for sand is also the same magnitude because of drained condition of the sand. When clay is in drained condition, it can be considered as ϕ-material like sand, and it will be able to provide sufficient confining pressure for sand column.

CONCLUSION

  1. Two failure mechanisms for composite ground element are possible. The mechanism in which sand and clay fail individually, is probable from results of laboratory tests.
  2. Anisotropy of composite ground element is clarified by theory

and laboratory tests.

3. To treat composite ground as anisotropic c-ϕ material, is proposed.

4. For effective improvement, composite ground should be located at the place where confining stress will increase by loading, and where active deformation is expected, because apparent friction angle of composite ground is higher in active state than in passive state, and because ϕ-material displays its strength under high confining pressure.

ACKNOWLEDGMENTS

The authors wish to thank Mr. Futagami, O. for his design of the apparatus used in this paper. The authors are grateful to Mr. Okuyama, K., Mr. Okabe, N. and Mr. Watanabe K. who assisted with this work.

REFERENCES

[1] Enoki, M., "What is Ground Improvement and What is Pile Foundation?," Proceedings of 8th Asian Regional Conference on Soil Mechanics and Foundation Engineerings, Vol.2, 1987, pp.333-334.

[2] Aboshi, H. et al., "The Compozer, a Method to Improve Characteristics of Soft Clays by Inclusion of Large Diameter Sand Columns," Proceedings of 1st International Conference on Soil Reinforcement, Vol. 1, 1971, pp.211-216.

[3] Fellenius, W., "Calculation of the Stability of Earth Dams," Proceedings of 2nd Congress on Large Dams, 1936, pp.445-462.

[4] Yagi, N. and Yatabe, R., "Simple Shear Apparatus with Arbitrary Stress Control," Journal of Japanese Society of Soil Mechanics and Foundation Engineerings, Vol.31, No.7, 1983, pp.23-28. (in Japanese)

[5] Enoki, M., Yagi, N., Yatabe, R., "Evaluation of Bearing Capacity Analysis Method of Improved Ground," Proceedings of International Conference on Geotechnical Engineering for Coastal Development, 1991. (Under submission)

[6] Enoki, M., Yagi, N., Yatabe, R. and Ichimoto, E., "Generalized Limit Equilibrium Method and Its Relation to Slip Line Method," Soils and Foundations, 1990. (Under submission)

[7] Broms, B.B., "Soil Improvement Methods in Southeast Asia for Soft Soils," Preprint of 8th Asian Regional Conference on Soil Mechanics and Foundation Engineerings, 1987, pp.1-36.

Hisao Aboshi, Yasuo Mizuno, and Mashahiko Kuwabara

PRESENT STATE OF
SAND COMPACTION PILE IN JAPAN

REFERENCE: Aboshi, H., Mizuno, Y., and Kuwabara, M., "Present State of Sand Compaction Pile in Japan," _Deep Foundation Improvements: Design, Construction, and Testing_, _ASTM STP 1089_, Melvin I. Esrig and Robert C. Bachus, Eds., American Society for Testing and Materials, Philadelphia, 1991.

ABSTRACT:The sand compaction pile method (hereinafter abbreviated as SCP method) has been developed in Japan as an improvement method for soft ground. One of the reasons of the above is that SCP method can be applied to both sandy and clayey soil. Complicated soil strata in Japan have also contributed to the development of this method. This paper evaluates and verifies the applicability of the design procedure of SCP method to sandy and clayey grounds by comparing the measured data with designed values. Moreover the latest technolgy of SCP method, that is the Mechatronic Consolidation System which can set up sand compaction piles in accordance with the variation of soil properties, is also introduced.

KEYWORDS:soil improvement, sand compaction pile, sandy soil, clayey soil, liquefaction, loading test

INTRODUCTION

The sand compaction pile method (hereinafter abbreviated SCP method) has been developed in Japan as a soft ground improvement method. It is made up of well compacted sand piles which are set up in soft ground using sand or similar materials through a vibrating casing pipe. In fact this method has undergone a greater progress and found a wider application in Japan than in any other countries because complicated soil strata with various characteristics of soft soil are mostly encountered in Japan and also the method itself is applicable to both sandy and clayey grounds using the same equipment and machines.

This paper evaluates and verifies the applicability of the design procedure to sandy and clayey grounds by comparing the measured data with predicted values. Moreover the Mechatronic Consolidation System which can set up sand compaction piles in accordance with the different properties of soil is introduced with a view to enhance the further development of SCP method.

HISTRY OF SCP METHOD[1]

Design Procedure

The design of SCP method was based on theoretical work presented by Dr.Murayama and Tanimoto who have published their reserch papers in 1957[2][3] and 1958[4]. Moreover the presentation of Dr.Murayama's paper in 1962[5] has established the method's applicability to clayey soil followed by a number of research and actual construction works by which their theories were proved further justified.

Dr. Aboshi is technical advisor. Mr. Mizuno is chief manager. Mr. Kuwabara is engineer. Authors belong to Fudo construction Co., Ltd.. 1-2-1, Taito, Taito-Ku, Tokyo, 110, JAPAN.

Development of Construction Procedures

The histry of SCP construction method is as follows. The hammering method was developed in 1957, followed by the development of a method which introduced the vibrating casing pipe to improve the efficiency of the construction. Moreover the use of SCP for offshore application was developed in 1967 in order to extend the applicability of the method. The automatically controlled SCP driving system was also invented in 1981 accomodating the variation of the soil properties.

The practical construction efforts seem to go far ahead of the formulation of a sound theoretical system.

THE APPLICATION OF SCP METHOD TO SANDY GROUND

Design Procedure for Sandy Ground

The effectiveness of improvement of the sandy ground by SCP method is generally checked by means of standard penetration tests conducted at intermediate points between piles (inter-pile). However, PS logging and/or dynamic tri-axial compression tests on undisturbed samples are employed in some specific cases. In order to check the increase in subgrade reaction, the reaction coefficient (of subgrade) is periodically measured. The design procedure for sandy ground proposed so far is estimated by the improvement effect through calculations using N-values of standard penetration tests, and it can be diversified into the following two methods.

Conventional Design Procedure Disregarding Fine Particle Content : In the design for sandy ground to be improved by SCP, filling effect of pile material is presumed to be primary and vibration effect to be secondary. Fig. 1 shows the relationship between N-Dr-e, through which the relative density of original ground "Dro" can be estimated to give N-value of original ground "No" and effective overburden pressure, "σ'_v" at the depth concerned. The void ratio of original ground, "e_0" is also estimated using the effective grain size, "D_{60}" and uniformity coefficient, "U_c" of the original ground. On the other hand, the superstructure load and other requirements specify the target N-value of inter-pile, "N_1", relative density "Dr_1" and target void ratio "e_1".

Consequently, pile pitch "x" can be determined through the following equation, using the pile diameter, 70cm and replacement ratio "a_s",

$$a_s =(e_0-e_1)/(1+e_0)=\triangle e/(1+e_0)$$
$$A=x^2=A_s/a_s \qquad \text{(a square arrangement)}$$
$$A= (\sqrt{3}/2)x^2=A_s/a_s \quad \text{(a regular triangular arrangement)}$$
$$\text{(see Fig. 2)}$$

An other method is available to determine the replacement ratio in accordance with the result of actual construction on sandy ground by SCP method. Fig. 3 shows the relationship between original N-value "No" and inter-pile N-value after improvement "N_1" with the replacement ratio "a_s" being the parameter. The required "a_s" can be easily determined through Fig. 3, which has been employed in the recent designs.

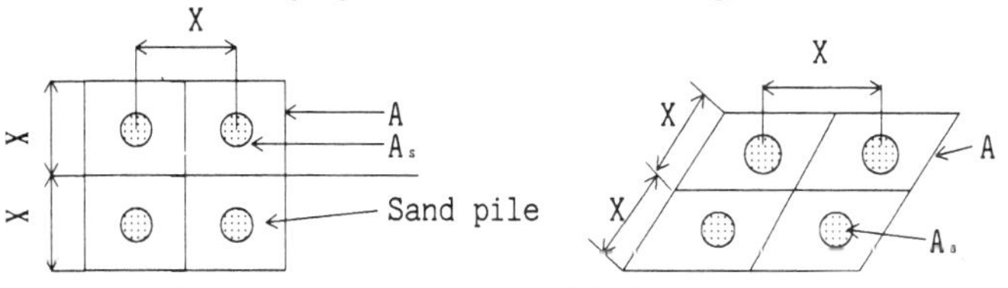

(a) Square Arrangement   (b) Regular Triangular Arrangement

FIG. 2 -- Spacing of sand piles

New Design Procedure with Due Consideration of Fine Particle Content[1] : Fig. 4 shows the relationship between fine particle content "Fc" and inter-pile N-value after improvement "N_1", through which it is observed that the higher the value of Fc is, the smaller the value of N1 will be given. Therefore, the earlier stated design procedure is meant for such sandy ground as having less than 20% Fc and is unable to estimate the reduction in improvement caused by fine particles.

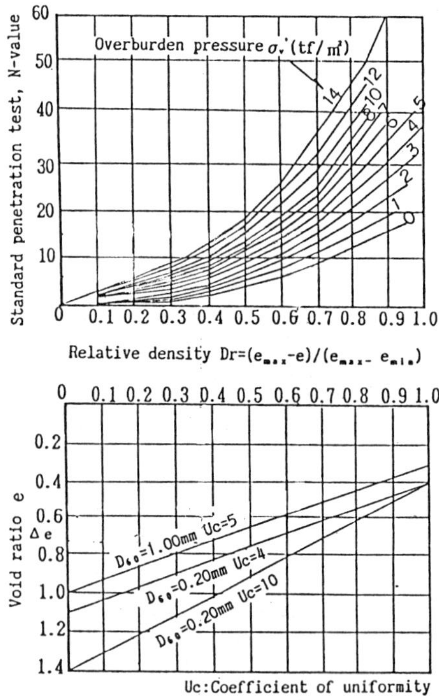

FIG. 1 -- Relationship between
N-value, relative density
and void ratio[6]

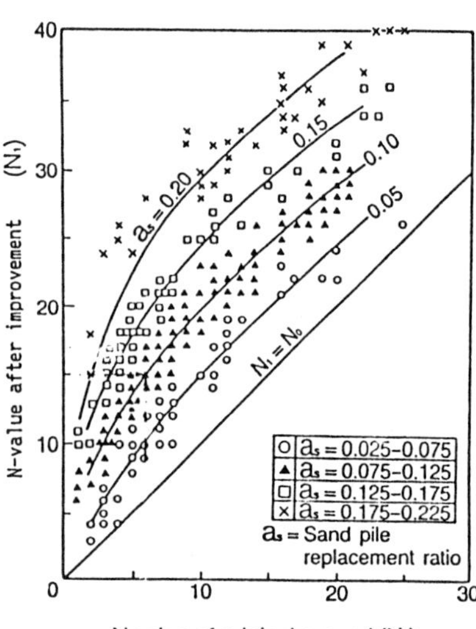

$$F_c = \frac{\text{Weight of particles finer than } 75\,\mu}{\text{Total weight of soil}} \times 100$$

FIG. 3 -- Relationship between
No-value, a, and N1-value
(Fc< 20%)

Here in order to estimate inter-pile N-value often the improvement of sandy ground containing fine particles, an equation was proposed formulating the relationship between governing factors such as fine particles content and effective overburden pressure etc. and the parameter such as relative density and void ratio. The relevant data were collected from actual constructions for sandy ground containing substantial amount of fine particles. Those factors governing inter-pile N-value after improvement of sandy ground with fine particles are assumed to be N-value of original ground "No", effective overburden pressure, "$\sigma_v{'}$", fine particle content "Fc", replacement ratio "a,", vibration effect and so on. However, the vibration effect is not taken into account in the said formulation because it is not considered in the conventional procedure. Fig. 5 shows the flow-chart for design calculation.

Assumption of Original Void Ratio:At first, original void ratio "eo" is assumed taking into consideration No, $\sigma_v{'}$ and Fc at the location and depth concerned, where the relationship between N-Dr-e in Fig. 1 is generally applied. However, since this relationship is not formulated, Eq 1 proposed by Meyerhof[8] and the relationship of Fc-e_max and Fc-e_min (see, Fig. 6) are employed.

$$Dr(\%) = 21 \cdot \sqrt{N/(0.7 + \sigma_v{'})} \tag{1}$$

where
$\sigma v'$:effective vertical stress (kgf/cm²)

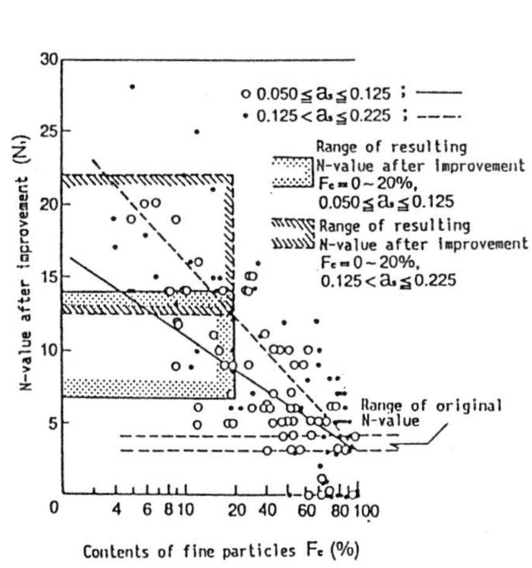

FIG. 4 -- Relationship between
F_c and N_1

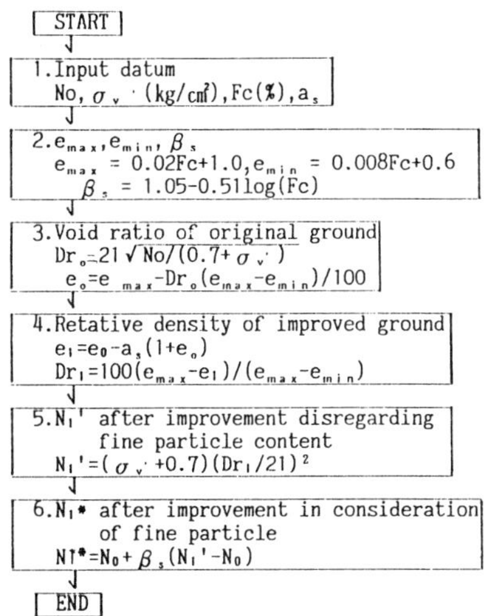

FIG. 5 -- Flow-chart for
estimation of N_1* [7]

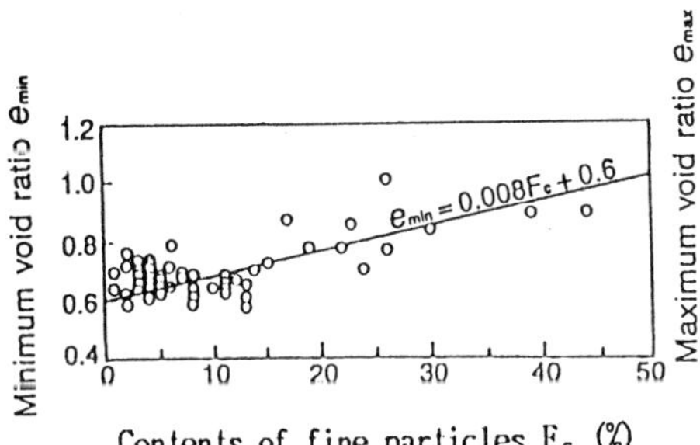

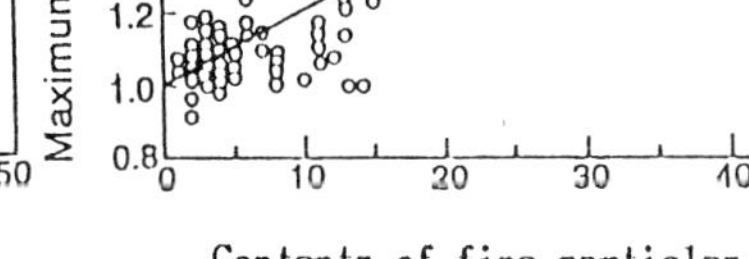

FIG. 6 -- Relationship between
F_c and e_{max}, e_{min} [9]

Calculation of Inter-Pile N-value, "N_1'" after Improvement Disregarding Fine Particle Content:Secondly, inter-pile void ratio, "e_1" and relative density "Dr_1" after improvement are assumed, using the replacement ratio "a_s", subsequently, inter-pile N-value after improvement "N_1'" can be obtained, disregarding the influence of fine particle content.

Assumption of Inter-Pile N-value, "N_1*" after Improvement in Consideration of Fine Particle Content:However, in case of high Fc, measured N_1* tends to be smaller than calculated N_1', which is because of the underestimation of Dr and its resultant overestimation of N_1'. This is also because of the upheaval and lateral flow of ground during the construction of sand piles which jeopardize the reduction of void ratio. Therefore, reduction in improvement effect shall be made by the decrement ratio, "β_s" given by Eq 2. The relationship between β_s and Fc show in Fig. 7 gives the regressive Eq 3 and the overestimation of N_1' can be rectified by Eq 4, which finally gives N_1*.

$$\beta_s = \triangle N/\triangle N' = (N_1{}^* - N_0)/(N_1' - N_0) \tag{2}$$

where
$\triangle N$:increase in N-value governed by fine particles
$\triangle N'$:increase in N-value without consideration of fine particles

$$\beta_s = 1.05 - 0.51 \cdot \log_{10} Fc \tag{3}$$
$$N_1{}^* = N_0 + \beta (N_1' - N_0) \tag{4}$$

The correlation between $N_1{}^*$ estimated by conventional and new design procedure and N_1 actually measured is shown in Fig. 8. $N_1{}^*$ estimated by new design procedure shows good correlation with N_1 actually measured. On the contrary, conventional design procedure gives larger N_1 than that by new design procedure, which leads to risky design. This is mainly because of fine particle content, and it is also the reason why conventional design procedure should not be simply applied to such sandy ground as containing substantial amount of fine particles.

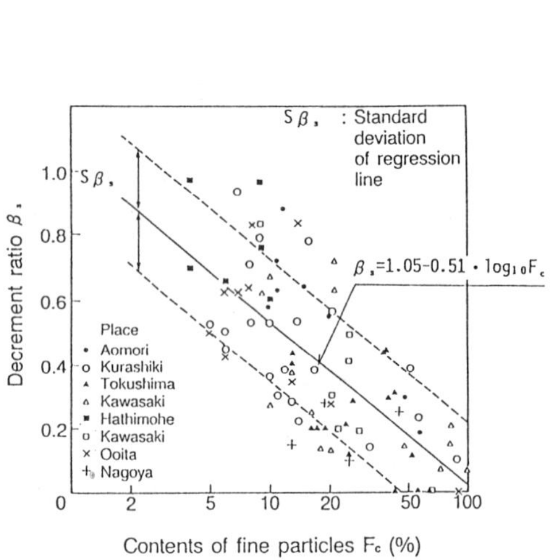

FIG. 7 -- Relationship between F_c and β_s

FIG. 8 -- Relationship between measured and estimated value

Application to Sandy Ground

As to improvement of loose sandy ground, the SCP is employed with an aim to increasing of bearing capacity and preventing of liquefaction. Some countermeasure against liquefaction of loose sandy ground is indispensable in Japan as it has earthquake. There are many kinds of countermeasures to prevent liquefaction due to earthquake. But SCP has been most often employed out of the various kinds of methods, which is entirely due to the fact that those grounds improved by SCP have been free from liquefaction even after they experienced severe earthquakes.[10]

When "the 1978 Miyagiken-oki Earthquake" took place, those storage tanks of petroleum built on the sandy ground improved by SCP suffered no damages due to liquefaction. Ishihara et al[11] elaborated on the above phenomenon by conducting dynamic tri-axial compression tests on undisturbed samples and simple response analysis. N-value at inner and outer areas improved by SCP, result of dynamic tri-axial compression test and result of simple response analysis are shown in Fig. 9,10 and 11, respectively.

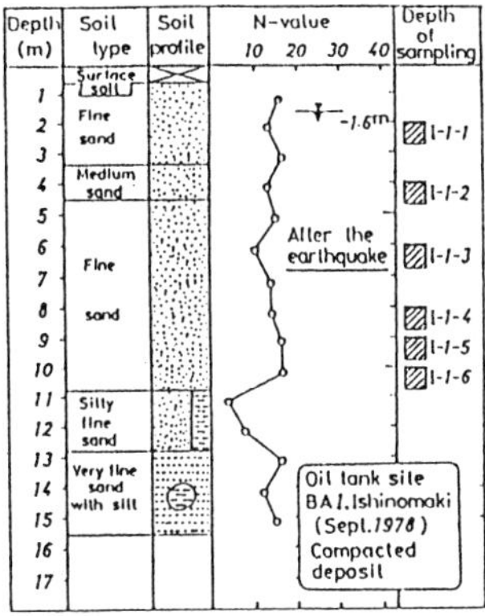

FIG. 9(a)-- Standard penetration resistance and depths Osterberg sampling at the compacted site after earthquake [11]

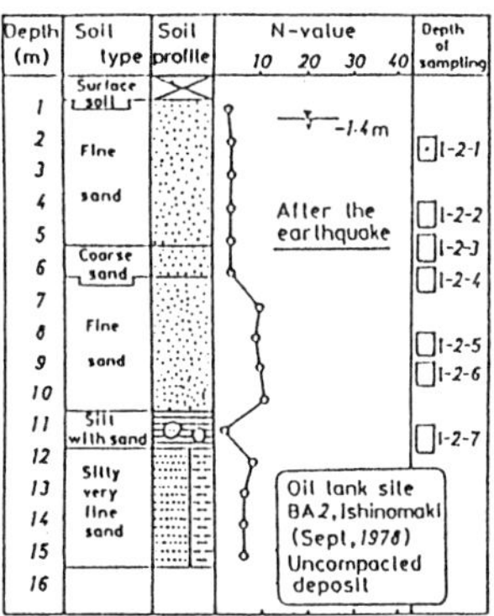

FIG. 9(b)-- Standard penetration resistance and depths Osterberg sampling at the uncompacted site after earthquake [11]

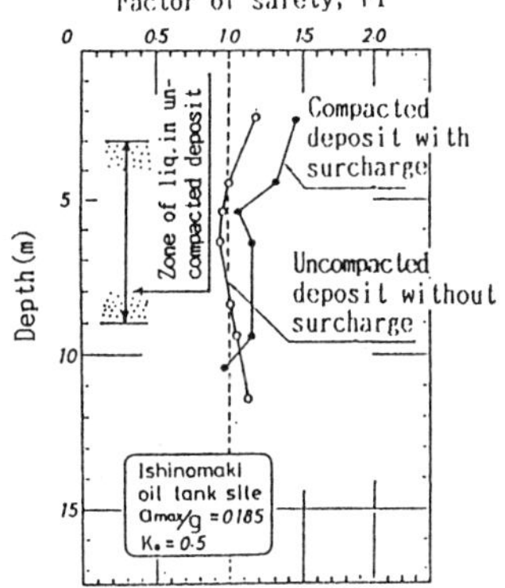

FIG. 10 -- Factor of safety against failure in the compacted and uncompacted deposits [11]

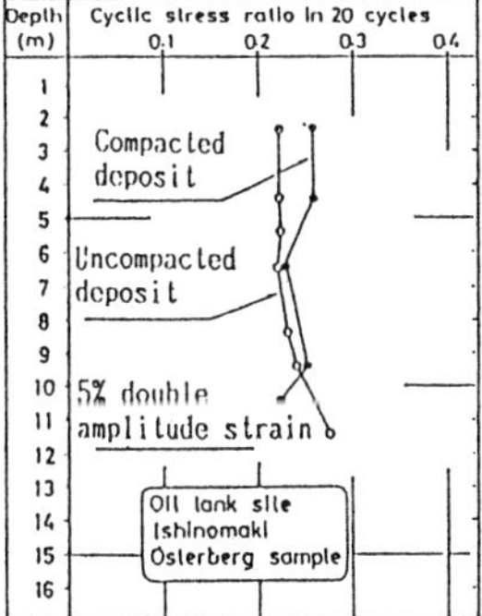

FIG. 11   Comparison of cyclic strengths between the compacted and uncompacted deposits [11]

APPLICATION OF SCP TO CLAYEY GROUND

Design Procedure for Clayey Ground

Murayama[5] named the clayey ground improved by a number of sand compaction piles as "the composite ground".

Sand piles driven in the composite ground bring the effects of stress concentration, increase of shear strength and acceleration of consolidation. Murayama[5] established the following equations which show that vertical stresses on sand pile "σ_s" and on surrounding clay "σ_c" are calculated through the equilibrium condition of vertical stresses acting on composite ground, where stress distribution ratio "m" and replacement ratio "a_s" work as parameters, as is illustrated in Fig. 12 and Eqs 5~12.

$$(A_s + A_c) \cdot \sigma = A_s \cdot \sigma_s + A_c \cdot \sigma_c \tag{5}$$
$$\sigma_h \geqq (1+\sin \phi_s)/(1-\sin\phi_s) \cdot \sigma_s \tag{6}$$
$$\sigma_h \leqq \sigma_c + \sigma_u \tag{7}$$

By the above Eqs 6 and 7

$$\sigma_s / \sigma_c = (1+\sin \phi_s)/(1-\sin\phi_s) \cdot (1 + \sigma_u / \sigma_c) \tag{8}$$
$$m = \sigma_s / \sigma_c \tag{9}$$
$$\sigma_c = \sigma / \{1 + (m-1)a_s\} \tag{10}$$

$$\sigma_s = m\sigma / \{1 + (m-1)a_s\} \tag{11}$$
$$a_s = A_s / (A_s + A_c) \tag{12}$$

Where

A_s : cross-sectional area of a sand pile
A_c : cross-sectional area of a clayey ground
σ : vertical loading intensity
σ_s : vertical stress on the sand piles
σ_c : vertical stress on the clayey ground
σ_h : lateral confining stress on the cylindrical
surface of the sand pile
ϕ_s : internal friction angle of sand pile
σ_u : upper yield value of the clayey ground
m : stress distribution ratio
a_s : replacement ratio

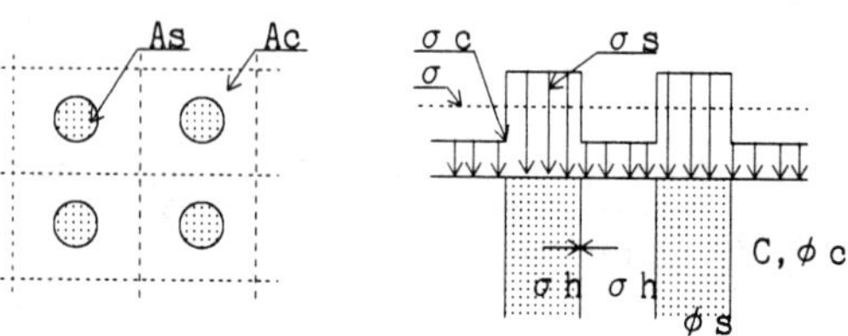

FIG. 12 -- Concept of composite ground

Estimation methods for improvement effect of composite ground can be classified into those for sand piles themselves, for inter-pile clayey soil and for entire composite ground. Sand piles themselves are generally checked by standard penetration tests. Inter-pile clayey soil can be tested by conducting laboratory tests on undisturbed samples to compare the test results with the design parameters and check the stability. Likewise, the entire composite ground also can be examined by comparing the outcomes of field measurements on displacements, water pressure, earth pressure and so on with the estimated values at the time of designing. Horizontal loading test is occasionaly conducted to verify the increase of horizontal subgrade reaction.

Meanwhile, non-displacement type method called "Non Flow Compozer" has been applied for improvement of ground in Japan. Design procedure of this method is same as that of SCP[12].

<u>Design Procedure for Slope Stability Analysis of SCP-improved Soils:</u>The shearing strength of composite ground that is required for bearing capacity and sliding resistance is formulated by Eqs 5~12 and expressed by Eq 13.

$$\tau_{sc} = (1 - a_s) \cdot c + a_s \cdot (\mu_s \cdot \sigma + \gamma_s \cdot Z_s)\tan\phi_s \cdot \cos^2\alpha \tag{13}$$

where

τ_{sc} : shearing strength of composite ground
a_s : a_s = As/(As+Ac), here called "replacement ratio"
c : cohesion of clay
γ_s : unit weight of sand pile
ϕ_s : internal friction angle of sand pile
σ : vertical loading stress intensity
μ_s : stress concentration coefficient of sand pile
μ_s = m / { 1+(m-1) $\cdot$ a_s }

As for consolidation of clay, the increased cohesion c, is expressed by Eq 14.

$$c = c_0 + \Delta c$$
$$= c_0 + \mu_c \cdot \sigma \cdot U \cdot c/p \tag{14}$$

Where

c_0 : initial strengh of clay
Δc : increment of cohesion by consolidation in case of SCP
U : consolidation degree
c/p : ratio of cohesion increase
μ_c : stress reduction coefficient of clay
μ_c = 1 / { 1+(m-1) $\cdot$ a_s }
m : stress distribution ratio

In the case of slip-circle failure calculation, a shearing phase at depth Z is assumed as shown in Fig. 13, where it is considered that the strengths of both sand pile and surrounding clay are mobilized against the vertical load.

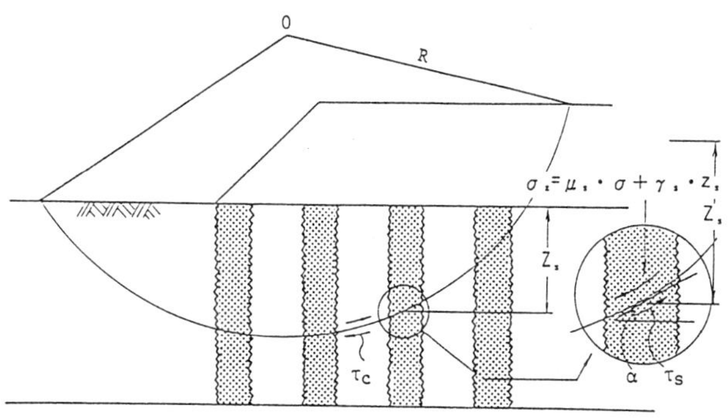

FIG. 13 -- Stability calculation of
soft ground

Comparison of each elememt of Eq 13. with actually measured result is made by using the results of unconfined compression tests on inter-pile clayey soil and N-value at pile core.

Disturbance and Its Recovery Due to Installation of Sand Piles:In Eqs 13, 14, original ground strength "c_o" is normally used as the strength of clayey ground around the sand pile is disturbed just temporarily. However, the disturbance is recovered before the construction of superstructurtes. Fig. 14 shows the recovery process of the disturbance of clayey soil after sand pile installation.

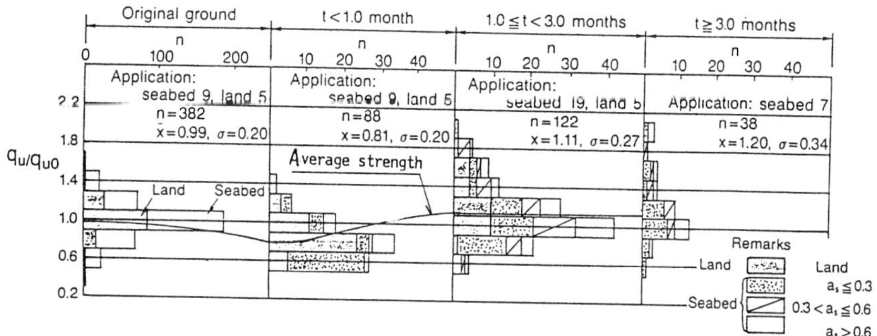

FIG. 14 -- Disturbance and recovery of strength
in the ground improved by SCP [1]

Increased Strength of Clayey Ground:In Eq 14, increase in strength of clayey ground around the sand pile, "Δc" is reduced by the stress concentration effect on sand pile. Fig. 15 shows the relationship between actually measured value of Δc at various working sites and stress distribution ratio "m".

Strength of Sand Pile:The strength of sand pile is assessed by evaluating its N-value. It is also affected by the strength of surrounding soil, and the actually measured results are shown in Fig. 16. In Eq 13, $\phi s= 30\text{-}35^\delta$ is generally used, and its changes depend upon the type of sand that is to be used in the sand pile.

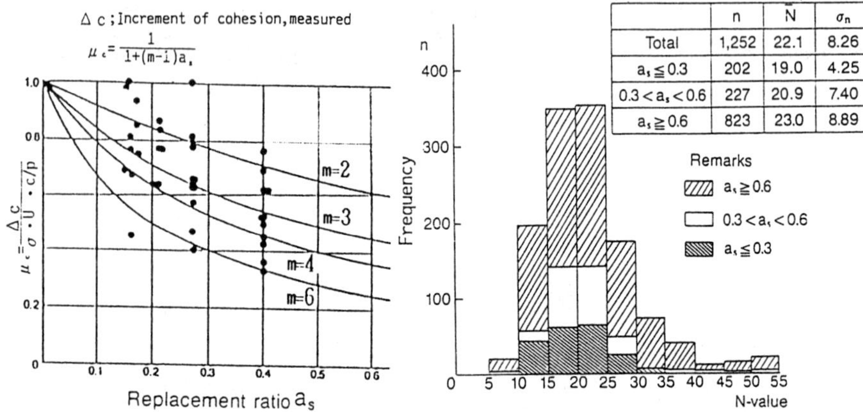

FIG. 15 -- Effect of stress
concentration ratio μ_c
on the increase of Cohesion [1]

FIG. 16 -- N-value at pile core [1]

Design Procedure for Settlement :Sand piles in composite ground also possess the ability accelerating the consolidation of clayey ground, in a similar manner to that of the sand drain method. However, the composite ground is characterized by its effect on the reduction of settlement due to consolidation. The settlement S_o of clayey subsoil can be calculated by the following equation.

$$S_o = m_v \cdot \sigma \cdot H \tag{15}$$

Where, m_v is the modulus of volume compressibility and H is the thickness of the layer. The settlement S of composite ground is estimated by the following equation, taking into consideration the effect of stress reduction expressed by Eq 10.

$$S = m_v \cdot (\mu_c \cdot \sigma) \cdot H \tag{16}$$

Comparing Eq 15 with Eq 16, settlement reduction ratio β_c equals μ_c.

$$\beta_c = S/S_o = \mu_c = 1/\{1+(m-1) \cdot a_s\} \tag{17}$$

From the viewpoint of consolidation theory and stress concentration, actual consolidation rate is roughly as fast as or even faster than the estimated value by Barron's Theory in the case of composite ground. However, taking various factors into account, it is obvious that the consolidation rate can be estimated, for practical purposes, using Barron's Theory.
The effect on settlement is evaluated by measuring the settlement of entire improved ground. Fig. 17 shows the relationship between estimated and actually measured values. Fig. 18 shows the relationship between settlement reduction coefficient "β_c", stress distribution ratio "m", and replacement ratio "a_s" and also shows the actually measured values.

Study on the Coefficient of Lateral Subgrade Reaction of Composite Ground [1] :The estimation of k_h (coefficient of lateral subgrade reaction) in the improved ground by SCP is studing as follows.
Relative K-value "k_{hs}", corresponding to 1 cm displacement of the pile with 1 cm diameter, based on the results of LLT conducted in sand pile is shown in Fig. 19.

The relation between unconfined compression strength and relative K-value "k_{hc}", is given by Eq 18.

$$k_{hc} = 5.116 \cdot q_u^{0.934} \quad [14] \tag{18}$$

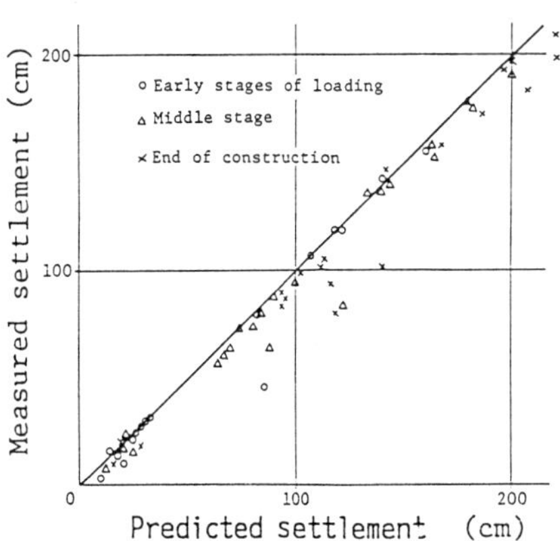

FIG. 17 -- Example of relation between measured and predicted settlement of composite ground [1]

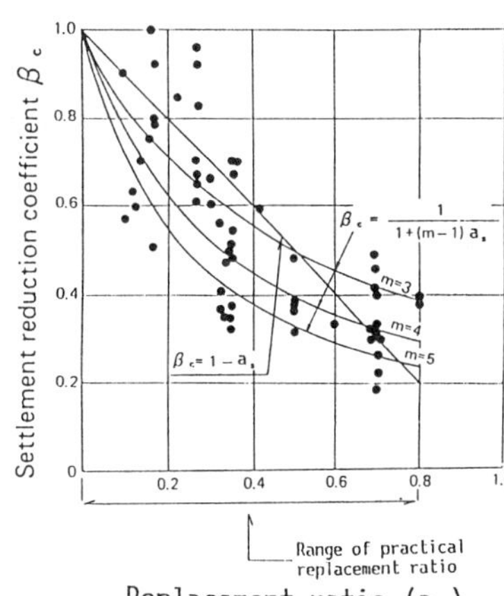

FIG. 18 -- Relationship between settlement reduction coefficient and replacement ratio [1]

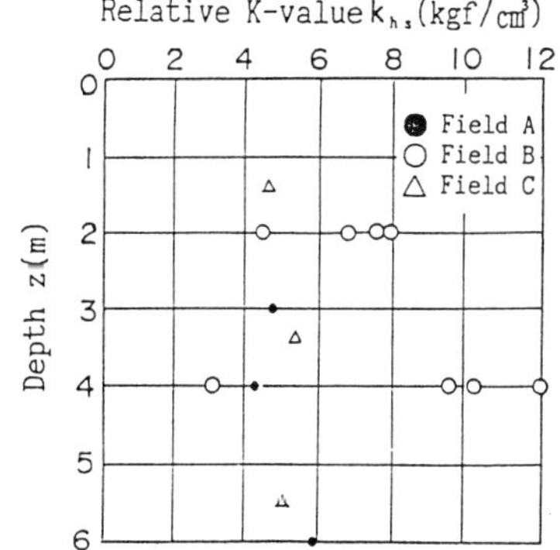

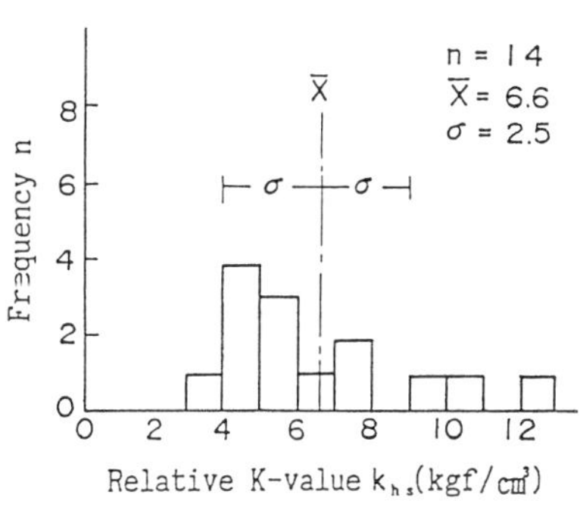

FIG. 19 -- Measured K-value in SCP by Lateral Load Test [13]

Thus, following calculation method for k_h is proposed in Eq 19.

$$k_h = k_{hs} \cdot a_s + k_{hc} \cdot (1 - a_s) q_u$$
$$= 6.6 \cdot a_s + 5.1 \cdot (1 - a_s) q_u \qquad (19)$$

where

k_h : K-value (kgf/cm^3) of the ground improved by SCP method (relative K-value)
a_s : replacement ratio
q_u : unconfined compression strength of inter-pile clay (kgf/cm^2)

Fig. 20 shows relative K-value estimatated by Eq 19 and the same obtained through field lateral loading test on a test pile, which hold good agreement with each other.

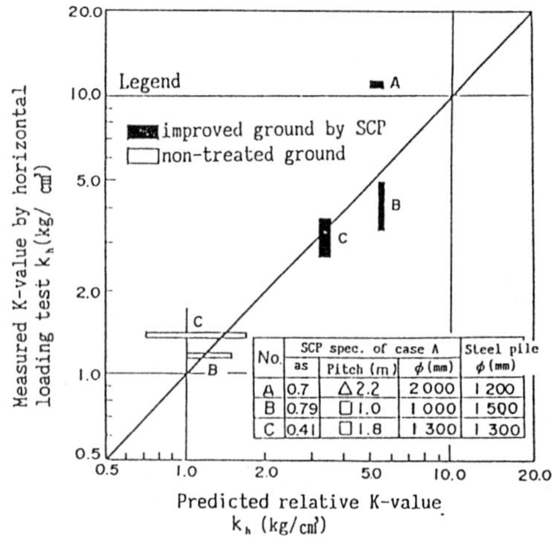

No.	SCP spec. of case A			Steel pile
	as	Pitch (m)	φ (mm)	φ (mm)
A	0.7	△ 2.2	2 000	1 200
B	0.79	□ 1.0	1 000	1 500
C	0.41	□ 1.8	1 300	1 300

FIG. 20 -- Relationship between measured
and Predicted K-value [13]

New Application to Clayey Ground

As far as land work is concerned, a number of testings inclusive of sliding failure tests carried out so far have led to the establishment of the practical design procedure for composite ground as stated earlier. As for marine work, relatively high replacement ratio has been often employed. However, low replacement ratio being not higher than 30% is occasionally superior in terms of economy, that depends upon the type and construction period of super-structure to be built there on (see Fig. 21). As such the on-site tests for offshore SCP method of low replacement ratio had been conducted for three years from February 1986 to March 1989[15],[16]. Objectives of the mentioned tests are as follows;

① to comprehend the consolidation process of composite ground.
② and to comprehend the shear characteristics of composite ground under ultimate loading condition.

The testing body and soil properties of original ground are illustrated in Fig. 22 and 23, respectively. Two-step loading was adopted with a view to assessing the consolidation behaviour at the first step and shear characteristics at the second. Inter-pile soil disturbance caused by SCP driving, its recovery process and strength gained by consolidation were confirmed through the results of soil investigation carried out prior to second step loading.

Furthermore, the soil-behaviour due to loading was carefully observed during consolidation process by first step loading, and this observation was continued until the stage of sliding failure by second step loading. The soil movements were measured and recorded by inclinometers, differential settlement gauges, and displacement pegs soil stresses by piezo-meters and earth pressure-cells as is shown in Fig. 24.
Through the above results, the validity of the design procedure for composite ground stated is now being confirmed.

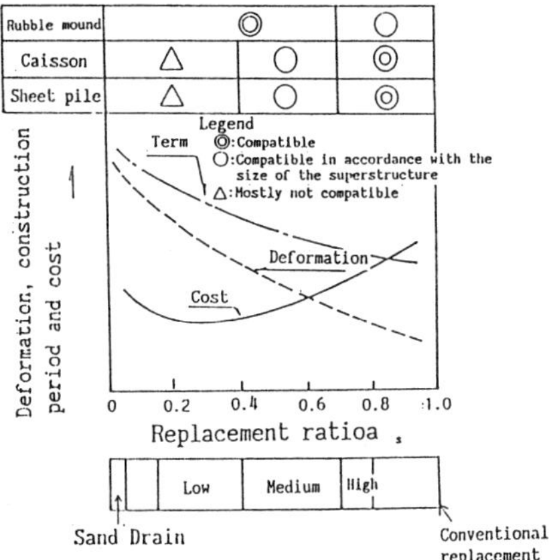

FIG. 21 -- Standing of low replacement ratio SCP method [15]

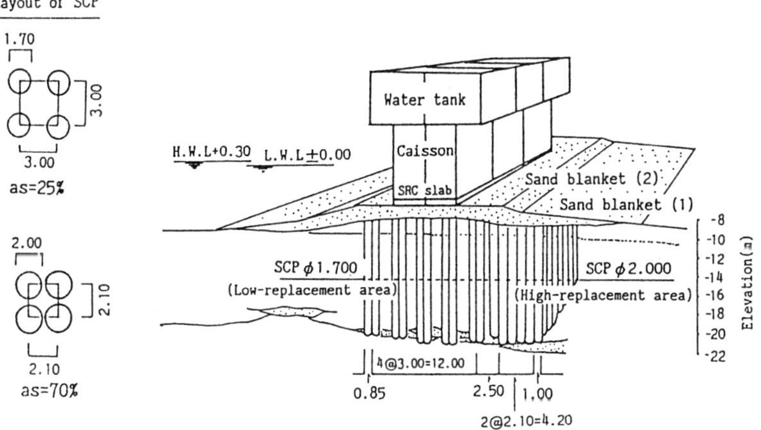

FIG. 22 -- Illustration of the loading test [15].

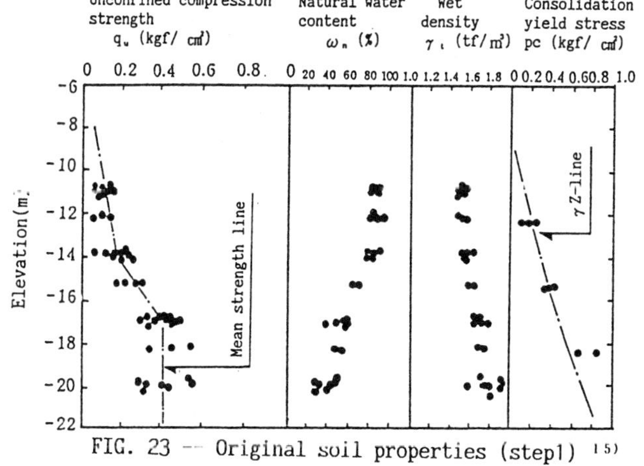

FIG. 23 -- Original soil properties (step1) [15]

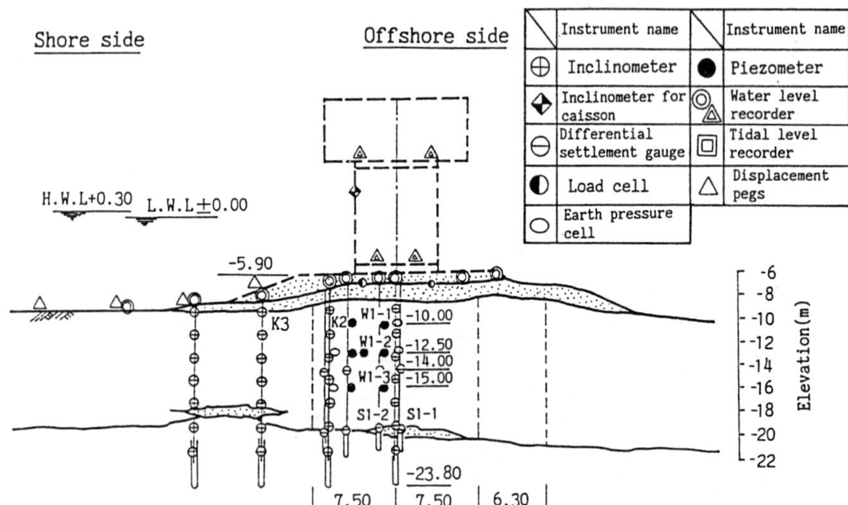

FIG. 24 -- Arrangement of instruments
(cross sectional view) [15]

FUTURE TRENDS OF SCP-METHOD

Mechatronic Consolidation System

Mechatronic Consolidation System has been developed incorporating the mechatoronics technology, such as auto-control technique, with the SCP method in order to ensure efficient construction and eliminate the uncertaintity caused by the variation of soil properties, thus resulting in increased design reliability as to the foundation of superstructures and also reduction of the construction cost.[17]

In this system the sand pile can be set up changing pile diameters and its strength along the depth to secure homogenious strength, as shown in Fig. 25.

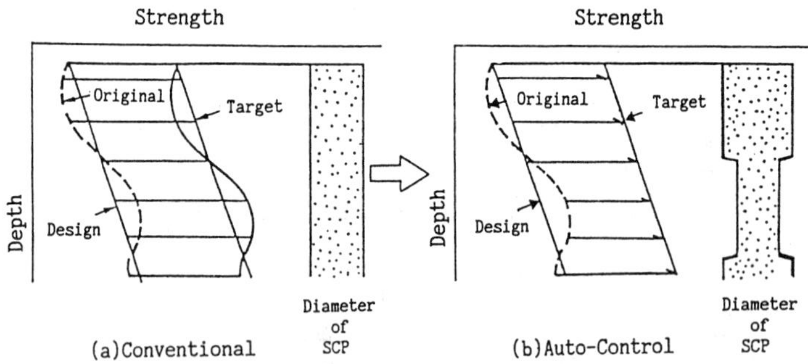

FIG. 25 -- Concept of reliability improvement

In applying the auto-control SCP method as a countermesure against liquefaction of sandy ground, the followings can be concluded from the results of standard penetration test.[18]

① The auto-control SCP method provides less variation in soil strength (N-value) distribution than the conventional SCP method dose, as shown in Fig. 26.

② The probability of N ≦ Ncr (the critical N-value) "P_{NG}", obviously drecreases.
③ This auto-control system reduces costs compared to the conventional system, yet achieves the same reliability of performance.

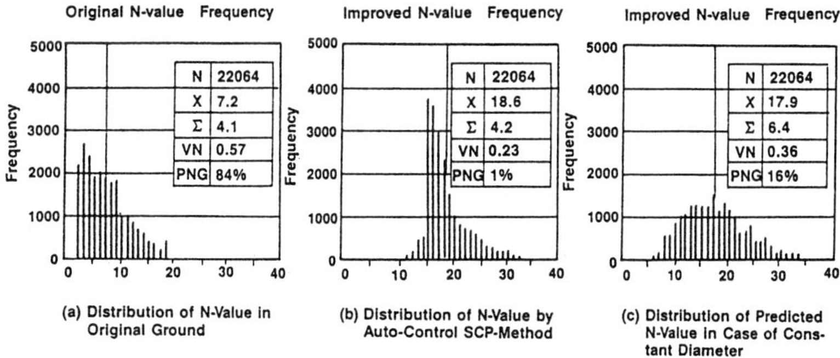

FIG. 26 -- Distribution of N-value
after/before improvement [18]

Robotization

Robotization of SCP is being developed with a view to reducing labor, and improving the work condition while includes the safety features.
The new SCP construction machine on the ground, which can be operated by wireless system, was already produced, and it achieved success in the trial construction.

CONCLUSION

In this paper, the followings as to the state-of the art SCP method in Japan are reported.
① The history of developement of SCP in Japan is summerized.
② It is pointed out that conventional design procedure of SCP has a limit when applied to the sandy ground which contains fine particle content. Therefore, new design method of SCP, cosidering fine particles content, is propoocd. It is also shown that SCP has been most popularly used as a countermeasure against liquefaction of sandy ground in Japan.
③ In addition, the design procedure of SCP for clayey ground is introduced and justified as viewed from construction experiences. It is also reported that, in the case of offshore construction, low replacement ratio method might be superior from an economic view point, depending upon the type of structures to be built thereon. As such, sliding falure test on on-site proto-type structure was carried out to investigate the consolidation process and the bearing capacity under ultimate loading condition.
④ Lastly, Mechatronic Consolidation System, which is the latest operation control system of SCP, and its robotization are introduced.

REFERENCES

1) Aboshi, H. and Suematsu, N.:"The State of the Art on Sand Compaction Pile Method", 3rd International Geotechnical Seminer, Soil Improvment Methods, 1985.
2) Murayama, S:"Soil Improvement by Sand Compaction Pile (Compozer Method)", Seminar Report of Osaka Constructers Association, pp.1-17, 1957 (in Japanese).
3) Murayama, S:"Soil Compaction Method and Equipment", Construction Equipment, Kansai Branch of Japan Society of Civil Engineering, pp.24-33, 1958 (in Japanese).
4) Tanimoto, K:"Sand Compaction Pile Method and Vibrating Pile Driving Method", New Method of Soil and Foundation, the Japanese Society of soil Mechanics and Foundation Engineering, 1960.

5)   Murayama, S:"Vibro-Compozer Method for Clayey Ground", _Mechanization of Construction Work_, No.150, pp.10-15, 1962 (in Japanese).

6)   Ogawa, M. and Ishidou, T.:"Application of Vibro-Comporzer Methods on Sandy Ground", _Tuchi-to-Kiso_, vol.13, No.2, pp.77-82, 1965 (in Japanese).

7)   Mizuno, Y., Suematu, N. and Okuyama, K.:"Design Method of Sand Compaction Pile for Sandy Soils Containing Fines", _Tuchi-to-Kiso_, vol.35, no.5, pp.21-26, 1987 (in Japanese).

8)   Meyerhof, G.G.:"Discussion of Session 1", Proc. of 4th International Conferencce on Soil Mechanics and Foundation Engineering, vol.3, 1957.

9)   Hirama, T.:"A Few Knowledge of Application of Relative Density", Proceedings of Japan Symposium on Geotechnical Properties and Relative Density of Sand, 1981 (in Japanese).

10) Suematsu, N., Yoshimi, Y. and Sakaki, Y.:"Contermeasures for Reducing Damages due to Liquefaction", _Tuchi-to-Kiso_, vol.30, no.4, pp.71-79, 1982 (in Japanese).

11) Ishihara, K. Kawase, Y. and Nakajima, M:"Liquefaction Characteristics of Sand Deposits at an Oil Tank Site during the 1978 Miyagiken-Oki Earthquake", _Soils and Foundation_, vol.20, no.2, 1980.

12) Imayama, T., Tateishi, S. and Nakajima, H.:"On Soil Stabilization Work of Higashi Omiya By-pass in Oyamadai District", 13th Japan Road Conference, 1979 (in Japanese).

13) The Japanese Society of Soil Mechanics and Foundation Engineering:"The Countermeasures against Soft Subsoil", pp.132-135, 1988 (in Japanese).

14) Nakajima, H., Ito, T., Takeuchi, T., and Imai, T.:"Mechanicals Propertie of Ground in around Osaka Area, based on the Results of LLT Measuement", _Geotechnical Magazine_, vol.78, No.4, 1972(in Japanese).

15) Takahashi, T., and Shiomi, M.:"On-Site Proto-Type Test for Low Replacement SCP Method in Maizuru port", _Civil Engineering echnique_, vol.43, No.10, pp.81-89, 1988 (in Japanese).

16) Okada, T., Yagyu, T and Yukita, Y.:"Field Rupture Test of Soil Improved by Sand Compaction Pile Method with Low Sand-Displacement Ratio", _Tsuchi-to-Kiso_, vol.37. No.8, pp.57-62, 1989 (in Japanese).

17) Kawakami, T., Katsuhara, M. and Isoda, T.:"New Automated Operation and Feed-back Control in Soil Improvement Method", The 5th internatiol symposium on robotics in construction, pp.801-810, Tokyo, 1988.

18) Kanatani, Y., Shono, H., Tsuboi, H. and Matsuo, M:"Soil Improvement Work as a Countermeasure against Liquefaction by Sand Compaction Pile using the Auto-control Method", Proc, of ICASP5, pp.870-877 Vancouver, 1987.

Masaaki Terashi, Masaki Kitazume, and Shogo Minagawa

BEARING CAPACITY OF IMPROVED GROUND BY SAND COMPACTION PILES

REFERENCE: Terashi, M., Kitazume, M., and Minagawa, S., "Bearing Capacity of Improved Ground by Sand Compaction Piles," _Deep Foundation Improvements: Design, Construction, and Testing_, _ASTM STP 1089_, Melvin I. Esrig and Robert C. Bachus, Eds., American Society for Testing and Materials, Philadelphia, 1991.

ABSTRACT: Bearing capacity of the ground improved by sand compaction pile method is investigated primarily by a series of centrifuge model tests. Bearing capacities obtained under the various loading conditions are well explained by the simple stability analysis which has been commonly employed in Japan. The effectiveness of a simple construction control diagram is confirmed based on the deformation of the ground. These findings are also verified by the full scale load test leading to failure of the improved ground.

KEY WORDS: soft clay, sand compaction pile method, bearing capacity, inclined load, centrifuge model test, full scale test, stability analysis

INTRODUCTION

Sand compaction pile method (SCP) has been applied to improve the soft ground often found in the Japanese coastal area. When the method is applied to improve clay deposits, the improved ground is often called a composite ground. The behavior of the composite ground under an external load is thought to depend upon many factors; (1) the shear strength of compacted sand piles, (2) the replacement area ratio, a_s, (3) geometric conditions such as the ratio of the width of improved area, W to the width of foundation, B, (4) ratio of the length of sand piles to the depth of the soft layer, (5) shear strength profile of the original soft clay, (6) external load condition (eccentricity and inclination), (7) loading rate, and other minor details.

Dr. Terashi is the chief of Soil Stabilization Laboratory and Mr. Kitazume is a senior research engineer at Port and Harbour Research Institute, 3-1-1 Nagase, Yokosuka, Japan; Mr. Minagawa is an engineer at Fudo Const. Co. Ltd., 4-10-40 Ohsu, Nakaku, Nagoya, Japan

The authors investigated the bearing capacity of the submarine composite ground with low a_s improvement under a combination of vertical and horizontal loads. This corresponds to a situation of breakwaters or revetments. The investigation is carried out by centrifuge model tests, finite element analysis and simple stability analysis. By the comparison of the bearing capacities with the calculations, the validity of the practical design method is confirmed. Usefulness of a failure control diagram is confirmed based on the deformation of the composite ground. These are also confirmed by a full scale test carried out separately.

In the present article, only the brief outline of the investigation is described due to the page limitation. The details of the investigation will appear in a separate publication [1].

CENTRIFUGE MODEL TESTS

General Descriptions

When the behavior of concern is not precisely known and is anticipated to depend on many factors as stated above, the rigorous idealization or simplification of the problem is inevitable and it is necessary to carry out the investigation by the parametric study for major influential factors. The author's approach to this problem primarily depends upon a series of scaled model tests by means of geotechnical centrifuge modeling technique which satisfies the similarity and is best suited for the parametric study.

In a preliminary study, the influence on the bearing capacity of the density of sand piles, ratio of W to B, and loading speed has been investigated under the strain controlled vertical loading test. Based on the preliminary study, conditions of model ground and test procedures are determined. Major characteristics of the model tests described in the present paper are as follows;

a) a two dimensional rigid foundation rests on the sand mound which is spread on the composite ground,
b) the replacement area ratio, a_s is 0.28 which would be a typical a_s value for the low a_s improvement expected in the marine works,
c) the improved area is symmetrical with respect to the center line of the foundation,
d) the sand piles penetrate through soft clay layer and reach the reliable sandy layer.

These conditions a) to d) may conform to the common practice of breakwater construction in Japan.

To carry out scaled model tests, PHRI geotechnical centrifuge is used. The radius of the centrifuge is 3.8 m measured to the surface of the swing platform. The maximum payload is 2.7 tons and the maximum acceleration is 115 g. Thus the capacity is 300 g-tons. The details of the PHRI geotechnical centrifuge and its ancillary equipments are reported elsewhere by Terashi [2].

Preparation of the Model Ground
____

The material used for the sand pile and sand mound is Toyoura
standard sand and the clay used in the tests is Kaolin clay. Both
materials are selected because their characteristics are well known
and also they are commercially available. The characteristics of these
materials are listed in Table 1.

TABLE 1 -- Engineering Properties of Model Materials

	Gs	Wl	Ip	Cu/p	Cv (cm^2/min)	C_c	C_s
Kaolin clay	2.692	59.0	42.2	0.314	0.15	0.49	0.12

	Gs	D_{50}	Uc	e_{max}	e_{min}	ϕ	
						γ_d	γ_d
						1.537	1.609
Toyoura sand	2.66	0.24	1.5	0.979	0.623	36.0	38.2

A thick normally consolidated clay layer is reduced in scale and
prepared in a strong specimen box which has the following inside
dimensions: 30 cm deep, 10 cm wide and 50 cm long. One side of the
strong box is made of glass to allow for photographic measurement.
All the model tests are carried out in the plane strain condition.

The Kaolin clay is thoroughly remoulded at a water content of 120
% which is sufficiently higher than its liquid limit. A drainage layer
of Toyoura sand with 50 mm thickness is placed at the bottom of the
strong box. Then the slurry of Kaolin clay is poured into the box.
The preliminary consolidation is conducted under a vertical pressure
of 10 kN/m^2 on the laboratory floor.
After the preliminary consolidation,
the glass plate of the strong box is
disassembled to place surface markers
on the side surface of clay which are
used later in the photographic measurement. The strong box is re-
assembled and brought onto a swing
platform for self weight consolidation
under 50 g in order to prepare the
normally consolidated clay ground.

The total thickness of the clay
layer is approximately 20 cm for all
the model tests. Due to the preconsolidation and the self weight consolidation, the completed model ground
has a thin layer of over-consolidated
clay underlain by the thick normally
consolidated clay. Figure 1 shows the
shear strength profile of the model
ground thus prepared.

After the self weight consolidation, the centrifuge is once stopped
for the preparation of improved ground

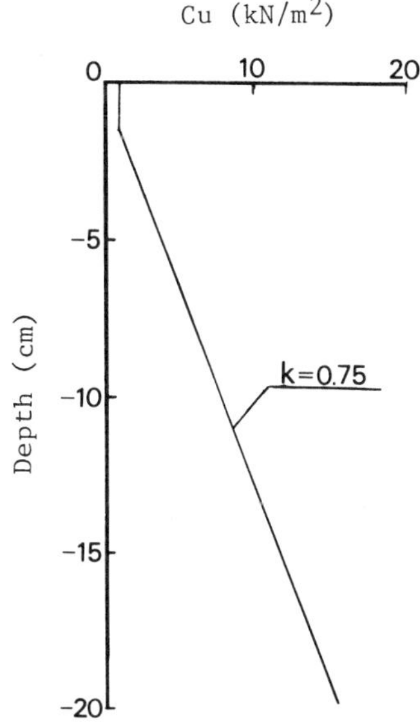

FIG. 1 -- Undrained   Shear
        Strength   Profile

on 1 g laboratory floor. Model compacted sand piles are manufactured following the procedure devised by Kimura et al.[3] and installed into the model ground. Saturated Toyoura sand is poured into water filled tubes whose inner diameter is 20 mm. The sand and tubes are subjected to vibration until the specified density is attained. The sand piles thus prepared are then slowly frozen and both ends are trimmed. The thin-walled tubes with a 20 mm outer diameter are inserted into the clay ground at a regular rectangular pattern with a distance of 33 mm which corresponds to a low a_s of 0.28. Then the clay inside the tubes are removed by a tiny auger to make holes. Finally frozen sand piles are inserted into the holes and left for gradual thawing.

After the soil improvement, the strong box is mounted again onto a swing platform of the centrifuge for the loading test.

All the model tests are carried out in the 50 g field. Therefore the prototype thus simulated in the strong box is an approximately 10 m thick alluvial clay deposit which is improved by large compacted sand piles with a 1 m diameter.

Test Procedure

The setup of the model is shown in Fig. 2. Model foundation is 10 cm by 10 cm and Toyoura sand is glued on the bottom of the foundation to simulate the rough base condition. The position of the model foundation is adjusted to the center line of the symmetric sand mound

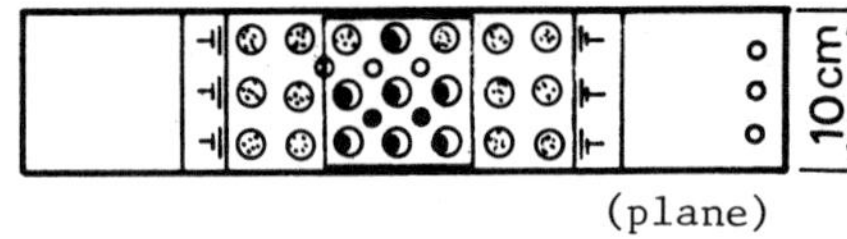

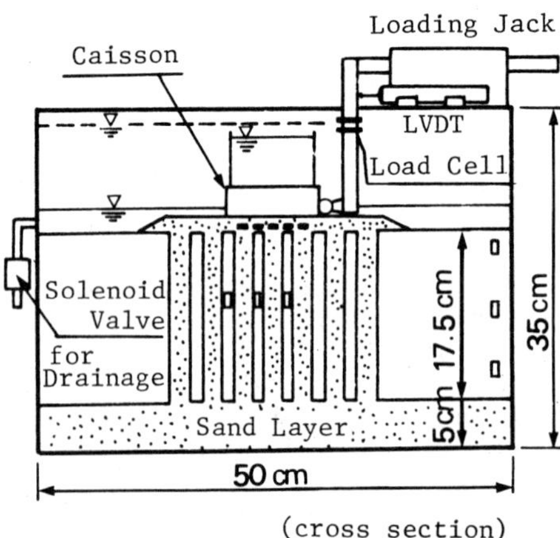

FIG. 2 -- Setup of Model for Inclined Load Test

and improved area. Thus the foundation rests on the sand mound over the central 9 compacted sand piles in 3 rows as shown in the figure. In addition to this central piles, 12 compacted sand piles in 4 rows are installed outside the foundation area. The ratio of W/B is therefore 2.3.

In order to reflect the prototype loading condition of the break-water or revetment to the model test, it is most appropriate to apply the vertical load component in advance to applying the horizontal load component. The vertical load component is given in stages; first stage drained loading and final undrained loading. The mound and part of the foundation load is given to the ground as a first stage load and left until the soil come to a new equilibrium in 50 g before the final undrained loading. The vertical component of the final load is applied by quick lowering of the water level. The horizontal component is applied immediately after that by means of the horizontal loading jack. The horizontal load is applied to a level very close to the foundation base so as to reduce the influence of load eccentricity.

During the loading test, earth pressure increments due to loading are measured at the surface of the improved ground just beneath the foundation. Transducers placed on the central sand piles are aimed to measure the pressure increment p_s at the top of the sand piles. Similarly, transducers are placed on the clay surface between piles to measure the pressure increment p_c on clay. Excess pore water pressures are also measured in the clay at the positions denoted by open squares in the figure. Photographs are taken intermittently during the loading. From the series of photographs, the displacement vector loci are determined based on the coordinates of markers. After the loading test, the strong box is disassembled and the deformation of the sand piles are directly observed.

Test conditions and results of a series of tests are summarized in Table 2. As explained above, the first stage load listed in the table is the load under which the improved ground is consolidated in 50 g before the second stage undrained loading. The vertical load component, V of the yield load is a total applied vertical load component including the first and second stage loads. Horizontal load component, H is the load where the improved ground comes to yield. Load inclination is the inclination of the resultant load at yield which is measured from vertical. Therefore test No. 1 is the vertical bearing capacity test with zero inclination and No. 3 to No. 5 are the tests for the inclined loading. Based on the vertical loading test results, the magnitude of the vertical load components in the inclined load test are decided so that the total vertical load

TABLE 2 -- Test Conditions and Major Test Results

test No.	1st Load (kN)	Total Load at Yield		Load Inclination (deg)
		V.Load (kN)	H.Load (kN)	
1	0.10	0.60	0	0
3	0.10	0.10	0.057	29.7
4	0.10	0.29	0.088	16.5
5	0.10	0.45	0.070	8.8

component at the yield may become one sixth, a half and three quarters of the vertical bearing capacity.

Test Results

Vertical load - settlement curve of test No. 1 is shown in Fig. 3. The load increases with increasing settlement and neither peak nor the final constant load is observed. The bearing capacity of the improved ground is determined as a yield of the ground. The yield load of the ground is defined by the intersection of the initial tangent line of the curve and the tangent line at the straight portion of the curve at the larger settlement. The arrow in the figure shows the position of the bearing capacity thus determined.

For the inclined load test, horizontal load H - horizontal displacement d_h curves are obtained and shown in Fig. 4. In test No. 3 for the largest load inclination, H increases with increasing d_h but the load becomes constant after d_h reaches a certain value. The bearing capacity for this particular case is therefore determined as this final constant value. Whereas the H - d_h curves for the smaller load inclination do not show either peak or the final constant value. The bearing capacity for these cases is determined by the intersection of two tangent lines. As shown in the figure, it is known that the horizontal component of the bearing capacity is highly dependent upon the load inclination.

In test No. 3, the foundation moves almost horizontally with a very little vertical settlement which is considered to be a sliding failure of the foundation on the surface of the

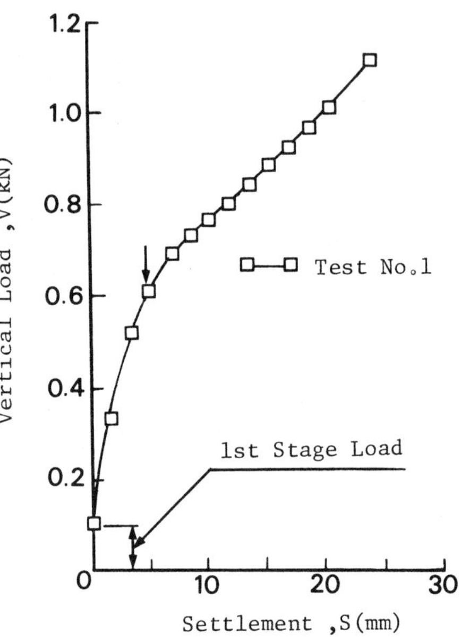

Fig. 3 -- V-S Curves for
Vertical Loading

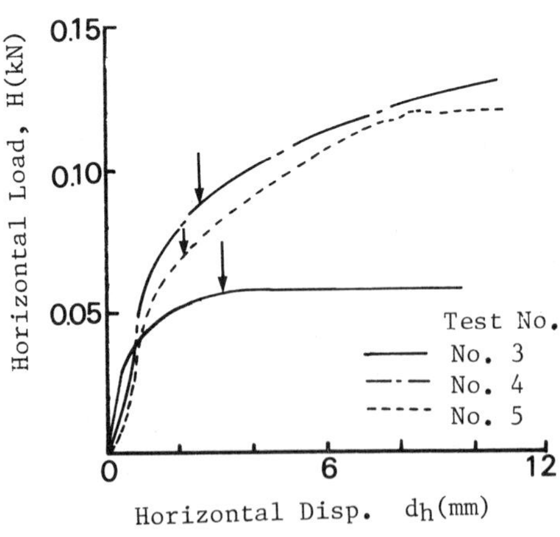

FIG. 4 -- H-d_h Curves for
Inclined Load Test

sand mound. In the case of sliding of the foundation, the improved ground is not involved in the deformation. At the smaller load inclinations (or at the larger V levels), the foundation movement under a horizontal load accompanies a relatively large vertical settlement and suggests that the improved ground is involved in the deformation. This movement of the foundation is characterized by the larger settlement of the toe compared to the heel. The term toe is used in the present article to denote the farther edge of the foundation from the horizontal loading point and the opposite edge of the foundation is called the heel.

In the vertical loading, the earth pressures p_s at the top of sand piles and settlement relation also showed a yielding at the settlement level where the composite ground yields as a whole. In the inclined loading, the stress concentration to the sand piles at the toe is significant. And p_s at the toe and horizontal displacement relation shows the yield at the time when the ground yields as a whole. These tendency implies that the yield of the improved ground is triggered by the yield of the compacted sand piles.

In all the cases, horizontal loading is continued to the larger displacement even after the yield of the ground. From the series of the photographs taken during the tests, a displacement vector is drawn for each marker and shown in Figs. 5 and 6 respectively for tests No. 1 and 5. In these figures, broken lines show the original position of sand mound, original and final position of foundation and the circles below are the positions of sand piles.

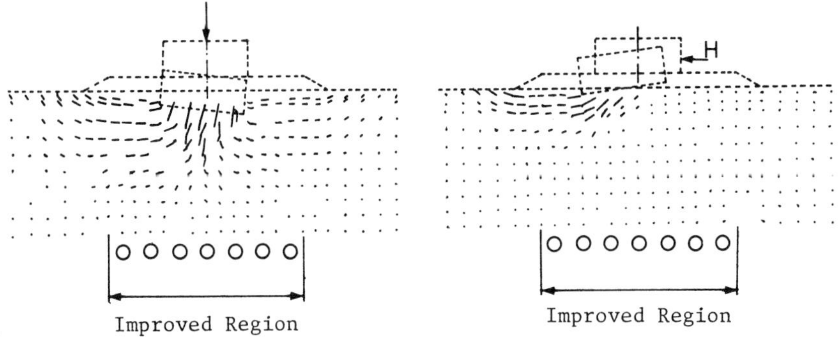

FIG. 5 -- Vector Loci (No. 1)   FIG. 6 -- Vector Loci (No. 5)

Figure 5 shows the vector loci of the loading test with no load inclination. It is known by the vertical vector loci underneath the foundation that the sliding wedge symmetrical with respect to the center line of the foundation penetrates into the ground. The soil outside the wedge do not show any vertical movement. All the vectors in the surrounding soil within the improved region are horizontal. Figure 6 shows the vector loci for the inclined loading test. It is known that the sand piles on the right hand side including the sand piles at the heel show almost no deformation. The direction of the displacement vectors at the center and toe of the foundation are inclined and suggest the formation of an asymmetric wedge with increasing horizontal load. The sand piles outside the foundation area

on the left hand side displace almost horizontally from the beginning
of loading.

Figure 7 is the photograph taken after the load test (Test No. 1)
to observe the deformation of sand piles directly. The deformation of
the sand piles agree well with the behavior found by the vector loci
in Fig. 5. Shear planes are clearly observed at the boundary of the
sliding wedge suggested by the vector loci. The piles outside the
wedges are deformed at their tops by the penetration of wedges but the
overall improved ground does not reach the general shear failure.

FIG. 7 -- Failure Mode of Compacted Sand Piles (No. 1)

The vertical load component, V and the horizontal load component,
H at yield or failure of the improved ground for all the tests are
plotted on Fig. 8 to obtain the bearing capacity envelope on the V - H
plane. The horizontal load which can be supported by the improved
ground increases with the increasing vertical load. However, when V
reaches around a half of the vertical bearing capacity, H becomes

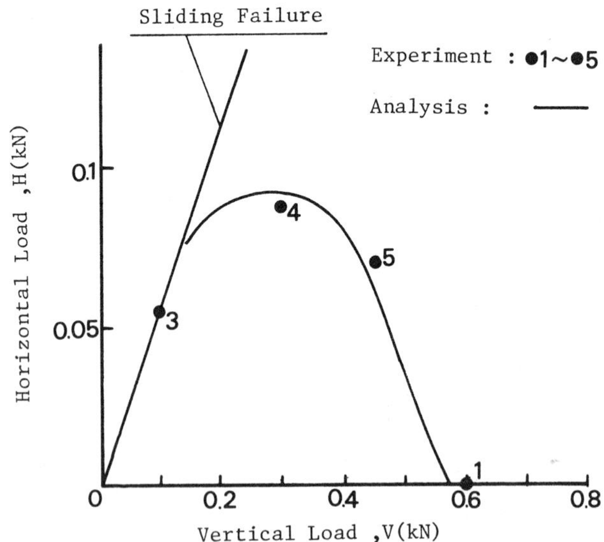

FIG. 8 -- Bearing Capacity Envelope in V-H Plane

maximum and decreases with further increase of V level. The data point on the V axis is the vertical bearing capacity. The yield load of the improved ground forms an envelope of a shape similar to a cigar. The failure envelope similar to this has been also found for the bearing capacity of sand under the inclined load previously [4]. The solid straight line and the curves shown together with test data in Fig. 8 are the estimated bearing capacity based on the simple stability analyses which will be discussed later.

Comparison with Practical Design Technique

Several failure criteria for the composite ground have been proposed and employed in the routine design for many years in Japan. Among others, following equation has been most frequently applied these days especially to low a_s improvement on land.

$$\tau = (1 - a_s) (c_o + kz + \mu c \, \Delta\sigma_z \, c_u/p \, U) + (\gamma_s z + \mu_s \, \Delta\sigma_z) \, a_s \, \tan \phi_s \, \cos^2\theta \qquad (1)$$

where,
τ : average shear strength of composite ground
a_s : replacement area ratio
$c_o + kz$: undrained shear strength of clay
z : depth
$\Delta\sigma_z$: average of the induced vertical stress due to an external load
c_u/p : rate of strength increase
U : degree of consolidation
γ_s : effective unit weight of sand pile
ϕ_s : internal friction angle of sand pile
θ : inclination of slip surface measured from horizontal plane
μ_s : coefficient of stress concentration
$\quad \mu_s = n / (1+(n-1)a_s)$
μ_c : coefficient of stress reduction
$\quad \mu_c = 1 / (1+(n-1)a_s)$
n : stress concentration ratio, $n = p_s/p_c$

In the equation, the shear strength of the composite ground is simply considered as a weighted average of the shear strengths of clay and sand. In order to take the stress concentration to sand piles and stress reduction to the clay into account, the formula has introduced the factors μ_s and μ_c respectively. As is easily understood, the formula is obviously not perfect to explain the complicated behavior of the composite ground. However, the simplicity of the formula is of practical use, if the calculations based on the formula give the reasonable estimates of the failure or yield of the improved ground.

For each test condition, the design constants are obtained by element tests. The bearing capacity is calculated by the Fellenius method of slip circle analysis combined by the shear strength expressed by the equation and already shown by a solid curve in Fig. 8. The curve gives the cigar shaped bearing capacity envelope in the V-H plane and is acceptable both qualitatively and quantitatively. The curve is the calculation based on the equation with stress concentration ratio n = 3. n = 3 is within the range of the present centrifuge test results and, at the same time, a value corresponds to the value which has been applied in the case histories.

Also calculated and shown by a solid line in Fig. 8 is the simple sliding failure of the foundation on the surface of sand mound. For this failure mode, the maximum horizontal load is calculated as a product of the effective weight of the foundation and the factor of friction of the sand mound, tan ϕ. As is observed in the figure, the results obtained both by calculation and experiment are in perfect accordance for the sliding failure.

From these comparisons, the validity and practical use of the current design procedure is confirmed as long as it is used with the adequate design constants.

FULL SCALE TEST

A full scale loading test of the improved ground has been carried out at Maizuru Port, Kyoto Prefecture for three years from fiscal years 1986 to 1988 by the Third Bureau of Port Construction, Ministry of Transport [5]. The purpose of the full scale test is to establish the design method for low a_s improvement suited for marine construction works. Stability of a breakwater to be constructed on soft marine clay is the selected problem for the full scale test.

The ground at the test site is composed of alluvial deposits. The thick normally consolidated clay layer appears at the surface which is underlain by a stiff sandy layer. As shown in Fig. 9, the superstructure is composed of three concrete caissons underlain by a

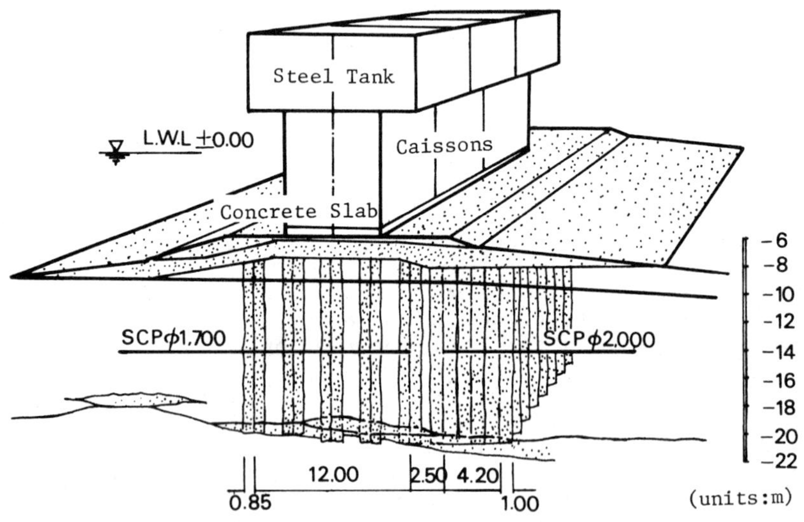

FIG. 9 -- Illustration of the Full Scale Load Test

concrete slab which is placed on the sand mound. The huge steel tank is placed on top of caissons to apply an additional load in the case where the ground does not fail by the load of caissons and fill material. As the breakwaters are typical two dimensional structures, the concrete slab is placed in order to restrict each caisson from independent three dimensional movement.

As shown in Fig. 9, the ground just underneath the superstructure and that in the left hand side in the figure is improved by a_s = 0.25. The right hand side of the foundation is heavily improved by a_s = 0.70. These layout of the improvement is planned to control the failure of the ground to left hand side which was necessary in order to reduce the cost for instrumentation.

The loading sequence in the full scale test is stage loading as in the case of centrifuge model tests. The sand mound, the concrete slab and the caissons partially filled with water are the first stage load. The second stage load is applied by the weight of slag and water supplied into caissons and the tank.

After the soil improvement work, the first stage load, 30 kN/m^2 altogether is placed on the improved ground and left for 10 months to allow consolidation. The second stage loading is started on July 17, 1988 with the filling of concrete caissons with slags by which the total load reached approximately 80 kN/m^2. Eight days are required for the placement of steel tank on top of the caissons. During this period the total load is kept almost constant by canceling the weight of the steel tank by dewatering the caisson. On July 26, the very final loading is carried out by supplying water into the concrete caisson and into the steel tank. Vertical load settlement curve of the undrained loading on July 26 is shown in Fig. 10.

Open circles in the figure show the measured load settlement relation. The time of the day is also shown together. At 6:45 water is filled up in the concrete caissons and the loading is once interrupted for changing pipelines from caissons to the tank. At 7:15 loading is re-started by supplying water into the steel tank. During this interruption, the load - settlement curve shows the continued settlement under the constant load. The solid circles are the corrected data for this interruption. From the corrected load - settlement curve, the yield of the ground is determined as 106 kN/m^2 which is around 7:45.

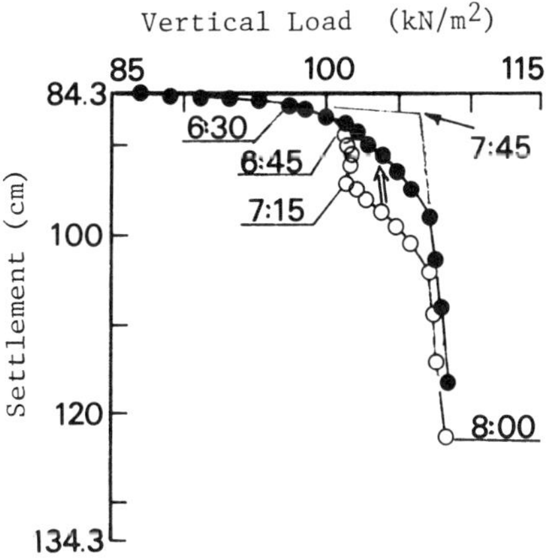

FIG. 10 -- Load Settlement Curve
(Full Scale Test)

Figure 11 shows the displacement of the superstructure and deformation of the ground together with the time of the day. From the

displacement of the superstructure, it is known that the superstructure begins to settle vertically with the increase of the weight
but does not show appreciable horizontal movement nor rotation up to
the yield of the ground at 7:45. After the yield of the ground, the
superstructure started to tilt. The tilt of the superstructure results
in the rapid increase of load eccentricity because the structure is
quite heavy at its top. This is considered to be a major reason for
the final drastic failure of the ground as shown in the figure at
8:30. And this is the reason why the load - settlement curve of full
scale test has the ultimate load as shown in the Fig. 10 while the
corresponding V - S curve by centrifuge modeling has not.

The bearing capacity of the improved ground is compared with the
simple stability analysis as described earlier with n = 3. When the
field test result is analyzed as two dimensional problem, the factor
of safety obtained by the calculation is 0.92. Although the length to
width ratio of the superstructure is taken as 4.5, the field test is
still three dimensional. A cylindrical slip surface is utilized to
take the three dimensional effect into account. Then the factor of
safety comes to almost agreeable value of 0.98.

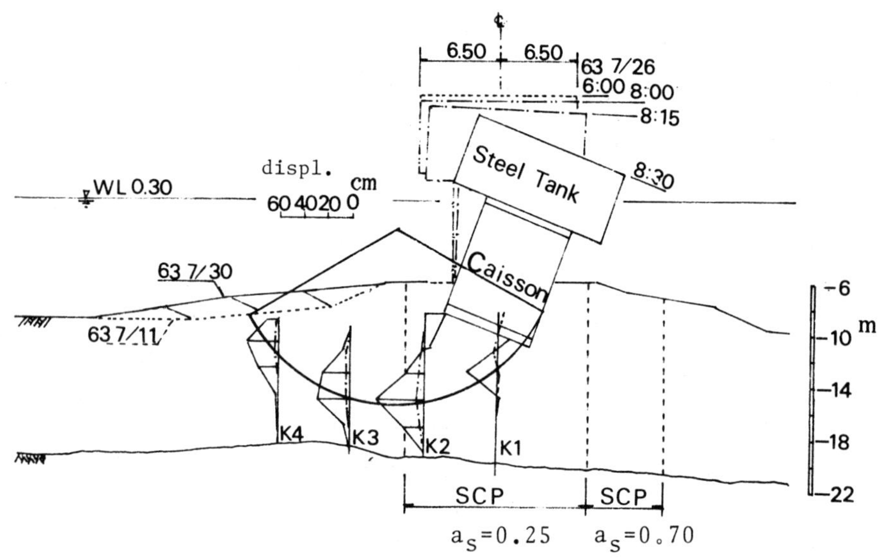

FIG. 11 -- Displacement of the Structure
and Deformation of the Ground

FAILURE PREDICTION DURING CONSTRUCTION

The failure of the improved ground is predicted in advance of the
construction, for example, by the simple stability analysis as
described earlier. However, the prediction always contains a certain

degree of error which comes from the inaccuracies in, for example, the soil's data, design method itself, and construction control. Therefore it is convenient to have a tool to predict and confirm the actual safety of the structure under construction. The effectiveness of one of such a tool proposed by Matsuo and Kawamura [6] has been investigated in the present study. The method is quite simple in which only a plot into a diagram of the displacement of the ground after each lift of the fill is required. Figs. 12 (a) and (b) are the diagrams for a model test and the full scale test respectively.

In the original diagram proposed, the vertical axis of the figure is a vertical settlement, S of the ground surface underneath the center line of the embankment and the horizontal axis is the horizontal displacement δ at the ground surface near the toe normalized by S. The meaning of the diagram is self-evident. If the plots with increasing fill height go straight up or go up toward the left in the diagram, the consolidation settlement is dominant over the shear deformation of the ground and the embankment is stable. Whereas, if these plots go up to right, the shear deformation or plastic lateral flow of the ground is increasing and the ground is approaching to large deformation or to failure.

The model test result for test No. 1 is plotted in Fig. 12 (a). The measurements of S and δ of the model ground are made on a series of photographs taken during the centrifuge flight. S is measured using the marker on the center line of the foundation at the clay surface. Measurements of δ are made on two different markers at the model ground surface. Therefore two series of plots are drawn on the diagram. As shown in the figure, the plots of solid circles are made based on δ measured at a point 1/2 B apart from the foundation edge where B is the foundation width. The plots of open circles are made based on δ measured at a point 1.0 B apart from the edge. In the diagram, first stage load is applied at the origin of the diagram. Due to the consolidation of the ground under the first stage load, plot moves to the point 0 where the second stage load is applied. The ground comes to yield at the point 3. The direction of the plots of solid circles is typical for the ground approaching to failure. In comparison to this, the plots of open circles seems to be less effective in predicting the failure.

Full scale test result is plotted in Fig. 12 (b). Here again two different plots are drawn on the diagram. For both plots, the same vertical settlement data is used. The plot, K2 uses the horizontal displacement data obtained at a point 3/10 B apart from the edge of the foundation. The other plot, K3 is based on the horizontal displacement data obtained at a point 1.0 B apart from the foundation edge. The K2 plot shows the tendency approaching to failure but the K3 plot is less effective.

The solid curves in the figure are the criteria for the safety of the embankment under construction. p_j/p_f attached to each line is a quantitative measure of the safety of the embankment. p_j is the magnitude of the fill pressure at j-th lift and the p_f is the fill pressure at failure. These factors are obtained by Matsuo and Kawamura [6] based on their numerical analyses together with the empirical data obtained from a number of embankments on land. Hence the failure of the improved ground is expected to occur near the $p_j/p_f = 1$ line. However the p_j/p_f of the point 3 of the centrifuge test result is around 0.75 and far less than unity. For the full scale test, the failure of the ground at 7:45, p_j/p_f is less than 0.9.

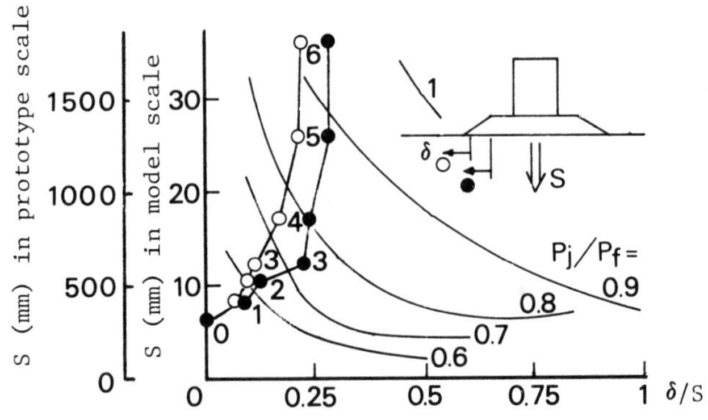

(a) Model Test in Centrifuge (No. 1)

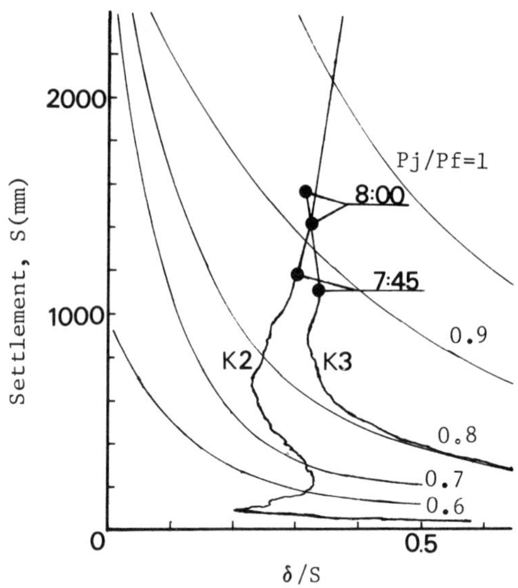

(b) Full Scale Test at Site

FIG. 12 -- Diagram for Prediction of Failure

From Fig. 12, it is known that the criteria established for the earth fill construction on land are inadequate for the quantitative prediction of the failure of the gravity type structure which rests on an embankment. However, it is confirmed that the construction control diagram is qualitatively useful for predicting the failure of the improved ground as long as the point for measuring horizontal displacement is selected not at the embankment toe but at a point closer to the foundation.

CONCLUDING REMARKS

The behavior of the composite ground with low a_s improvement is studied. The major findings by the current study are;

1) Influence of various factors on the bearing capacity is revealed.
2) The local failure governs the yielding of the composite ground in most cases of undrained loadings.
3) Bearing capacity of the composite ground with low a_s improvement can be estimated by a simple slip circle method of stability analysis combined with the practical formula for the shear strength of composite ground.
4) The prediction of the failure of the composite ground can be carried out by the simple construction control diagram at least qualitatively at the moment.

No description is given due to page limitation regarding the numerical analysis for the model tests which are carried out using the elasto-visco-plastic finite element analysis . The input data for the analysis are determined by the element tests on the model materials. The obtained load - settlement relations by FEM are in good accordance with centrifuge model test results. In order to utilize the construction control diagram, adequate criteria for the breakwater construction must be developed. This may be carried out with the aid of FEM analysis and centrifuge modeling in the near future.

REFERENCES

[1] Terashi, M. and Kitazume, M. (1990)"Bearing capacity of clay ground improved by sand compaction piles of low replacement area ratio," Report of Port and Harbour Research Institute, Vol. 29-2.
[2] Terashi, M. (1985)"Development of PHRI Geotechnical Centrifuge and Its Application," Report of the Port and Harbour Research Institute, Vol. 24-3, pp. 73-122.
[3] Kimura, T. et al. (1983)"Centrifuge Tests on Sand Compaction Piles," Proc. 7th Asian Regional Conference on Soil Mechanics and Foundation Engineering, Vol. 1, pp. 255-260.
[4] Terashi, M. and Kitazume, M. (1987)"Bearing capacity of a foundation on top of high mound subjected to eccentric and inclined load," Report of Port and Harbour Research Institute, Vol. 26-2, pp. 3 - 24
[5] Yagyu, T. and Yukita, Y. (1989)"Field Rupture Test of Soil Improved by Sand Compaction Piles with Low Sand-replacement Ratio," Proc. of The 24th JSSMFE Conference
[6] Matsuo, M. and Kawamura, K. (1977)"Diagram for Construction Control of Embankment on Soft Ground," _Soils and Foundations_, Vol. 17, No. 3, pp. 37 - 52.

Heinz J. Priebe

VIBRO REPLACEMENT - DESIGN CRITERIA AND QUALITY CONTROL

REFERENCE: Priebe, H. J., "Vibro Replacement - Design Criteria and Quality Control," <u>Deep Foundation Improvements: Design, Construction, and Testing</u>, <u>ASTM STP 1089</u>, Melvin I. Esrig and Robert C. Bachus, Eds., American Society for Testing and Materials, Philadelphia, 1991.

ABSTRACT: Stone Columns are applied to improve cohesive soils. Although the equipment and procedure are the same as for the vibro compaction of non-cohesive soils, the inherent principles with regard to the improvement are completely different. The fact that the effects of both techniques overlap each other must not lead to wrong design criteria and/or wrong interpretations of the results.
Stone Columns are not independent structural members like piles. Theoretical approaches for the design are facilitated if the improvement is related to the conditions of the surrounding soil. However, approximations are indispensable, especially with regard to the interpretation of load tests which are the only reliable method of in-situ testing.

KEYWORDS: Ground improvement, vibro compaction, stone columns, foundation design, load test

INTRODUCTION

The development of the deep vibratory compaction technique began in the early thirties and is inseparably linked with the Keller company of Germany. The history is well documented by Kirsch [1, 2] and Jebe and Bartels [3].

Dipl.-Ing. H. J. Priebe is Senior Soils Engineer with Keller Grundbau GmbH, Kaiserleistr. 44, D-6050 Offenbach, Federal Republic of Germany.

The original intention of the new technique of ground improvement was to densify compactible non-cohesive soils only, this being called "Vibro Compaction". From experience it was noted that with an increasing content of fines the compaction effects were reduced. However, the hole created by the vibrator did not immediately collapse. This led to the concept of installing into the hole created by the vibrator, load bearing columns consisting of well compacted material as a kind of reinforcement to the existing soil. The beginning of this development by the Keller company of Germany cannot be defined precisely but can be placed around the end of the fifties.   This variation is called "Vibro Replacement" or "Vibro Displacement" depending whether jetting water is used or not. The various aspects of the technique are described by Barksdale and Bachus [4].

CHARACTERISTICS OF THE TECHNIQUES

The equipment and the procedures of the deep vibratory compaction techniques are comprehensively described in literature, e.g. by Greenwood and Kirsch [5]. Although they are basically the same for vibro compaction and vibro replacement, the inherent principle in each technique with regard to the achievable improvement is completely different. This part has to be emphasized for a better understanding and for a more reliable evaluation of both techniques as the effects overlap in most cases of practical application.

The effect of vibro compaction is a direct improvement of the existing soil by densification. It depends firstly on the compactibility of the soil and secondly on the efficiency of the vibrator being used. Due to the complexity of the parameters involved, it is not possible to establish accurate design criteria in advance. Without preceding large scale compaction trials, a design depends mainly on the experience of the contractor who may be influenced by competitive considerations. The drawback of the technique in advancing an optimum design is compensated by the fact that in-situ quality control tests exist like soundings which are simple to perform and unequivocal in results. In particular, Static Cone Penetration Tests have proved to be economical and reliable.

Stone columns installed with the aid of vibrators exert a significant lateral pressure on the treated soil. Nevertheless, in saturated cohesive material the increase in density is only marginal. However, it has to be emphasized that, even if a measurable densification of the surrounding soil is indicated by in-situ tests like soundings, no conclusions can be drawn from this about the reinforcing effect of the stone columns. Only full scale loading tests, preferably on groups of stone columns, are reliable in-situ measures for quality control, i.e. to prove the effiencency of this soil improvement technique. Since they are time and cost consuming, it is significant that reliable criteria for a foundation design exist on the bases of theoretical approaches. Self-evident that such a design requires a good performance control.

Unlike piling, vibro replacement is an improvement technique, even if the properties of the treated soil remain more or less unchanged. Since the stone column behaviour depends directly on the lateral support of the surrounding soil, the improving effect can be related to the performance of the soil without columns. Thereby, a relative improvement is established and the parameters of the soil being difficult to determine, are more or less eliminated in theoretical approaches as demonstrated by Priebe [6]. Furthermore, errors in the soil investigation affect evaluations of the performance before and after treatment likewise. The remaining parameters like the stone column geometry and the properties of the backfill material can be obtained easily and with sufficient accuracy.

CRITERIA FOR THE APPLICATION

At the design stage of a project, it is of vital importance to decide which technique has to be adopted. The choice depends on the compactibility of the soil concerned which depends on many parameters. However, the main feature to estimate the compactibility is the grain size distribution of the soil and particularly the content of fines. Figure 1 shows a grain size diagram with a shaded zone. If the content of fines of a soil is less than some 5 % and its distribution curve entirely to the right of the shaded zone in the diagram, the soil will be generally well compactible. If the content of silt is approximately between 5 % and 15 % or if the distribution curve runs into the shaded zone elsewhere, it is advisable to use imported coarser backfill material to improve the contact between soil and vibrator. In the many remaining cases where the shaded zone is crossed by the distribution curve or where the soil is totally fine grained, the design has to be based entirely on the reinforcing effect of installed stone columns.

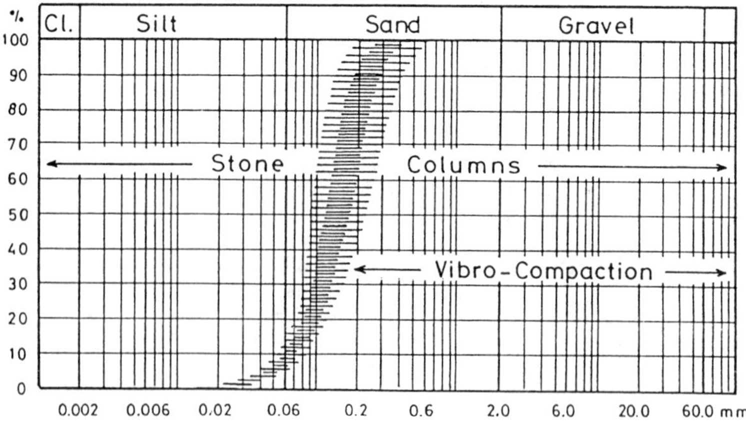

FIG. 1 -- Application Ranges of Deep Vibratory Compaction Technique

In applying vibro replacement, in many cases the surrounding soil is considerably compacted although the   reinforcing   effect of the stone columns is more essential than the densification of the existing soil. Particularly in these cases, it is often   difficult to convince engineers of the double effect. Too often, consideration is given only to the densification achieved but not to the additional support provided by the columns consisting of stiffer material.

Typical for above-mentioned one-sided consideration of vibro replacement are most evaluations where vibro replacement is applied to   reduce the   liquefaction potential of a   ground. Most practical methods of evaluating the liquefaction   potential of soils like the one proposed by Seed and Idriss [7] are based on the in-situ density measured by soundings or similar. In applying these methods, vibro replacement is required to densify the   existing soil to a level beyond the risk of liquefaction. The stabilizing effect of the   stone   columns is   not taken   into   consideration at all.   A proposal that   considers also   the effect   of the stone columns is given by Priebe [8]. There it is   recommended, to use the methods derived for untreated ground, but to   reduce the Cyclic Stress Ratio calculated for the expected   earthquake, by a factor equal to the ratio of the remaining pressure on the soil between the stone columns and the total overburden. This pressure ratio as a function of the area ratio which is the ratio of the grid size and the cross section of a stone column, is shown in Figure 2.

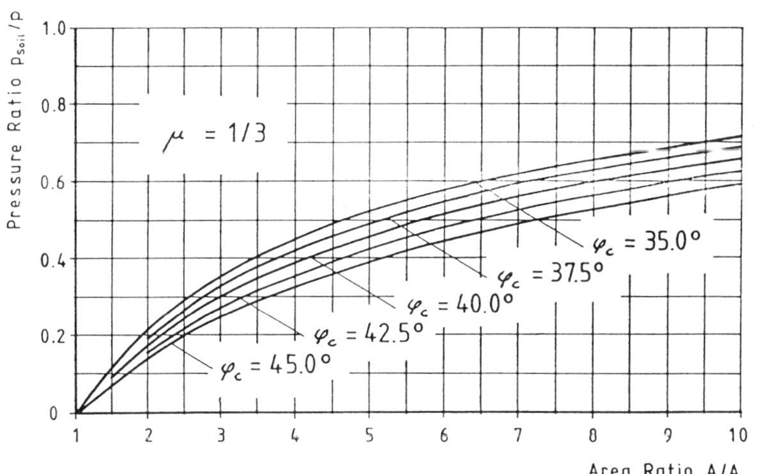

FIG. 2 -- Ratio of the Pressure on the Soil between Stone Columns
          and the Uniform Overburden

μ = Poisson's ratio of the treated soil
φ_C = Friction angle of the column material

GENERAL ASPECTS OF FULL SCALE LOAD TESTS

As already mentioned, vibro replacement is a specific system of soil improvement. The bearing capacity of the installed stone columns depends predominantly on the lateral support of the surrounding soil, i.e. the columns are not independent structural members like piles. Depending on the intensity of the treatment, the performance of foundations on stone columns is comparable to that of foundations on a suitable and somewhat homogeneous soil.

Isolated footings on piles settle relatively little up to the ultimate bearing capacity, particularly in cases where the load performance relies mainly on skin friction. Therefore, admissible loads are derived from the ultimate bearing capacity rather than from the settlement performance. Conversely, footings on stone columns settle more from the beginning and exceed generally with increasing load acceptable settlements well before the ultimate bearing capacity is arrived at. Therefore, admissible loads of stone columns will be derived from the settlement performance rather than from the ultimate bearing capacity. Accordingly, the arrangements and specifications for large scale load tests should correspond to those of Static Load on Spread Footings (ASTM D 1194) rather than to those of Piles under Axial Compressive Load (ASTM D 1143). In either case, there is no reason to impose specifications tighter than in these standards.

As outlined before, the conditions for large scale load tests on stone columns are quite different from those on piles. With increasing load, the settlements are influenced by local shear failures below the edges of the test footing. For this reason, conclusions with regard to the settlement performance of a raft on a great number of stone columns, should be based on the more or less linear range of the load-settlement curve. It seems to be adequate to consider the range up to 2/3 of any observed final load or, at maximum, the range up to the working load.

LOAD TEST EVALUATION AS QUALITY CONTROL

Settlement Performance

Summaries of several useful approaches to evaluate the performance of stone columns are given by Barksdale and Bachus [4] and by Soyez [9]. With regard to the settlement performance, the theoretical approaches predominantly refer to an infinite grid of columns. The practically executable load tests with footings on few columns, do not fulfil the assumptions. Accordingly, all modifications to evaluate the settlement performance of a footing on a limited number of stone columns are more or less rough approximations.

Very practical design charts which consider load distribution as well as minor lateral support on marginal columns are presented by Priebe [10]. They allow to estimate the settlement of a rigid foundation on a limited number of stone columns as a function of the settlement of an infinite raft supported by an infinite grid of columns. The method presupposes that the footing area attributable to a stone column as well as the foundation pressure are identical in both cases. There exists an optimum layout for a given number of stone columns beneath a footing. However, in practical applications it is sufficient to determine the required grid size by dividing the footing area by the number of columns.

The main diagram, useful for the evaluation of load tests, is shown in Figure 3. The application is relatively simple because the relevant settlement ratio depends on the number and the diameter of the stone columns and the treatment depth considered. For layered soil, it is necessary to use differences as shown in the following settlement calculation since the settlement ratio relates always to the full depth from the foundation level.

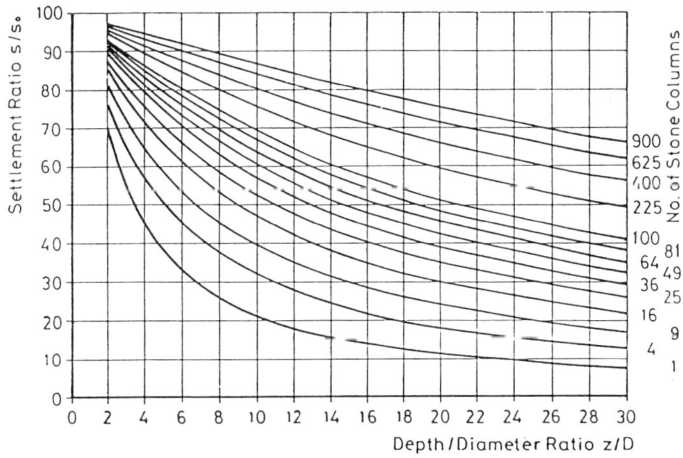

FIG. 3 -- Settlement Ratio of a Rigid Footing on Stone Columns and
an Infinite Raft on an Infinite Column Grid

Although originally developed in connection with the theoreti-cal approach of Priebe [6] to determine the general effect of vibro replacement, the diagram can be used in connection with any other approach, i.e. incorrect estimates depend on the general prediction of the stone column performance used in determining the reference settlement of the infinte raft rather than on the diagram.

An example of estimating the settlement of a footing on a group of stone columns is given by Barksdale and Bachus [4]:

```
Footing size:        a  = b  =  13.0 ft
Found. pressure:     б  = 2367 psf
Residual pressure on soil after the installation of stone columns:
                     б' = 1775 psf  (acc. to Barksdale and Bachus)
No. of columns:      n  =    4
Column diameter:     D  =    3 ft
```

Soil properties: 1. Layer   2. Layer

$$\gamma = 120 \text{ pcf} \qquad \gamma_{sat} = 125 \text{ pcf}$$
$$e_o = 0.9 \qquad e_o = 1.0$$
$$C_c = 0.06 \qquad C_c = 0.08$$
$$\sigma_o = 960 \text{ psf} \qquad \sigma_o = 1810 \text{ psf}$$
$$H = 10 \text{ ft} \qquad H = 8 \text{ ft}$$

Settlements of the footing (acc. to Barksdale and Bachus):

$$s_1 = 1.52 \text{ in} \qquad s_2 = 0.44 \text{ in}$$

The settlement of an infinite load area on stone columns is easily calculated using above-given residual pressure:

$$s_{o1} = 0.01436 \cdot H \qquad s_{o2} = 0.01187 \cdot H$$

Settlement calculation for a rigid footing acc. to Figure 3:

$$z_1/D = 10/3 = 3.33 \qquad z_2/D = 18/3 = 6.00$$
$$s/s_o = 0.62 \qquad s/s_o = 0.46$$

$$s_{1ri} = 0.01436 \cdot (10 \cdot 0.62 - 0) \cdot 12 = 1.07 \text{ in}$$
$$s_{2ri} = 0.01187 \cdot (18 \cdot 0.46 - 10 \cdot 0.62) \cdot 12 = 0.30 \text{ in}$$

The discrepancy to the values of Barksdale and Bachus is due to the consideration of a rigid footing. As calculations for the given footing size show, in case of flexible conditions the settlement ratio between the centre and the characteristic point is approximately 1.4.

$$s_{1fl} = 1.50 \text{ in} \qquad s_{2fl} = 0.42 \text{ in}$$

The agreement between the calculations is now really good.

Bearing Capacity

With regard to the bearing capacity, the conclusions of theoretical approaches are predominantly based on the performance of a single isolated column. Load tests on single columns are frequently performed, but in most cases they do not fully apply to the assumptions because the footing size exceeds the cross section of the stone columns. Furthermore, an uncontrollable tilting of the test footing is likely to occur which is not evidence of a faulty stone column. This is due to the fact that unlike a pile a stone column does not provide a rigid cross section.

In most cases it is desirable to predict the behaviour of column groups. It is questionable whether theoretical approaches based on the performance of an isolated column then apply at all. A
reasonable but somewhat complicated proposal for estimating the
bearing capacity of footings on a limited number of stone columns
is given by Barksdale and Bachus [4].

It seems preferable to evaluate the bearing capacity of a
footing on a group of stone columns on the basis of established
bearing capacity factors. This requires the graph of an approximate
ground failure line as shown in Figure 4, using a calculated composite shear strength within the zone of treated ground and outside
of this zone the shear strength of the surrounding soil.

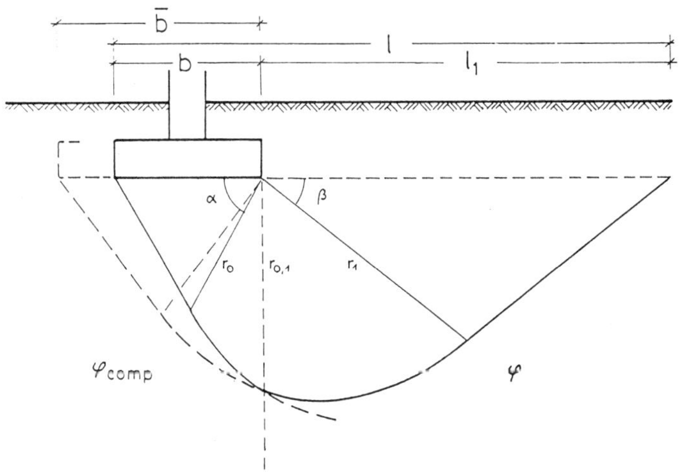

FIG. 4 -- Approximate Ground Failure Line

Now there are two methods available. According to the first
one, an average friction value is calculated or drawn from Figure 5
which delivers a comparable failure line. With this friction value
and an average cohesion along the failure line, the bearing capacity is calculated as normal.

According to the second method, the failure line of the
untreated ground is extended below the footing as also shown in
Figure 4 by the dashed line, arriving at an assumed footing width.
With this assumed width the bearing capacity is calculated as
normal, but using the friction value of the untreated ground. This
approach is surely on the safe side.

The advantages of the recommended methods are firstly that
they provide the possibility to easily consider any friction value
of the surrounding soil and secondly that they are also applicable
where the treated zone extends over the edge of the foundation.

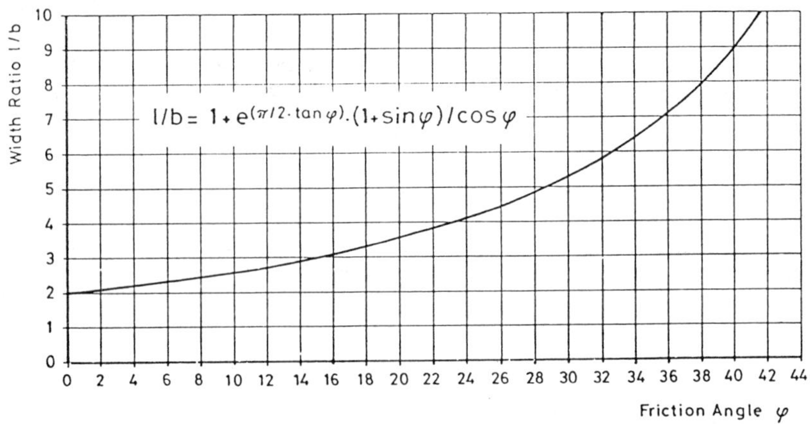

FIG. 5 -- Extent of a Ground Failure at Homogeneous Conditions Dependent on the Friction Angle of the Soil

An example of estimating the bearing capacity of a footing on a group of stone columns is also given by Barksdale and Bachus [4]:

Footing size: a = 13.5 ft = 4.1 m,   b = 10.5 ft = 3.2 m
                    d = 3.0 ft = 0.9 m

Existing soil: γ = 115 pcf = 18 kN/m^3
                    φ = 0.0 °,   c   = 1.000 ksf = 48 kN/m^2

Improved soil: φ_{comp} = 24.9 °, c_{comp} = 0.654 ksf = 31 kN/m^2

Bearing Capacity: P_{ult} = 1418 k = 6310 kN

The recommended methods require the determination of the approximate failure line according to Figure 4 (German Standard DIN 4017):

$$\alpha = 45 + \varphi_{comp}/2 = 57.45 °$$

$$\beta = 45 - \varphi/2 = 45.00 °$$

$$r_o = b \cdot \sin /\sin(90° - \varphi_{comp}) = 2.97 \text{ m}$$

$$r_{o,1} = r_o \cdot e^{[arc(90° - \alpha) \cdot \tan\varphi_{comp}]} = 3.87 \text{ m}$$

$$r_1 = r_{o,1} \cdot e^{[arc(90° - \beta) \cdot \tan\varphi]} = 3.87 \text{ m}$$

$$l_1 = 2 \cdot r_1 \cdot \cos\beta = 5.47 \text{ m}$$

$$l = l_1 + b = 8.67 \text{ m} = 2.71 \text{ b}$$

The following calculations are based on the German Standard DIN 4017. A similar procedure should be possible on the basis of other standards.

1. Method φ_{avg} = 11.8 °   (calculated acc. to Fig. 5)
               c_{avg} = 42 kN/m^2 (appr. (1.00·31+1.71·48)/2.71)

$$N_d = 2.92 \qquad \nu_d = 1.16$$
$$N_c = 9.18 \qquad \nu_c = 1.16$$
$$N_b = 0.40 \qquad \nu_b = 0.77$$

Bearing Capacity:
$$V_b = 3.2 \cdot 4.1 \cdot (42 \cdot 9.18 \cdot 1.16 + 18 \cdot 0.9 \cdot 2.92 \cdot 1.16 + 18 \cdot 3.2 \cdot 0.40 \cdot 0.77)$$
$$= 6821 \text{ kN}$$

2. Method $\bar{b}$ = 1.71·b   (calculated acc. to Fig. 4)

$$N_d = 1.00 \qquad \nu_d = 1.00 \qquad \text{(untreated ground)}$$
$$N_c = 5.14 \qquad \nu_c = 1.16 \qquad\qquad "$$
$$N_b = 0.00 \qquad\qquad\qquad\qquad\qquad "$$

Bearing Capacity:
$$V_b = 1.71 \cdot 3.2 \cdot 4.1 \cdot (48 \cdot 5.14 \cdot 1.16 + 18 \cdot 0.9 \cdot 1.00 \cdot 1.00)$$
$$= 6784 \text{ kN}$$

Both methods give similar results which reasonably correspond to the value of Barksdale and Bachus.

CONCLUSIONS

The main features of deep vibratory compaction techniques for soil improvement are the densification of compactible soils which is called vibro compaction, and the reinforcement of non-compactible soils with load bearing columns of coarse granular material which is called vibro replacement. Although the effects overlap each other in practice, design and interpretation conform to different principles. The decision depends mainly on the grain size distribution of the treated soil. Accordingly, a grain size diagram is submitted for a more reliable definition of the application ranges.

For vibro replacement design criteria exist. This is very important since the in-situ testing of stone columns is rather involved. Only full scale load tests allow reliable interpretations of the actual performance. However, conclusions on the general improvement provided by stone columns are difficult because the test conditions generally do not correspond to assumptions on which the theoretical approaches are based. Simple procedures are outlined to predict the performance of a footing on a limited number of stone columns. Vice versa, these procedures are suitable for interpreting load test results.

REFERENCES

[1] Kirsch, K., "Erfahrungen mit der Baugrundverbesserung durch Tiefenrüttler", Geotechnik 1/1979, pp. 21-32.

[2] Kirsch, K., "Over 50 Years of Deep Vibratory Compaction", Jubilee Presentation of Geotechnik 1985 to the 11th Intern. Conference on Soil Mechanics and Foundation Engineering, San Francisco 1985, pp. 41-45.

[3] Jebe, W. and Bartels, K., "The Development of Compaction Methods with Vibrators from 1976 to 1982", Proc. of the 8th European Conf. on Soil Mechanics and Foundation Engineering, Helsinki 1983, pp. 259-266.

[4] Barksdale, R.D. and Bachus, R. C., " Design and Construction of Stone Columns", Final Report SCEGIT-83-104, Federal Highway Administration, Washington, D.C. 20590, 1983.

[5] Greenwood, D. A. and Kirsch, K., "Specialist Ground Treatment by Vibratory and Dynamic Methods", Advances in Piling and Ground Treatment for Foundations, Thomas Telford Ltd, London 1983, pp. 17-45.

[6] Priebe, H., "Abschätzung des Setzungsverhaltens eines durch Stopfverdichtung verbesserten Baugrundes", Die Bautechnik 53, H.5, 1976.

[7] Seed, H. B. and Idriss, I. M., "Simplified Procedure for Evaluating Soil Liquefaction Potential", Journal of the Soil Mechanics and Foundations Division, ASCE, Vol.97, No. SM9, September 1971, pp. 1249-1273.

[8] Priebe, H., "The Prevention of Liquefaction by Vibro Replacement", Proc. of the Intern. Conference on Earthquake Resistant Construction and Design, Berlin 1990.

[9] Soyez, B., "Méthodes de dimensionnement des colonnes ballastées", Bulletin liaison Laboratoire central des Ponts et Chaussées, Paris, 1985, pp. 35-51.

[10] Priebe, H., "Zur Abschätzung des Setzungsverhaltens eines durch Stopfverdichtung verbesserten Baugrundes", Die Bautechnik 65, H.1, 1988, pp. 3-6.

Timothy D. Stark[1] and B. Michael Yacyshyn[2]

SPECIFICATIONS FOR CONSTRUCTING AND LOAD TESTING STONE COLUMNS IN CLAYS

REFERENCE: Stark, T. D. and Yacyshyn, B. M., "Specifications for Constructing and Load Testing Stone Columns in Clays," _Deep Foundation Improvements: Design, Construction, and Testing_, _ASTM STP 1089_, Melvin I. Esrig and Robert C. Bachus, Eds., American Society for Testing and Materials, Philadelphia, 1991.

ABSTRACT: The purpose of this paper is to provide guidelines for writing a performance specification for the construction and load testing of stone columns in cohesive soils. Specifications from five different sources, three specialty contractors and two agencies, were studied to develop the guidelines described herein. The two primary components of a performance specification are the performance criteria for the stone column foundation and the load testing procedures used to verify that the desired ground improvement has been achieved.

KEY WORDS: specifications, soil stabilization, vibro-replacement, cohesive soil, load tests

There are a number of construction techniques available to stabilize or improve soft clays. These methods include staged construction with and without prefabricated strip drains, geosynthetics, deep foundations, removal and replacement, and stone columns. Stone columns have been successfully used for a wide variety of projects and are becoming widely accepted as a stabilization technique for large area loads, such as embankments and fills. Stone columns may also be used to support spread footings. Stone columns are used to:

1.) reduce the total and differential settlement of the clay due to the applied load.

2.) reduce the time required for consolidation settlement to occur.

[1] Asst. Professor, Dept. of Civil Engineering, San Diego State University, San Diego, CA

[2] Senior Engineer, Kleinfelder, Inc., Irvine, CA

3.)   increase the bearing capacity of the clay.

4.)   increase the shear resistance of the clay which reduces the potential for slope instability.

The two major types of specifications used for the construction of stone columns are performance and procedural specifications.   In a performance specification, the specialty contractor is required to improve the clay to: 1) provide a specified average bearing capacity, 2) limit the total and differential settlement to a specified value, or 3) provide a minimum factor of safety against slope instability.   The specialty contractor then determines the most economical procedure to construct the stone columns, e.g. wet top-feed, dry top-feed or dry bottom-feed, that will provide the desired performance.   The performance specification then specifies a load testing program to verify the ground improvement. If the testing program shows that the objectives have not been met, then it's the contractors responsibility to perform the additional work necessary to meet the specifications unless changed conditions were encountered.

A procedural specification provides the specialty contractor with a detailed description of the construction method, process, and the equipment that must be employed to complete the project. If the desired ground improvement is not achieved and the specialty contractor has followed the specifications, the owner/designer assumes the responsibility for this failure. Therefore, this type of specification requires that the designer possess an extensive knowledge of the nature and distribution of the cohesive soil that is to be improved, stone column construction techniques, and the anticipated performance of stone columns in the native soil. Developing this knowledge usually requires a test program where several stone columns are constructed using different vibrators, spacings, and construction techniques. The columns are then load tested to determine which equipment, procedures, and spacings will provide the desired ground improvement. Because this testing can be expensive, procedural specifications are usually only used on very large ground improvement projects where the cost of the initial load testing can be justified. However, procedural specifications may result in savings on stone column construction costs because the majority of the uncertainties associated with the construction are eliminated prior to contract bidding.

The authors feel that developing procedural specifications requires an intimate knowledge of stone column construction that most practicing geotechnical engineers do not possess. More importantly, procedural specifications often limit the specialty contractor's ability to use a new or unique construction method that may be capable of meeting the project goals at a reduced cost and/or time. Therefore, a performance specification is currently recommended and this paper provides the background required to develop a comprehensive set of performance specifications for stone column construction in cohesive soils. Procedural and performance specifications from five different sources, three specialty contractors and two agencies, references [1-5], were used to develop the "guide" performance specification described herein.

Specifications from successful and unsuccessful projects were reviewed during this study. Both the successful and unsuccessful specifications had formats similar the guide specification presented herein. However, the successful specifications contained more site specific information which facilitated the bidding and construction processes. Therefore, each section of the performance specification should contain as much site specific information as possible. In addition, any site specific laws or ordinances should be clearly identified.

Guide Performance Specification

The guide performance specification consists of the following eleven sections: 1.) soil improvement objectives, 2.) specialty contractor qualifications, 3.) scope of work, 4.) requirements of regulatory agencies, 5.) submittals, 6.) construction of stone columns, 7.) materials, 8.) obstructions, 9.) quality control and assurance, 10.) payment, and 11.) load tests and insitu testing. The following paragraphs describe the information that should be included in each section.

Soil Improvement Objectives

This section should clearly describe the proposed project, the area covered by the project, and the known subsurface conditions. The subsurface information should include representative cross-sections, soil properties, and boring logs or insitu test results. Any laboratory test results, e.g. undrained shear strength, preconsolidation pressure, water content, etc., should be clearly presented on the cross-sections. This section should also clearly describe the expected performance of the stone column foundation and the major responsibilities of the contractor. The performance criteria could require the stone columns to provide any combination of the following: 1.) an average allowable axial capacity, typically 20 to 30 tonnes per column in soft to medium stiff clays, 2.) an average bearing pressure, typically 150 to 200 kN/m^2, 3.) a total settlement that is less than a specified value, typically 25 to 100 mm, 4.) a limiting differential settlement, a typical angular distortion for soft clays is 1/300 to 1/500, and/or 5.) a minimum factor of safety, usually 1.5, against slope instability. Discussion of these typical values is provided by Barksdale and Bachus [6] and Mitchell [7]. The stone column spacing and layout required to achieve the specified performance is determined by the specialty contractor. However, if there are structural concerns that override the geotechnical concerns, limiting column spacing and diameter can be specified by the geotechnical engineer.

Specialty Contractor Qualifications

This section should specify that the stone column construction must be performed by a contractor that has a history of specializing in this type of construction. The "specialty contractor" should be required to submit proof of three or more projects of a similar nature on which they have successfully installed stone columns within the last two to three years. With the proof, the specialty contractor should submit the names, addresses and telephone numbers of previous clients who can be contacted, and are familiar with the project and the contractor's performance. A list of specialty contractors who meet the aforementioned qualifications can also be included in this section. This pre-qualification clause may limit the number of respondents to the project, but the complexity and desired quality of the work should justify this. It should be noted that this type of prequalification may not be permitted on federal projects.

Scope of Work

This section should outline the entire scope of work that will be performed by the specialty contractor. The principal items of work may include some or all of the following:

1.) Preparation of construction drawings showing specific stone column locations, identification numbers and approximate depths.

2.) A detailed description of the equipment and procedures to be used to achieve the desired construction performance criteria.

3.) Furnishing crushed stone (or gravel) as required for the stone columns and, if necessary, the working pad.

4.) Control and disposal of surface water resulting from stone column construction operations.

5.) Site access and construction of gravel working platform, if necessary.

6.) Control and disposal of surface water resulting from stone column construction operations.

7.) Construction and removal of silt settling ponds or similar facilities as required, and site restoration.

8.) Load testing of stone columns prior to and during production as specified.

On most projects, the installation of all stone columns should be the responsibility of one specialty contractor and no part of the contract should be sublet without prior approval. Multiple specialty contractors may be necessary on very large projects to achieve the desired schedule. The specialty contractor should also furnish a qualified supervisor, who is on the job-site at all times during construction, all labor, equipment, materials, and related engineering services necessary to perform all ground improvement work.

Requirements of Regulatory Agencies

This section should describe any laws, ordinances, or any other regulatory requirements that the specialty contractor must comply with during the project. For example, the specialty contractor could be required to comply with all requirements pertaining to the prevention of nuisance to the public and adjacent property owners by noise, impact, vibration, dust, dirt, water, and other causes. Another example would be specifying that the specialty contractor must comply with all laws and regulations pertaining to surface runoff, siltation, pollution, and general disposal of the effluent from the construction of the stone columns and general site work.

Submittals

This section should describe all the submittals that the specialty contractor will be required to provide throughout the contract. The submittals that may be required include:

1.) proof of expertise in constructing stone columns.

2.) construction drawings showing stone column locations, approximate depths and identification numbers.

3.) a description of the equipment and construction procedures to be used.

4.) certification that the installation crew has had experience in performing the work specified.

5.) printed copies of manufacturers recommendations for installation or use of special equipment. For example, the installation procedure, specifications and/or a sample of the filter fabric which might be used in the working platform.

6.) the source and material properties of the backfill material.

7.) daily reports of the progress which include stone column locations, start and stop time, tip depth below grade, and stone quantity per location.

Construction of Stone Columns

A section describing the appropriate construction methods and techniques used for stone column construction should be provided. This is especially important if site constraints or project objectives preclude the use of one or more of the methods. The site constraints should be clearly stated in the section entitled Soil Improvement Objectives. It should be emphasized that a good performance specification will not unnecessarily limit the construction method and equipment that can be employed. The implications, if any, of selecting one method over the other on the performance of the finished product should be discussed in this section. However, the specialty contractor should ultimately decide on the method to employ. The minimum equipment necessary to construct the stone columns, and the column layout and minimum spacing can also be recommended in this section. There are currently three basic methods for constructing stone columns in cohesive soils:

1. Wet top-feed.

2. Dry top-feed.

3. Dry bottom-feed.

The wet top-feed method, also referred to as the vibro-replacement method, entails the use of water as a jetting fluid to aid probe penetration to the desired depth, maintain hole stability, and to facilitate gravel/stone distribution. After the hole is flushed out, stone is added in 0.3 - 1.0 meter increments and densified with a vibrator near the bottom of the probe. The wet top-feed method is usually used in very soft soils with a high ground water table where borehole stability is questionable.   The wet, top-feed method is usually the fastest of the three methods, it typically results in the largest diameter stone columns (typically 0.7 to 1.1 meters in diameter), is capable of supporting the highest design load per column, and allows the use of the widest range of stone/gravel material gradations. However, this method requires a large quantity of water, 2000 to 4000 gallons per hour per probe, which may affect site trafficability and may require special handling to avoid polluting local watercourses.

The dry top-feed method, also referred to as the vibro-displacement method, is essentially the same as the wet top-feed method, except air is used as a jetting fluid. Thus, this method is much cleaner than the wet top-feed method and does not require disposal of the jetting fluid. However, this method can only be used where the borehole can stand open when the probe is extracted so the stone can be inserted into the hole. This usually requires cohesive soils with a minimum undrained shear strengths of approximately 50 - 60 kN/m^2, Barksdale and Bachus [6], and/or a fairly deep ground water table. The dry top-feed method is slower than the wet top-feed method and, if the probe must be kept in the ground to maintain hole stability, the maximum particle dimension of the stone/gravel material may be limited to 2.5 cm by the probe/hole clearance.

The dry bottom-feed method is similar to the dry top-feed method except the stone/gravel material is conveyed to the tip of the probe using an eccentric tube attached adjacent to the probe. Therefore, the vibrator prevents caving of the hole and as a result this method can be used in very soft soils with a high ground water table. Air is used to aid initial penetration of the probe and to facilitate movement of the stone/gravel through the tube to the probe tip. The air pressure should be limited to no more than 275 to 415 kN/m^2 to prevent fracturing of the clay mass during stone column construction (this limiting value tends to be site specific and should be evaluated on a case-by-case basis). Due to the absence of a jetting fluid, the resulting stone columns have diameters that are approximately 15 to 25 percent smaller than the wet top-feed columns. The dry bottom-feed method is slower and requires more equipment than the wet method. However, this method is much cleaner, does not require the disposal of a jetting fluid and results in stone columns with fairly consistent diameters. The dry method also does not introduce additional water into the soft cohesive soils.

Materials

i.) Stone/Gravel Requirements

The construction method usually influences the gradation of the stone/gravel. Stone having a maximum particle dimension of 5 to 10 cm can be used with the wet top-feed and the dry top-feed methods. The size of the tube that transports the stone to the probe tip will limit the maximum particle dimension to no larger than about 2.5 cm in the dry bottom-feed method. In all methods, the stone should be angular, hard, unweathered, and free from organic or other deleterious materials. The fines content for any of the construction methods typically ranges from 0 to 10 percent of the minus No. 4 fraction. Two gradations, adapted from Barksdale and Bachus [6], which would be acceptable for the wet and dry top-feed methods arepresented in Table 1.

A gradation for the dry bottom-feed method can be obtained by reducing the maximum size particle in the above gradations such that it corresponds to the diameter of the tube transporting the stone. The designer should verify that the specified material gradation is locally available.

TABLE 1 -- Acceptable Backfill Gradations for Wet and Dry Top-Feed Methods.

Sieve Size (cm)	Alternative 1 Percent Passing	Alternative 2 Percent Passing
8.9	100	---
7.6	90 - 100	100
5.0	40 - 90	90 - 100
2.5	---	50 - 90
1.9	0 - 10	35 - 70
1.3	0 - 5	---
0.95	---	0 - 10

The specialty contractor should furnish the geotechnical engineer with a gradation curve (ASTM D422), a specific gravity (ASTM C127), and the loose and compacted densities (ASTM C29) of the proposed backfill material. The percent weight loss of the stone should not be more than 12 percent when subjected to the sulfate soundness test (ASTM C88). When subjected to the Los Angeles Abrasion test (ASTM C131), the stone should have a maximum loss of 40 percent at 500 revolutions. The latest version of the specified standards should be used for these tests.

ii. Working Pad Material

When treating soft cohesive soils, a working pad may be required to:

1. provide adequate support for the construction equipment.

2. to better distribute the working loads from the structure or embankment to the stone columns.

3. serve as a drainage blanket during subsequent consolation of the cohesive soil.

The thickness and material gradation of the working pad is a function of its eventual use. Many designers have used geosynthetics and fabrics to provide tensile reinforcement and filtering as necessary. The gravel used for the working pad should be hard, unweathered, and free of organics or other deleterious material. The gradation of the working pad material may be similar to the material required for the stone columns. However, the working pad material should not be large enough to hinder probe penetration.

Obstructions

The vibratory probes can be misdirected or meet refusal during penetration on in-situ debris that has a maximum particle dimension of 15 to 20 cm. Pre-drilling is usually required through dense or hard soil zones to provide probe access to other layers requiring treatment. Pre-drilling can also be used successfully through debris-laden zones. The author's experience indicates that the increased rate of the stone column production typically compensates for the pre-drilling costs. When pre-drilling is not appropriate, and obstructions are encountered, the obstruction may be removed or the effected stone column may be relocated. If the obstruction is removed, the void should be backfilled with gravel.

Quality Control and Assurance

This section should detail the requirements of a quality control and assurance program. The program could consist of the following items.

i.) Construction Records:

Detailed records regarding the construction and load testing of each stone column should be required. This information typically includes:

- stone column identification number.

- date.

- elevation of top and bottom of each stone column.

- quantity of stone placed in each stone column.

- estimate of ground heave or subsidence.

- vibrator power consumption during penetration and compaction.

- time to penetrate and time to form each stone column.

- jetting pressure (air or water).

- details of obstructions, delays, and any unusual ground conditions.

- as-built drawings showing specific stone column locations, identification numbers, and estimated depths.

- load test results and calculations.

ii.) Workmanship:

The specialty contractors workmanship can be evaluated in a qualitative manner by full-time observation of the procedures, methods, equipment, and construction rates during stone column production. If an initial pre-production load test program was performed and accepted, then the specialty contractor must employ similar construction techniques for the production phase of the stone column construction.

iii.) Tolerances:

The authors feel that specifying allowable tolerances for horizontal control, verticality, average stone column diameter, and maintenance of previous subgrade elevations should be included in a performance specification. These factors may impact the ultimate performance of the stone columns, but are difficult to evaluate through load tests or insitu testing. The intent of the stone column construction plays a key role in selecting the appropriate tolerances. Some typical tolerances are listed below and are applicable to sites where there are not significant soil variations:

Horizontal Control:	1/3 to 1 diameter
Verticality:	1 to 5 percent deviation
*Stone Column Diameter:	-10 percent
Subgrade Elevation:	7.5 to 15 cm

*Note: Oversized stone column diameter is only a concern when there is a separate pay item.

Physical measurements in the field are required to maintain horizontal control and subgrade elevation. Verticality is usually judged by observing the tilt of the probe as it penetrates into the ground. If the tilt is excessive and may result in a stone column exceeding the specified vertical tolerance, an additional stone column may be required or the pattern locally altered to provide the proper support. The average stone column diameter may be estimated from the volume of the stone/gravel material delivered into a single stone column and the assumed relative density of the in-place material.

Payment

A lump sum basis of payment is typically used with a performance specification. The lump sum basis of payment allows the specialty contractor to select the most efficient method of stone column construction to satisfy the performance criteria. However, the area and depths of improvement and the performance criteria must be clearly defined if a lump sum price is used. The lump sum should provide full compensation for furnishing all labor (including a qualified supervisor), materials, tools, supplies, equipment, and incidentals necessary to design, install, and proof test the stone columns constructed during the production phase of the construction. The effluent handling and disposal, and the initial load testing can be covered in a lump sum price or as a separate pay item depending on the project.

Load Tests and Insitu Testing

One of the most important parts of a performance specification is the load test program that should be used to verify the performance of the stone column foundation. A combination of load tests on stone columns constructed before, during, and after production should be specified to verify the design assumptions and the performance specification. There are three major types of load tests: (1)

short-term tests which are used to evaluate the ultimate bearing capacity, (2) long-term tests which are used to measure the consolidation settlement characteristics, and (3) horizontal or composite shear tests which are used to evaluate the composite stone-soil shear strength for use in stability analyses. The most common of these tests is the short-term load test on a single column. The five specifications reviewed during this study all specified short-term load tests that were generally in good agreement with the ASTM Standard D1194-87, entitled "Bearing Capacity of Soil for Static Load and Spread Footing." Table 2 shows the variations observed in the short-term load test procedures reviewed.

The short-term load tests should be performed after all excess pore pressures induced during construction have been dissipated. The load increment should closely correspond to the actual loading. For example, if the actual foundation load will be applied very slowly a load increment of approximately 10% should be used. A rapid loading may result in immediate settlement as well as consolidation settlement. If the actual load will be applied rapidly, a load increment of 20 to 25% should be used. A final acceptance criteria of 2.5 cm of settlement at 150 to 200% of the allowable/design load appears to be a reasonable criterion.

The ultimate or long-term settlement of the stone column foundation is usually estimated from the results of short-term load tests on single stone columns. Mitchell [7] reported that the ultimate foundation settlement due to a uniform loading of a large area was 5 to 10 times greater than the settlement measured in a short-term load test on a single column. However, there is very little field data available to confirm this behavior. Therefore, it is recommended that long-term load tests on a group of columns be conducted in conjunction with short-term load tests to develop an estimate of the ultimate settlement of the stone column foundation. The long-term load tests should be conducted on a minimum of three to four stone columns located within a group of 9 to 12 columns having the proposed spacing and pattern. The load should be applied over the tributary area of the columns and left in place until the cohesive soil reaches a primary degree of consolidation of 90 to 95%. The applied load could consist of column backfill material, native material, and/or the dead weight from the short-term load tests. The results of these tests will provide valuable information for estimating the ultimate settlement of the stone column foundation.

During the production phase of construction, one short-term load test is usually performed for every 5 to 10% of the stone columns installed. These tests are referred to as proof tests and are used to verify quality control during production. The load applied in the proof test is usually only 100 to 125% of the allowable/design load.

Insitu testing to evaluate the affect of the stone column construction on the native cohesive soil can be also specified. However, the specified test method should be selected on the basis of its ability to measure changes in lateral pressure in cohesive soils. The electric cone penetrometer (CPT), the flat plate dilatometer (DMT), and the pressuremeter (PMT) appear to provide the best means for measuring the change, if any, in lateral stress due to stone column construction. Due to the limited amount of information that will be obtained from CPT, DMT or PMT testing after column construction, it is recommended that long-term load tests on groups of stone columns be conducted instead of insitu tests. However, extensive insitu testing should be conducted during the initial subsurface investigation to reliably estimate the soil profile and the stone column design parameters.

TABLE 2 -- Variations Observed in Short-Term Load Test Procedures.

Parameter	Variation
Number of columns tested	3 - 5 (approximately one for each 100-150 columns)
Column configuration	single column to center column of a group of 9
Maximum load applied	100, 125, 150, and 200% of allowable/design load
Load increment	10 - 25% of allowable/design load
Load increment criterion	settlement less than 0.025-0.05 cm per hour
Final load increment criterion	0.013 - 0.025 cm per hour
Final acceptance criterion	total settlement less than 0.5 - 2.5 cm under 150 - 200% of the allowable/design load.

Summary

A procedural specification requires an intimate knowledge of stone column construction and performance that most practicing geotechnical engineers do not possess. In addition, a procedural specification may limit the contractor's ability to use a new or unique construction technique. As a result, a performance specification is currently recommended for the construction of stone columns in cohesive soils. The purpose of this paper is to provide guidelines for writing a performance specification. Specifications from five different sources, three specialty contractors and two agencies, were studied to develop the guidelines described herein. The performance specification consists of two main parts: 1.) The performance criteria for the stone column foundation, and 2.) the load testing program that should be used to verify that the ground improvement has been achieved.

The performance criteria clearly states the expected performance of the stone column foundation. This could require the stone columns to: 1.) provide a specified average axial capacity or bearing pressure, 2.) limit the total and differential settlement to a specified value, and/or 3.) provide a minimum factor of safety against slope instability.

The load test program should be specified such that the level of ground improvement can be evaluated. At present, it is very difficult to extrapolate the results of short-term load tests on single stone columns to the long-term behavior of the stone column foundation. As a result, a load test program involving long-term load tests on stone column groups is recommended to measure the ultimate settlement and capacity of the stone column foundation. The long-term tests could be conducted on three to four stone columns located within a group of 9 to 12 columns having the proposed spacing and pattern.

Acknowledgements

The authors express their appreciation to Geocon Inc. of San Diego for sponsoring the research described herein. The paper benefitted substantially from the critical review of Mr. Robert F. Hayden of GKN Hayward Baker.

References

1.   Engelhardt, K., "Specification for Soil Stabilization by the Vibro-Replacement Method," Sept., 1989, personal communication.

2.   Drumheller, J.C., "Specifications for Soil Stabilization (Vibro-Replacement)," Sept., 1989, personal communication.

3.   Institution of Civil Engineers, "Specification for Ground Treatment," Thomas Telford Ltd., London, 1987, 32p.

4.   Institution of Civil Engineers, "Specification for Ground Treatment: notes for guidance," Thomas Telford Ltd., London, 1987, 18p.

5.   Welsh, J.P., "Specifications for Soil Improvement by Vibro-Replacement," June, 1989, personal communication.

6.   Barksdale, R.D. and Bachus, R.C., "Design and Construction of Stone Columns," Report FHWA/RD-83/026, FHWA, Dec., 1983, 194p.

7.   Mitchell, J.K., "Soil Improvement: State-of-the-Art," <u>Proceedings of the Tenth International Conference on Soil Mechanics and Foundation Engineering</u>, Stockholm, 1981.

Barry C. Slocombe, and Michael P. Moseley

THE TESTING AND INSTRUMENTATION OF STONE COLUMNS

REFERENCE: Slocombe, B. C. and Moseley, M. P., "The Testing
and Instrumentation of Stone Columns," <u>Deep Foundation Improvements: Design, Construction, and Testing</u>, <u>ASTM STP 1089</u>, Melvin
I. Esrig and Robert C. Bachus, Eds., American Society for
Testing and Materials, Philadelphia, 1991.

ABSTRACT: The vibro replacement technique for constructing
stone columns was first used in Europe in the late 1950's and
has been used extensively for ground improvement work in the
United Kingdom for over twenty five years. Testing of ground
improvement work is a vital ingredient of the construction
process, and this paper sets out current British developments.

KEYWORDS: Vibro replacement, ground improvement, testing,
instrumentation.

Compaction of clean sands using depth vibrators has been practised
for over fifty years and commenced with the development of the key
tool, the depth vibrator, by Johann Keller GmbH, Germany, in the
1930's. The concept of adding stone during compaction, first
performed in the late 1950's, greatly extended the range of soils
capable of being improved. This technique, commonly described as
vibro replacement, is the predominant type of ground improvement used
in the United Kingdom.

METHOD OF CONSTRUCTION

The depth vibrator, electrically or hydrostatically driven, and
hung from a mobile crane, enters the ground under the combined effect
of weight, vibration and air or water jetting. On reaching the design
depth, a charge of imported stone is placed in the ground and the
vibrator is used to compact the stone (and surrounding ground if

Barry Slocombe is Senior Geotechnical Engineer and Michael Moseley
Business Development Director, Keller Group Limited, Oxford Road,
Ryton-on-Dunsmore, Coventry CV8 3EG, England.

granular).   By repetition, a dense stone column, tightly interlocked with the surrounding ground, is constructed to the surface.

The use of air or water jetting is dependent on soil conditions and in particular on the need to maintain the hole open during placement of the stone and to avoid contamination of the stone column with surrounding soil.   Water jetting is commonly used on sites with high water tables and generates slurry.   With many sites underlain by industrial wastes in the United Kingdom, the effluent arising from the water jetting has become increasingly unacceptable on environmental and practical grounds.   These difficulties have been overcome in the British market by the introduction of specially designed vibrators that enable stone to be fed direct to the nose of the vibrator, thus obviating the need for water jetting and ensuring a stable environment for the construction of high integrity stone columns.

TESTING

In considering testing procedures, for vibro replacement, it is important to recognize the differences between the British ground improvement market and those elsewhere.   The major factors are the short distances between sites and the ready availability of suitable hard stone at reasonable price.   This permits the easy movement of men and machines between sites and enables even small sites to be treated economically.

There are few areas of clean sandy soils and as such the support of multi-storey developments on densified sand is rare in Britain. The majority of sites require improvement to the engineering properties of inert man-made fills and/or weak natural strata for the support of low-rise housing or light industrial development.   Even when the grading of these soils is suitable for densification to take place during treatment, it is normal for stone columns to be constructed in view of their superior engineering performance at similar overall cost.   Depths of treatment are normally within the range of 3 to 6m but have attained almost 30m.

The main aim of the testing of this larger number of smaller sites is therefore to provide reassurance, quickly and economically, that the treatment scheme will meet the requirements of adequate bearing capacity and suitable settlement control for the proposed structure.

The testing of soils reinforced by stone columns must recognise the different response of the ground when testing granular in comparison to predominantly cohesive soils.   In-situ tests are therefore more appropriate where the soils respond to densification effects.   Surface loading tests are applied to these as well as mixed or cohesive constituents.   Table 1 offers the authors' opinion on how useful certain commonly performed methods are for testing treated soils.

Short duration tests on metal plates of 600mm diameter (small plates in Table 1) are the most common form of testing stone columns in Britain.   This is due to their speed and low cost.   However, such

TABLE 1
Suitability for testing stone columns

Test	Granular	Cohesive	Comments
McIntosh probe	*	*	Before/after essential. Can locate obstructions prior to treatment.
Dynamic cone	**	*	Too insensitive to reveal clay fraction. Can locate dense layers and buried features.
Mechanical cone	***	*	Rarely used.
Electric cone	****	**	Particle size important. Can be affected by lateral earth pressures generated by treatment. Best test for seismic liquefaction evaluation.
Boreholes + SPT	***	**	Efficiency of test important. Recovers samples.
Dilatometer	***	*	Rarely used.
Pressuremeter	***	*	Rarely used.
Small plate	*	*	Does not adequately confine stone column. Affected by pore water pressures.
Large plate	**	**	Better confining action.
Skip	**	**	Can maintain for extended period.
Zone loading	****	****	Best test for realistic comparison with foundations.
Full-scale	*****	*****	Rare.

*least suitable
*****most suitable

tests can only stress the soils to shallow depths and have been susceptible to anomalous readings, particularly when residual porewater pressures are present in the ground.

To overcome these limitations, and to provide more realistic simulation of applied building loads, zone loading or dummy footing tests are occasionally performed. Here, loadings of up to 3 times the the design bearing pressure are applied over a group of stone columns, typically of 4 to 9 in number. Significant expense is involved, a large part of which is the delivery of the ballast to provide sufficient reaction. As a result, these tests tend to be performed on the larger contracts or where the soil profile is variable in combination with plate tests to permit correlation between individual stone column and group performance.

It is important that the test base be of sufficient dimension and applied loading to induce significant stress into the "critical layer". This stratum is normally the weakest cohesive layer of significant thickness present on an individual site. This layer determines the allowable safe capacity of the stone column that is then utilised in the design of the treatment scheme. The most common theory used in such analysis is that of Hughes and Withers, 1974, (ref 1).

The majority of the plate and zone loading tests are now performed in accordance with the Institution of Civil Engineers Specification for Ground Improvement, 1987, (ref 2).

INSTRUMENTATION

It will be appreciated that having constructed the stone columns there is a degree of selectivity in the type of test and, more particularly, the test location. The instrumentation package has been designed to monitor the construction of every stone column on any project. The computer-generated records can then be assessed to determine the optimum test locations.

The equipment consists of a microprocessor-based system designed to be unaffected by the range of temperature and moisture normally occurring in Europe. The computer monitors and processes industry standard signals from four main sensor sources:

1.   Depth of vibrator
2.   Power consumption
3.   Weight of stone
4.   Skip movement

Information derived from these signals is displayed via a series of instructions to the plant operator on the LCD unit within the machine cab, see Fig 1. A printer, Fig 2, again inside the cab, provides permanent records whilst the main details of each stone column are stored on disc.

The power consumed by the vibrator motor is a good indicator of the degree of compaction of the stone column being achieved and response of the surrounding soils. The Keller vibrators are electrically driven and as such each lift during construction is aimed at compacting up to a site-specific amperage. In the event of the records showing that this had not been quickly achieved, this would suggest the possible presence of a localised weak zone, unsuitable stratum or operator error. This situation would be investigated and additional stone columns constructed or unsuitable materials removed as necessary.

One of the major advantages of properly constructing the stone columns to a required resistance is that localised looser or weaker zones are compensated by larger stone consumption. The weight of stone is sampled several hundred times on the first rise of the full hopper to give an average reading since a single point measurement would suffer major errors from rig vibration. It is common for more than one hopper load of stone to be consumed per stone column. This provides more accurate information on the stone consumption to be compared with the soils profile.

The print-out plots 3 parameters against time increments of 20 seconds:

1. Power consumption
2. Weight of stone
3. Depth of vibrator

Fig 3 illustrates the computer print-out of the construction of a stone column through about 4.0m of inert granular fill underlain by stiff clay.

During the initial 30 to 40 seconds of the plot, the vibrator is resting just into the site surface whilst the stone is being loaded into the delivery tube. The next 10 to 15 seconds are spent penetrating the ground to terminate the stone column within stiff clay. This is confirmed by the increase in amperage indicated by point A.

Construction of the stone column is achieved by raising the vibrator a short distance to allow the stone backfill to run out of the delivery tube and then repenetrating each lift up to ground level. In this example, high amperage readings have been consistently achieved.

The intervals between stone column construction occur as the bottom-feed vibrator is being charged with stone and the central plot states the weight of stone at each charge. This permits the calculation of the stone consumption for each of the various lifts. In this example the use of 1.8 tonnes of stone between about 3.5 and 5.0m depth would imply the fill was very loose towards its base and is substantially higher than the general average of 3.22 tonnes in 5.06m total depth.

The shape and continuity of the depth/time plot confirms that a continuous high-integrity column has been constructed in a series of short controlled lifts, each of which was properly repenetrated and compacted.

Fig. 1 -- Display unit.

Fig. 2 -- In-cab printer.

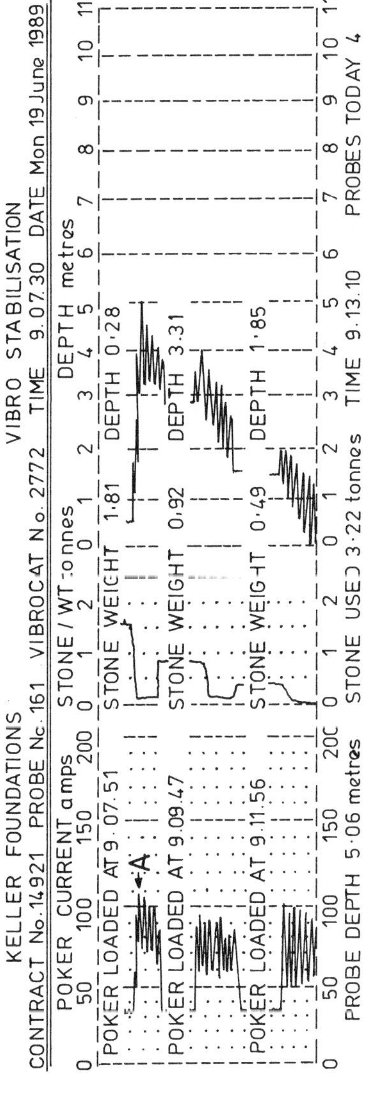

Fig. 3 -- Typical print-out.

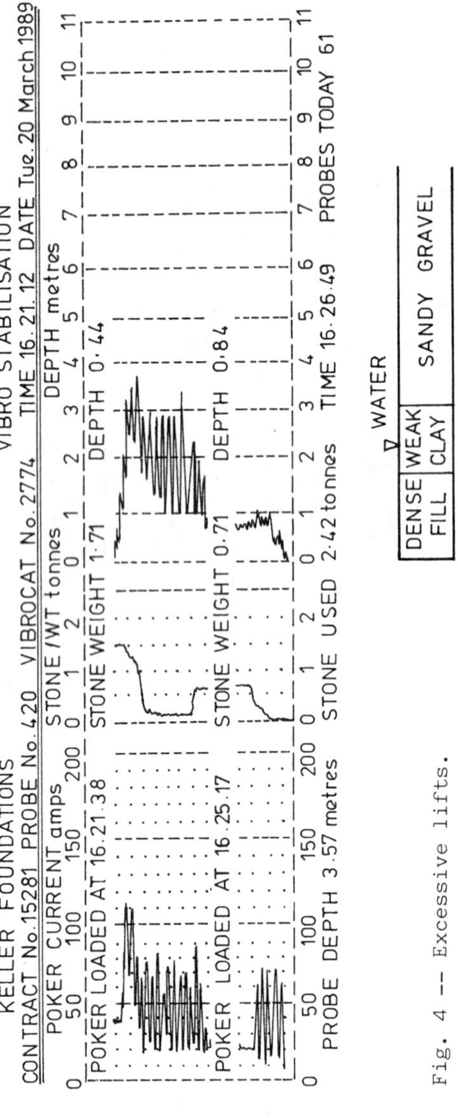

Fig. 4 -- Excessive lifts.

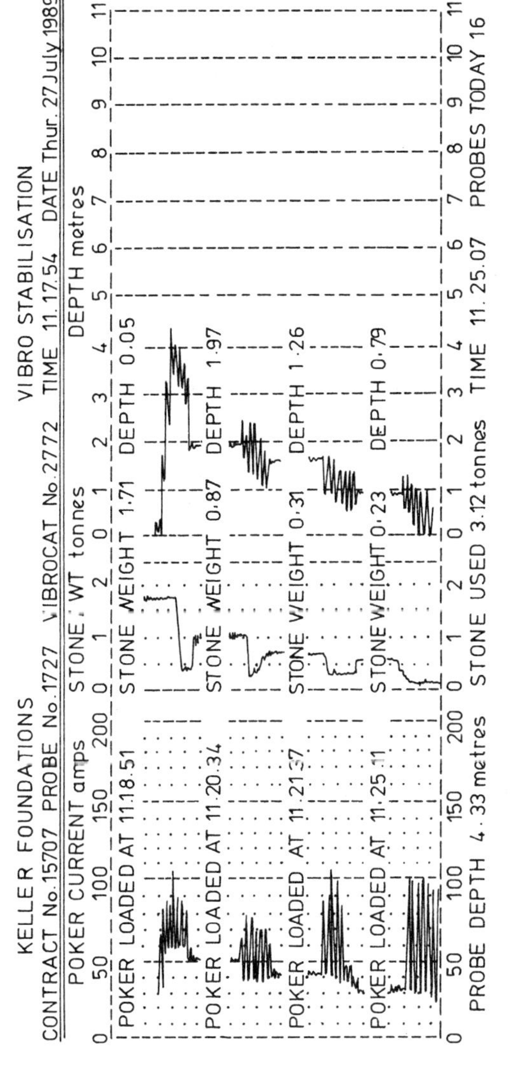

Fig. 5 -- Section not compacted.

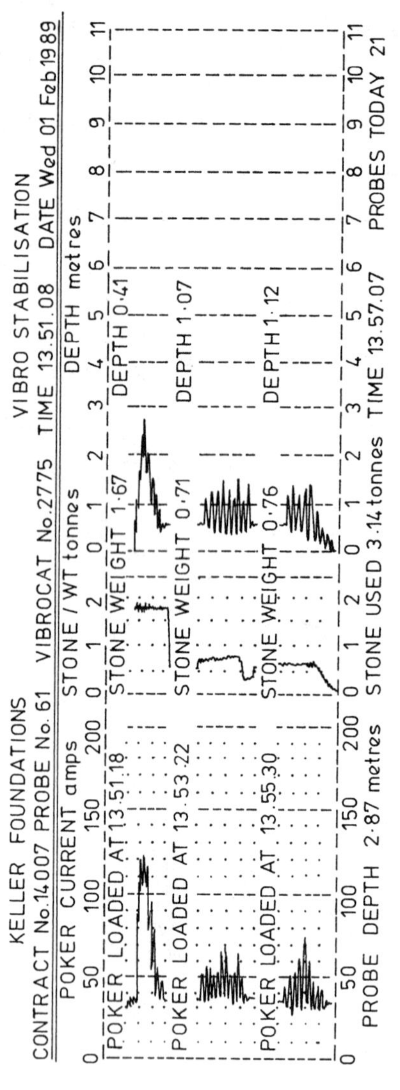

Fig. 6 -- Non-responsive condition.

Fig 4 illustrates a poorly constructed column where the distance that the vibrator was pulled back during each lift was excessive. This could have given rise to collapse of the bore during construction and contamination of the stone column.

Although lower amperage values are normally expected in weak clays the amperage achieved during each lift in this example was relatively low and inconsistent. The amperage values achieved within the surface 2.0m of fill could also have been anticipated to have been higher. It is of interest to note that this column was subjected to a short duration plate test that recorded almost 50% higher deflection than the other tests on the site.

In Fig 5, a section of stone column between about 2.2 and 3.3m depth was not compacted, possibly due to operator error. The amperage readings immediately above this section are also slightly low for the soil type and confirm that insufficient repenetration effort was being applied. This column had to be reconstructed.

Fig 6 illustrates the presence of a zone at about 1.0 m depth where in spite of repeated repenetration, proper amperage readings could not be achieved. An area of the site was identified that would not respond to treatment. This was excavated to remove the unsuitable material and reinstated using clean granular soil which was then treated to the Engineer's specification.

CASE HISTORIES

Three case histories are offered to illustrate the different response of the ground when treating granular and cohesive soils, the effect of confining action to the stone columns and an example of the instrumentation package.

Case history 1

A very large facility in the north of England required stone column treatment for seismic, load bearing and settlement purposes. The seismic aspect was confirmed by electric cone tests where soils were expected to respond to densification effects, and the other parameters by 21 large plate and 9 zone loading tests.

The soils were of glacial origin and very mixed, comprising interbedded clays, silts and sands with occasional gravel content. The cohesive deposits were either reasonably firm or locally stiff and the granular materials generally loose to medium dense.

The treated ground was required to support either raft foundations or conventional footings for a design bearing pressure of 175 kN/m^2 and stanchion loads of up to 150 tonnes with allowable settlement of 25mm. The "wet" vibro replacement process was adopted using very powerful vibrators in order to construct large stone columns that would act as drains through the cohesive deposits during a seismic event and because larger area replacement of the soils produces superior settlement performance. Treatment extended to depths of typically 10 to 15m and locally up to 27m below ground level.

The large plate tests were performed using a rigid steel plate of 0.9m diameter and loaded in 10 increments to loads of between 20 and 45 tonnes. The load was applied directly on top of individual stone columns. These tests revealed markedly different results between predominantly granular and cohesive areas.

The zone loading tests were performed on bases 2.7 or 3.0m square to twice working load, in a procedure similar to that of the Institution of Civil Engineers, and maintained for at least 24 hours. Groups of 4 stone columns were tested at stanchion locations and 2 columns for raft foundations. Recoveries of 30 to 50% were recorded on release of the loading. Typical average cone values within the granular soils were 120 to 150 kgf/cm^2. Recorded settlements (mm) were:

	Test	Number of columns	20T	30T	40T	150T	300T
					Settlement (mm)		
a)	Granular						
	Plate	1	3–4	3–5	4–11	...	...
	Zone	2	...	...	1	3–5	10–15
	Zone	4	...	...	1	1–3	5–8
b)	Cohesive						
	Plate	1	10–15	17–23	22–25	...	...
	Zone	2	...	...	2–4	10–17	35–45
	Zone	4	...	...	2–4	9–12	35–40

Comparison between large plate and zone loading tests clearly illustrates the enhanced rigidity of the column in both soils conditions when subjected to the increased confining action of the area loading tests.

Case history 2

Weak river flood plain soils were treated for a warehouse development in the south-west of England. The area of the 15000m^2 warehouse was bisected by a change in ground level. To the lower side, a succession of 1.0m thick dessicated crust, 2.0m thickness of soft silty clays including peat, sands and gravels and marl bedrock was present. The higher level was underlain by 2.0 to 3.0m thickness of firm red clay overlying the granular and marl deposits.

Landscaping requirements resulted in between 0.5 and 1.5m of upfill of rolled marlstone being placed over the site.

The specification required a design bearing pressure of 100 kN/m^2 for structural foundations up to 2.75m square and 50 kN/m^2 beneath floor slabs with differential settlements not greater than 1 in 500. In recognition of the differing soils and upfilling conditions, treatment grids were varied across the building area and all stone columns taken down to the sand and gravel stratum.

Ten 600mm diameter plate tests and 4 zone loading tests were performed to confirm that the specification had been achieved. 3.0m square bases on a single stone column were used for testing the floor slab areas and a 2.5m base on 4 columns for the structure. The zone loadings were raised at 16 tonne increments at hourly intervals to working load of 48 tonnes which was maintained for 24 hours. Similar increments were then applied up to twice working load. This was maintained for 24 hours for the column base and 1 floor test where movements had ceased for at least 6 hours. The other 2 tests were maintained for 8 days. These recorded creep settlement of about 15% in the firm soils and 25% in the softer zone during the 1 week longer period. All tests recorded between about 5 and 10mm at working load with recoveries of 20 to 40% on release of the load. Recorded settlements (mm) were:

| | | | | Settlement (mm) | |
Test	Soils	Grid	9T	48T (24 hr)	96T (24 hr)
Plate	All	...	10-20	...	...
Zone	Soft	2.2m	...	5-8	13-22
Zone	Firm	2.8m	...	5-10	32-38

Case history 3

A stone column scheme incorporating instrumentation and a zone loading test was performed in southern England for large retail units. Foundation bases of up to 3.5m dimension, designed to a bearing pressure of 150 kN/m^2, and floor slabs of 75 kN/m^2 loading, were treated and the test performed on a 2.5m square base.

The soils comprised a thin layer of made ground, shallow impersistent areas of weak organic clay and loose to medium dense fine clayey silty sand which became dense at about 3.0m depth. Groundwater was present at about 1.0m below ground level.

The zone loading test was performed on 6 stone columns with 4 increments to working load, held for 12 hours, and then 2 further increments to 150% working load which was maintained for 6 days. A recovery of 20% and creep between 1 and 6 days maximum load of 16% was

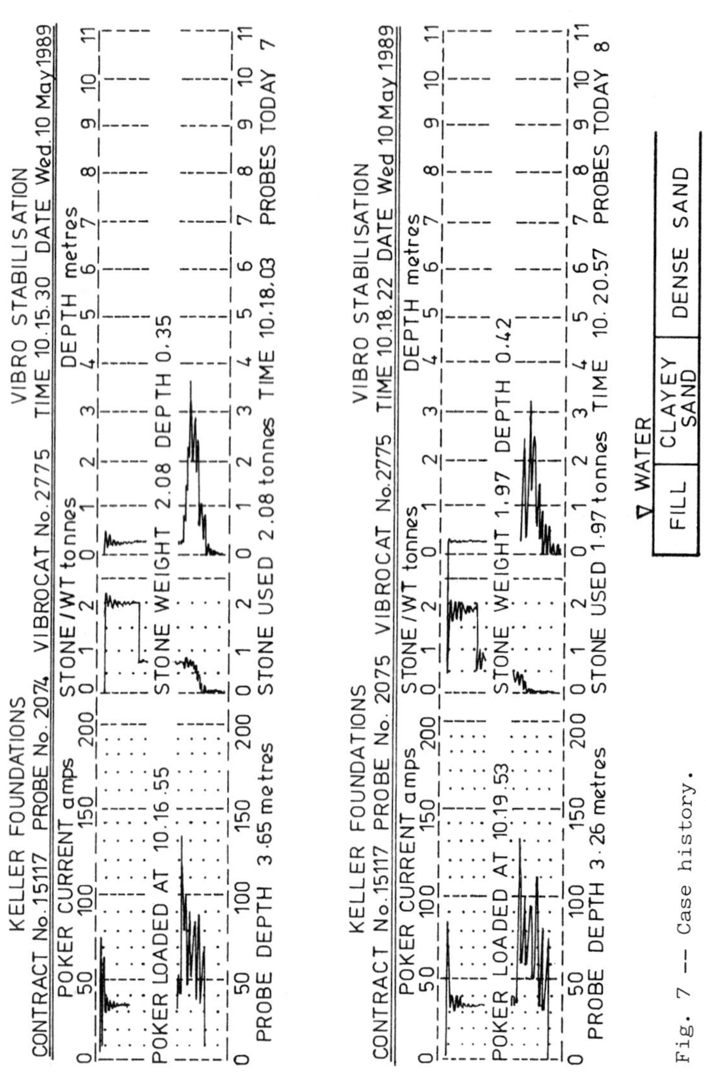

Fig. 7 -- Case history.

recorded. Recorded settlements (mm) were:

Test	Settlement (mm)		
	9T	94T (12 hr)	141T (24 hr)
Plate	6-10	...	...
Zone	...	8.4	17.9

Instrumentation records of two of the stone columns beneath the zone loading test are given in Fig 7.

CONCLUSIONS

The development of the ground improvement market in Britain has necessitated radical improvements in equipment, construction techniques and site records. The instrumentation package described above provides confirmation of the stone column integrity, documentation of every location and a record of any anomalies encountered. The computer generated information provides another step to improved site monitoring, ensures good workmanship and an assurance of quality to the client.

REFERENCES

(1)   Hughes, J.M.O. and Withers, N.J., "Reinforcing of soft soils with stone columns," Ground Engineering, Vol 7, No3, May 1974, pp42-49.

(2)   The Institution of Civil Engineers, Specification for Ground Improvement, Thomas Telford, London, 1987.

(3)   Mitchell, J.K., "Soil improvement-State of the Art Report," Proc.10th Int. Conf. on Soil Mechanics and Foundation Engineering, Stockholm, Vol.4, 1981, pp509-565.

(4)   Greenwood, D.A. and Kirsch, K., "Specialist ground treatment by vibratory and dynamic methods," Proc. Conf. on Piling and Ground Treatment, London, Thomas Telford, 1984, pp17-45.

(5)   Mitchell, J.K. and Huber, T.R., "Performance of a stone column foundation," Journal of Geotechnical Engineering, ASCE, Vol. 111,No2, February, 1985, pp205-223.

(6)   Priebe, H., "Assessment of the settlement of a soil improved by vibro replacement," (German), Bautechnik 65, H1, 1988, pp23-26.

(7)   Brauns, J., "The initial bearing capacity of stone columns in cohesive soils," (German), Bautechnik 55, H8, 1978, pp263-271.

(8)   Kirsch, K., Heere, D. and Schumacher, N., "Settlement performance of storage tanks and similar large structures on soft alluvial soils improved by stone columns," Proc. Conf. on Marginal and Derelict Land, Glasgow, Thomas Telford, 1987, pp651-663.

(9)   Bell, A.L., Kirkland, D.A. and Sinclair, A., "Vibro replacement ground improvement at General Terminus Quay," Proc. Conf. on Marginal and Derelict Land, Glasgow, Thomas Telford, 1987, pp697-712.

(10)   Bell, A.L., Slocombe, B.C., Nesbitt, A.M. and Finey, J.T., "Vibro compaction densification of a deep hydraulic fill," Proc. Conf. on Marginal and Derelict Land, Glasgow, Thomas Telford, 1987, pp791-797.

Tony M. Allen, Todd L. Harrison, John R. Strada, and Alan P. Kilian

USE OF STONE COLUMNS TO SUPPORT I-90 CUT AND COVER TUNNEL

REFERENCE: Allen, T. M., Harrison, T. L., and Strada, J. R., and Kilian, A. P., "Use of Stone Columns to Support I-90 Cut and Cover Tunnel," Deep Foundation Improvements: Design, Construction, and Testing, ASTM STP 1089, Melvin I. Esrig and Robert C. Bachus, Eds., American Society for Testing and Materials, Philadelphia, 1991.

ABSTRACT: A case history is presented describing the use of stone columns to support a portion of a cut and cover tunnel. The tunnel was intended to be supported on spread footings founded on dense granular glacial soil. Inspection and testing of the bottom of the footing excavation revealed the presence of soil below the footing level inadequate for the design loads. A number of alternatives were evaluated. Stone columns were chosen, resulting in a savings of over $350,000 when compared to the next lowest cost alternative. Performance measures were used to ensure that the stone columns improved ground performance sufficiently to result in less than 5.1 cm (2.0 in.) of footing settlement under a maximum load of 290 kN/m^2 (6 ksf). Two plate load tests were performed on the stone columns, and instrumentation was placed in the ground and on the structure. Data are presented indicating structure induced pressures in the foundation soil, settlement at four depths below footing level, and rotation of the footing.

KEYWORDS: Spread footing, silty sand, plate load tests, settlement, soil improvement, instrumentation.

Interstate 90 (I-90) in Seattle, Washington is currently undergoing major reconstruction and expansion by the Washington State Department of Transportation (WSDOT). The new configuration

T. M. Allen and T. L. Harrison are geotechnical engineers at the Washington State Department of Transportation (WSDOT), P.O. Box 167, Olympia, WA 98504; J. R. Strada is the State Materials Engineer, WSDOT; A. P. Kilian is the Chief Geotechnical Engineer, WSDOT.

consists of three general purpose lanes in each direction, with two center reversible lanes built to carry high occupancy vehicles (HOV), light rail transit, or automobiles.

The subject project area is a portion of a 790 m (2600 ft) long cut and cover tunnel. The tunnel consists of three cells: one each for the three westbound, three eastbound, and two HOV lanes. The tunnel lid supports landscaped park areas , bike paths, and crossing roadways.

The foundation improvement described is for the south wall of the eastbound roadway cell of tunnel Units 9 and 10 . Support for the tunnel structure consists of spread footings, as shown in Figures 1 and 2. The design spread footing for these units is 87.8 m (288 ft) long by 9.4 m (31 ft) wide. The footing was designed for maximum footing loads of 290 kN/m^2 (6.0 ksf). Maximum allowable settlements were 5.1 cm (2.0 in.) total and 2.5 cm (1.0 in.) differential.

During construction it was discovered that the existing soil at footing level was not capable of supporting the design footing loads within the settlement constraints. Alternative designs were considered to support the structure. Excavation and structural backfill, steel "H" piles, drilled shafts, and in-situ soil improvement were the alternates investigated in depth. The latter was chosen based on costs and construction impacts.

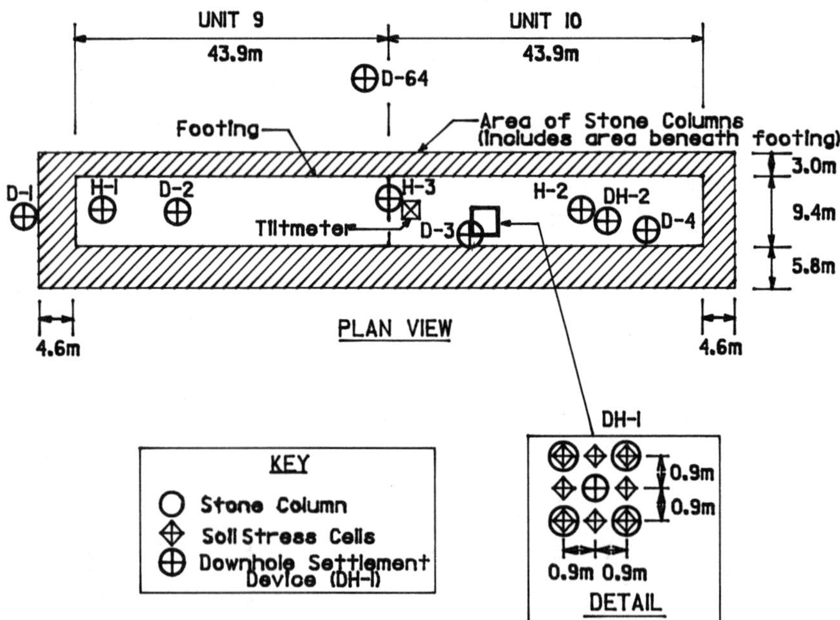

Figure 1: Plan View of Footing and Stone Columns.

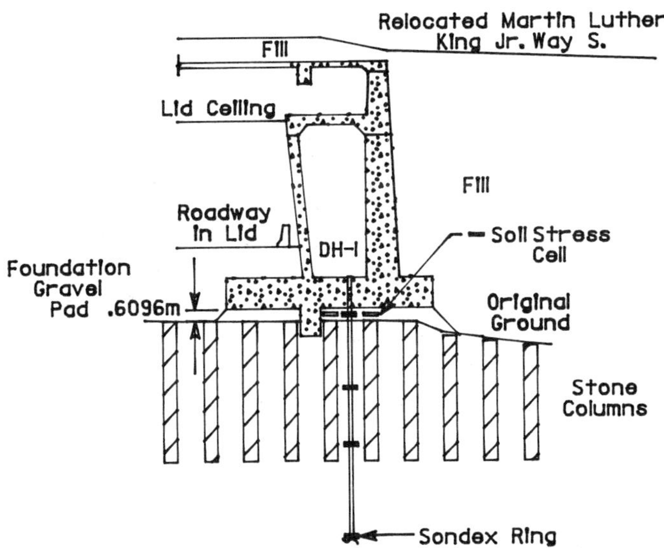

Figure 2: Cross Section of Footing and Stone Columns.

GEOLOGIC AND SOIL CONDITIONS

The project area is located within the Puget Sound lowland in
the western half of Washington State. The Puget Sound lowland is a
north-south trending basin which has been subjected to several
cycles of continental glaciation over the last 2.5 million years.
These glacial events and the processes occurring during interglacial
periods are responsible for the sediments which underlie the general
project area.

The most recent advance and retreat of glacial ice is known as
the Vashon Stade of the Fraser Glaciation.   Geologic units
encountered at the project site are interpreted to be sedimentary
deposits associated with the Vashon ice. A geologic cross section
beneath Units 9 and 10 is shown in Figure 3.

The deepest sedimentary unit encountered in test borings below
Units 9 and 10 is interpreted to be Vashon Till. The Vashon Till
consists of very dense, well graded, very silty, gravelly sand.
Standard penetration test (SPT) results in this unit average over
100 blows/0.3 m (blows/ft).

The Vashon Recessional unit overlies the Vashon Till. This
unit is interpreted to be sediments from glacial meltwater,
deposited as the glacial ice was receding. The thickness of this
deposit varies from 1.5 to 12 m (5 to 40 ft) between the west end of
Unit 9 and the east end of Unit 10. SPT results and design soil
properties for the Vashon Recessional unit are listed in Table 1.
SPT and grain size test results were used to determine engineering
properties for design.

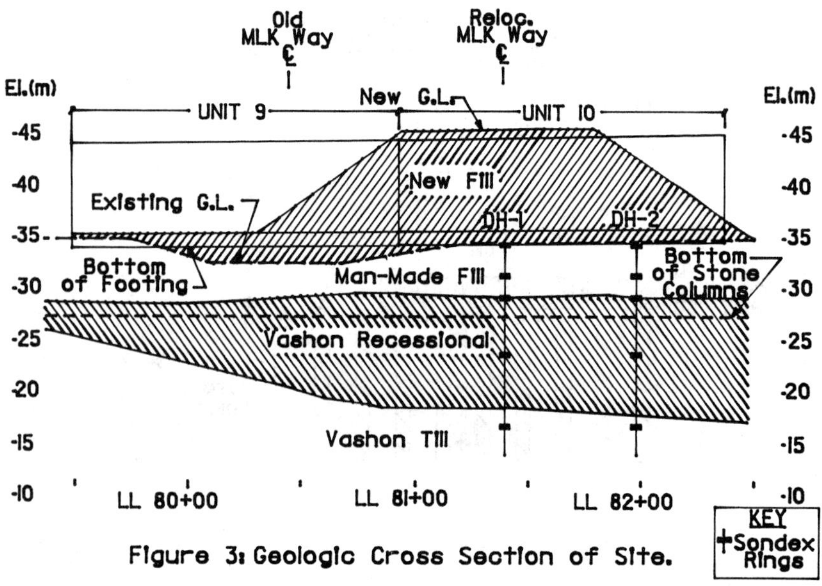

Figure 3: Geologic Cross Section of Site.

Man-made fill placed during previous construction overlies the
Vashon Recessional unit. SPT results and design soil properties
derived from the SPT data and grain size analysis for this unit are
also listed in Table 1. During the design phase of the project, the
thickness of the man-made fill unit was estimated to be
approximately 3.7 to 4.3 m (12 to 14 ft). However, additional test
holes drilled during construction found the thickness to be more in
the order of 6.1 m (20 ft).

A perched water table was present above the relatively
impermeable Vashon Till deposit. The depth to the phreatic surface
below the spread footing foundations for Units 9 and 10 varies from
4.3 to 5.2 m (14 to 17 ft).

TABLE 1: SPT Results and Design Soil Properties

Soil Unit	Soil Classification	N_{spt} Avg.	N_{spt} Range	% Pass. #200 Avg.	Range	Phi deg.	Cc	Unit Wt. g/cm^3
Man-Made Fill	SM	12	5-21	22	11-30	30	0.018	1.8
Vashon Recessional	SP/SM, GP/GM	40	7-78	7	6-8	39	0.009	2.0

FOUNDATION ALTERNATIVES

Foundation design in the contract consisted of spread footing support founded on Vashon Recessional soil. Routine footing site review revealed the presence of unsatisfactory foundation material for the design footing loads. The unsatisfactory material extended to a depth up to 6.1 m (20 ft) below footing level based on additional site exploration.

Revised settlement estimates for the footing were in the order of 10 to 13 cm (4 to 5 in.) based on SPT correlations. This settlement was greater than tolerable for the cut and cover tunnel structure and necessitated redesign of the foundation system.

The following alternatives were considered: overexcavation and replacement with structural fill, dynamic compaction, vibroflotation, drilled shafts, steel "H" piles, and stone columns. The cost for overexcavation was estimated at $1,100,000 and would result in unacceptable construction impacts, e.g. time delays, and conflict with adjacent projects. Dynamic compaction was not feasible due to the potential for damage to adjacent structures as a result of excessive vibration and lateral deformation. Vibroflotation as a method of soil compaction was not feasible due to the high silt content of the existing fill. Drilled shafts were the second lowest cost alternative at an estimated cost of $800,000 plus construction impact costs. Steel "H" piles were estimated at a cost of over $1,000,000 with major design and schedule impacts. Stone columns were determined to be the lowest cost alternative with actual construction costs of $446,365. Additional benefits included only minor modification of the contract spread footing design and minimal impact to the schedule and contractor activities.

STONE COLUMN DESIGN AND ACCEPTANCE CRITERIA DEVELOPMENT

Structural requirements for the footings and lid dictated a maximum allowable settlement of 5.1 cm (2.0 in.) total and 2.5 cm (1.0 in.) differential at the design footing load of 290 kN/m^2 (6 ksf). Two approaches were taken to establish the stone column design. One approach was to determine the required stone column diameter and spacing based on prediction of footing settlement utilizing available methods. The other approach was to establish a settlement criteria for a plate load test that would yield a footing settlement of less than 5.1 cm (2.0 in.). Using the performance method, the contractor would select the column spacing and diameter in this second approach to ensure that the plate load test deflection is less than the load test criteria.

The first approach required that footing settlement be predicted with confidence based on the methods available to predict the settlement of the soil unit improved with stone columns as well as the soil unit below the stone column treated soil. Conventional semi-empirical methods such as the Hough Method [1] were used to estimate the settlement of the soil below the stone columns.

Methods specifically developed for stone column treated soil were
used to estimate the settlement of the soil unit treated with stone
columns (see Ref. #2, pp. 47 to 62). Based on these methods, total
footing settlement with the use of stone columns was estimated to be
10 to 13 cm (4 to 5 in.) at a load of 290 kN/m^2 (6 ksf). This
estimate was approximately the same as the estimate for a footing
on unimproved soil. Common sense made it difficult to believe that
in reality the stone columns would provide no soil improvement as
indicated by these calculations. Additionally, contractor load test
data from similar sites indicated real improvement should be
expected.

The second design approach dictated that the plate load test
settlement at a 290 kN/m^2 (6 ksf) load must not exceed 1.9 cm (0.75
in.) if the total footing settlement is not to exceed 5.1 cm (2.0
in.). This 1.9 cm (0.75 in.) criteria was determined by
back-analysis of the properties of the stone column improved soil
necessary to obtain a footing settlement of 5.1 cm (2.0 in.) using
conventional semi-empirical settlement prediction techniques. The
contractor for the project stated that based on his experience this
criteria could be met. It was therefore decided that the second
design approach would work best.

It was expected, based on the prevous load test data provided
by the contractor, that 0.9 m (3 ft) diameter stone columns
installed at a spacing of 1.8 m (6 ft) center to center would be
needed to ensure that the settlement criteria would be met. The
plate size for the load test was set at 1.8 m by 1.8 m (6 ft by 6
ft) based on this expectation, and would be centered over a single
column. Stone columns were placed beneath and outside the limits of
the footing as shown in Figures 1 and 2 to ensure that improved soil
would be present throughout the area stressed by the footing.

Because the 6.1 m (20 ft) deep loose fill soil unit was
contributing to the majority of the estimated settlement for the
footing, it was decided that the stone columns need only penetrate
through this soil unit to bear on the more competent Vashon
Recessional.

STONE COLUMN CONSTRUCTION

Stone columns were installed using the "vibro-replacement"
(wet) method. The "vibro-displacement" (dry) method was initially
considered; however, the presence of the water table and the
potential for caving soils in the hole made this method too risky.

A 165 horsepower vibrator probe was jetted into the ground to
form the hole for each stone column. Stone was placed into each
hole with a loader while the probe was moved up and down to compact
the stone. The stone used was a subrounded gravel with a size range
of 0.64 to 3.8 cm (0.25 to 1.5 in.). Water from the jetting process
was allowed to flow into a retention pond.

Installed at the site were 537 columns totalling 2871 lineal meters (9,419 lineal feet). Average stone column depths were 4.3 m (14 ft) at the Unit 9 footing and 6.7 m (22 ft) at the Unit 10 footing. The bottom elevation of the stone columns was established by the occurrence of high penetration resistance of the probe. This high resistance was consistently encountered with 0.9 to 1.5 m (3 to 5 ft) of penetration into the Vashon Recessional (see Figure 3). Production rates for stone column installation averaged approximately 120 lineal meters per day (400 lineal feet per day). Stone column installation took a total of 6 weeks to complete.

PLATE LOAD TESTS AND QUALITY CONTROL

Two plate load tests were conducted. One plate load test was conducted after 10 percent of the columns were installed. This test was located at the east end of Unit 10 at Station LL 82+26. The other load test was conducted within Unit 9 at LL 80+40 after all of the stone columns were completed.

The plate load testing apparatus consisted of a 1.8 m (6 ft) square by 2.5 cm (1.0 in.) thick steel plate stacked beneath a 1.5 m (5 ft) diameter by 5.1 cm (2.0 in.) thick steel plate. The upper plate was welded to stiffeners extending up to a bearing plate for the jack. The plates were centered over a single stone column and were placed in intimate contact with the column and surrounding soil. The stone column tested was completely surrounded by other stone columns. The reaction frame consisted of two 6.1 m (20 ft) long W36x160 beams attached to six vertical augered-in-place 320 kN (72 kip) tiedown anchors. The tiedown anchors were 23 cm (9 in.) in diameter and were installed to a depth of 7.3 m (24 ft) below the plate. The anchors were located 2.7 m (9 ft) away from the center of the plates.

The load was applied to the plate with a single hydraulic jack. The plates were loaded to 968 kN, or 290 kN/m^2 (216 kips, or 6 ksf), in increments of 10 percent of the maximum load. The load was measured with an electronic load cell calibrated to 1800 kN (400 kips). Deflection was measured with five dial gauges. Four of the gauges were evenly distributed around the outside edge of the 1.5 m (5 ft) diameter plate, and one gauge was placed near the center of the plate. Both plate load tests met the acceptance criteria of 1.9 cm (0.75 in.) of deflection at the maximum test load.

Quality control for the columns not load tested was maintained by observation of current drawn from the vibrator power source, volume of stone placed in each column, and the details of the construction procedure used. When the vibrator is turned on but not placed in the ground, a current draw of 60 amps was typically measured. The current drawn typically increased to 150 amps during probe penetration into the fill, and increased to 200 amps once the dense Vashon Recessional was encountered at the tip of the probe. These observations were correlated with the depth of the probe and test hole information to ensure that the stone columns were placed

to the bottom of the loose fill soil. Proper compaction of the
stone was ensured by measuring the number of buckets of stone placed
in each column, and by observing the compaction procedure and the
power consumption of the vibrator. Typically, a current draw of 200
amps indicated that maximum compaction was obtained. Visual
observation at the ground surface indicated the column diameter was
0.9 to 1.1 m (3 to 3.5 ft).

INSTRUMENTATION

     Soil stress and settlement beneath the Unit 10 footing were
measured to determine how well the stone column method of soil
improvement was working. Settlement at the footing level and at
several depths below the ground surface was measured using downhole
settlement devices. The devices consisted of Sondex rings embedded
and anchored in the soil and attached to compressible tubing. The
devices were installed directly between a group of four stone
columns using a soil auger. SPT's were performed at 1.5 m (5 ft)
intervals to determine the improvement in soil density which
occurred due to the stone columns. Once the downhole settlement
device was inserted into the augered hole, the void outside the
device was filled with a weak grout. A tiltmeter installed in the
footing base was used to determine differential settlement across
the footing width.

     Soil stress was measured by eight pneumatic total stress cells
placed around downhole settlement device DH-1, as shown in Figure 1.
Four of the stress cells were placed directly on stone columns, and
four were placed in native soil between the stone columns.
Thermistors were placed beside four of the stress cells to measure
temperature for stress cell calibration needs. The stress cells
were placed 1 ft below the footing adjacent to the footing shear
key. Compacted sand fill was placed over the stress cells. The
locations of all of the instrumentation are shown in Figure 1.

RESULTS OF LOAD TESTING AND INSTRUMENTATION

     A direct comparison of the SPT data obtained before ("H" and
"D" holes) and after ("DH" holes) the stone columns were installed
is made in Figure 4, showing the improvement in soil density. The
average penetration resistance of the existing man-made fill before
the stone columns were installed was 12 blows/0.3m (blows/ft). The
average penetration resistance after the stone columns were
installed increased to 30 blows/0.3m (blows/ft). It was expected
based on empirical correlations [2] and contractor experience that
the standard penetration resistance of the existing fill would only
increase to 20 blows/0.3m (blows/ft) due to the densification caused
by the stone columns. The density of the upper 0.9 to 1.5 m (3 to
5 ft) of the Vashon Recessional increased markedly due to the
penetration of the stone columns into this soil deposit, with before

and after standard penetration resistances of 40 and 78 blows/0.3m
(blows/ft), respectively. No increase in the SPT resistance after
stone column installation was expected in the Vashon Recessional
unit. Therefore, based on the SPT data, the stone columns were more
effective than expected.

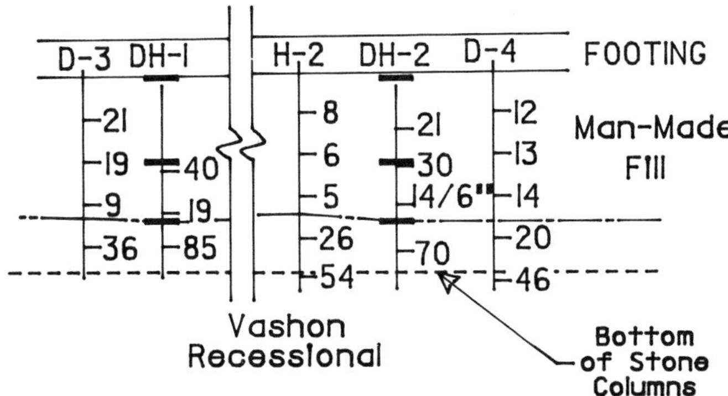

Figure 4: Comparison of SPT Results Before and After Installation of Stone Columns

The results of the plate load tests are shown in Figures 5 and
6. The total settlement measured at the maximum test load of 968
kN, or 290 kN/m^2 (216 kips, or 6 ksf) was 1.3 cm (0.5 in.) or less
for both plate load tests. Some time dependent settlement was
observed for these tests at the maximum test load, as shown in
Figure 6. The rate of settlement appeared to be decreasing with
time.

Figure 5: Load Test Results

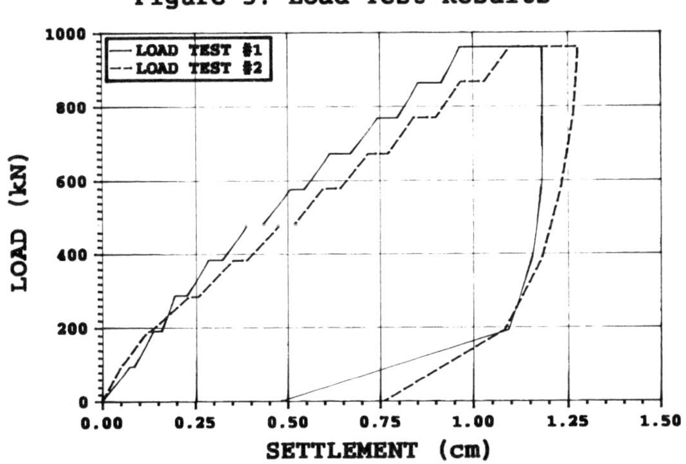

Figure 6: Settlement vs. Time At Maximum Test Load

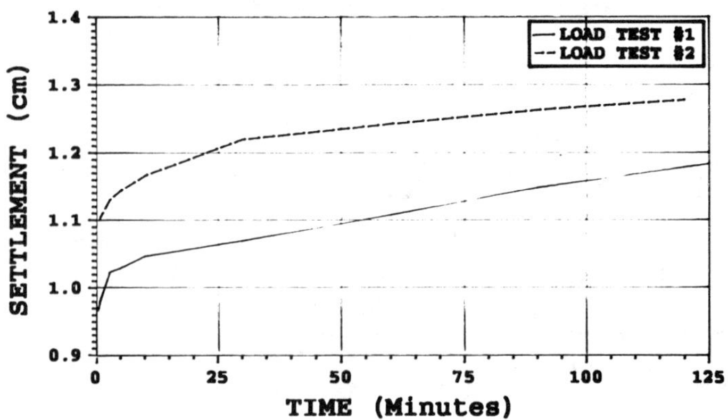

The measured soil stress beneath the Unit 10 footing is shown in Table 2. This table shows that the predicted load is approximately twice the measured load. Predicted loads were determined from unfactored dead loads. The predicted load shown is the average dead load across the footing width, not the peak load resulting from eccentric loads.

A 1.2 m (4 ft) deep shear key for the footing was located adjacent to the stress cells, as shown in Figure 2. Less soil deflection is required to mobilize shearing resistance than to mobilize compressive resistance. It is therefore likely that much of the vertical stress in the soil below the footing is concentrated along the sides of the shear key. The load that was transferred by compression to the soil in the vicinity of the stress cells was distributed equally between the stone columns and the soil, as shown in Table 2.

It was concluded that the measured soil stress did not reflect the full stress applied to the soil. Therefore, settlement measurements were correlated to calculated stress. The calculated footing load at the August 15, 1989 instrumentation reading (i.e., the average dead load) was 170 kN/m^2 (3.55 ksf). When the landscape fill is placed on the lid, the average dead load will increase to its maximum value of 210 kN/m^2 (4.3 ksf). The 290 kN/m^2 (6 ksf) design load for the footing is actually a peak load at the edge of the footing which also includes live load.

The measured settlement at the ground surface is also shown in Table 3. It was not possible to take initial readings for the downhole settlement devices until the 1.4 m (4.5 ft) thick footing slab was poured. Therefore, some settlement occurred before the initial readings were taken. The measured settlement at the ground surface as reported in Table 3 was increased by 0.56 cm (0.22 in.). to account for this unmeasured settlement. The magnitude of the increase was determined using the Hough Settlement Method [1] and an iterative process of adjusting the Hough Factor of the stone column

TABLE 2: Summary of Soil Stress Measurements

DATE	ELAPSED TIME (DAYS)	STAGE OF CONSTRUCTION (REMARKS)	PREDICTED LOAD (KN/m^2)	AVERAGE MEASURED LOAD (KN/m^2) (corrected to 20 deg. Celcius)			STANDARD DEVIATION OF AVG. MEAS. LOAD (ALL CELLS)
				CELLS ON STONE COL.	CELLS ON NATIVE SOIL	AVERAGE FOR ALL	
10/25/1988	0	INIT. STRESS CELL READINGS.	0.00	4.42	3.67	4.31	1.89
11/03/1988	9	REBAR IN PLACE.	15.37	2.92	3.29	3.25	1.54
11/16/1988	22	CONCRETE FOR FOOTING IN PLACE (1.37 M DEPTH).	37.53	27.32	34.50	35.00	2.78
11/30/1988	36	CONCRETE FOR FOOTING IN PLACE (1.37 M DEPTH).	37.53	19.89	24.37	25.50	2.33
12/20/1988	56	CONCRETE FOR 6.1 M HT. OF SHORT WALL AND 12.2 M HT. OF TALL WALL IN PLACE	46.19	26.40	32.19	33.51	2.75
01/18/1989	85	65% OF WALLS AND 50% OF SOFITS POURED.	58.78	32.25	39.52	40.76	4.27
02/28/1989	126	FOOTINGS AND WALLS COMPLETED.	76.59	26.86	32.78	33.85	4.91
06/07/1989	225	FOOTINGS, WALLS, AND APPROACH FILL COMPLETED.	124.46	48.28	64.27	64.27	11.60
07/07/1989	255	GIRDERS, WALLS, FOOTINGS, AND APPROACH FILL COMPLETED.	144.23	53.62	69.72	69.96	12.03
08/15/1989	294	FOOTINGS, WALLS, APPROACH FILL, AND SUPERSTRUCTURE COMPLETED.	169.94	78.13	77.64	77.92	14.66

Table 3: Summary of Settlement and Tiltmeter Measurements

DATE	ELAPSED TIME (DAYS)	(DH-1) TOTAL FOOTING SETTLEMENT AT EL. 40.5 (cm)	(DH-2) TOTAL FOOTING SETTLEMENT AT EL. 34.4 (cm)	MEASURED TILT (degrees)	DIFFERENTIAL SETTLEMENT AT DH-1 (cm)
10/25/1988	0	0.00	0.00		
11/03/1988	9	0.00	0.00		
11/16/1988	22	0.56	0.56		
11/30/1988	36	0.56	0.56		
12/20/1988	56	0.86	0.56		
01/18/1989	85	1.17	0.86		
02/28/1989	126	1.17	1.47		
06/07/1989	225	3.30	2.08	.014 so.	0.23
07/07/1989	255	3.30	2.08	.006 so.	0.10
08/15/1989	294	3.91	2.39	.011 so.	0.18

treated soil. The iterations were considered complete when the actual settlement measured plus the settlement predicted for the footing slab load by the Hough Method matched the total predicted settlement for the footing as currently loaded (i.e., 170 kN/m^2 or 3.55 ksf). The magnitude of this adjustment was also reasonable based on visual inspection of the measured settlement curves. The settlement correction at elevation 29 m (96 ft) and below determined by the method just described was considered to be negligible.

The ground surface settlement tabulated in Table 3 for DH-1 is plotted in Figure 7. The two lower plots show settlement measured at other selected depths below the ground surface. This figure also summarizes on the right the amount of settlement which occurred within each soil unit at the maximum load (i.e., 170 kN/m^2 or 3.55 ksf). The measured settlement occurring within each soil unit, ΔHm, is the difference between settlement readings obtained from the Sondex rings at the top and base of the unit. ΔHp is the predicted settlement for each soil unit determined as described later.

The total settlement measured at the Unit 10 footing plus the 0.56 cm (0.22 in.) correction previously discussed was 3.91 cm (1.54 in.) at an average footing load of 170 kN/m^2 (3.55 ksf). The differential settlement measured along the length of the footing was 1.5 cm (0.60 in.) in 17 m (56 ft) of length, and less than 0.25 cm (0.10 in.) in 9.4 m (31 ft) across the footing width. The total measured settlement was less than the allowable settlement of 5.1 cm (2.0 in.) total and 2.5 cm (1.0 in.) differential.

Figure 7 shows that 3.00 cm (1.18 in.) of this settlement occurred in the existing fill treated with stone columns, whereas 0.91 cm (0.36 in.) of settlement occurred in the underlying Vashon Recessional soil. Settlement prediction methods available for stone columns in sandy soil (see ref. #2, pages 47 to 62) indicate that settlement of the stone column treated fill soil should be approximately 5.1 cm (2.0 in.) at a footing load of 170 kN/m^2 (3.55 ksf). The settlement predicted by these methods is almost twice as much as what actually occurred. Other settlement prediction methods must be used to predict settlement of the soil below the stone columns. If the Hough Method is used, the settlement of the underlying Vashon Recessional soil would be estimated at 2.5 cm (1.0 in.), resulting in a total predicted settlement of 7.6 cm (3.0 in.). Considering that the total measured settlement was only 3.91 cm (1.54 in.), this estimate is obviously quite conservative. It appears that some refinements in the available stone column settlement prediction methods are warranted to make these methods more useful, especially when dealing with structural foundations.

Alternatively, the plate load test data can be extrapolated to give a more accurate settlement prediction for the footing. Properties for settlement calculation in the upper soil stratum (i.e., the stratum of soil treated with stone columns) can be back-analyzed with accuracy using the plate load test settlement data. The settlement characteristics of the soil below the stone columns must be estimated independent of the load test data.

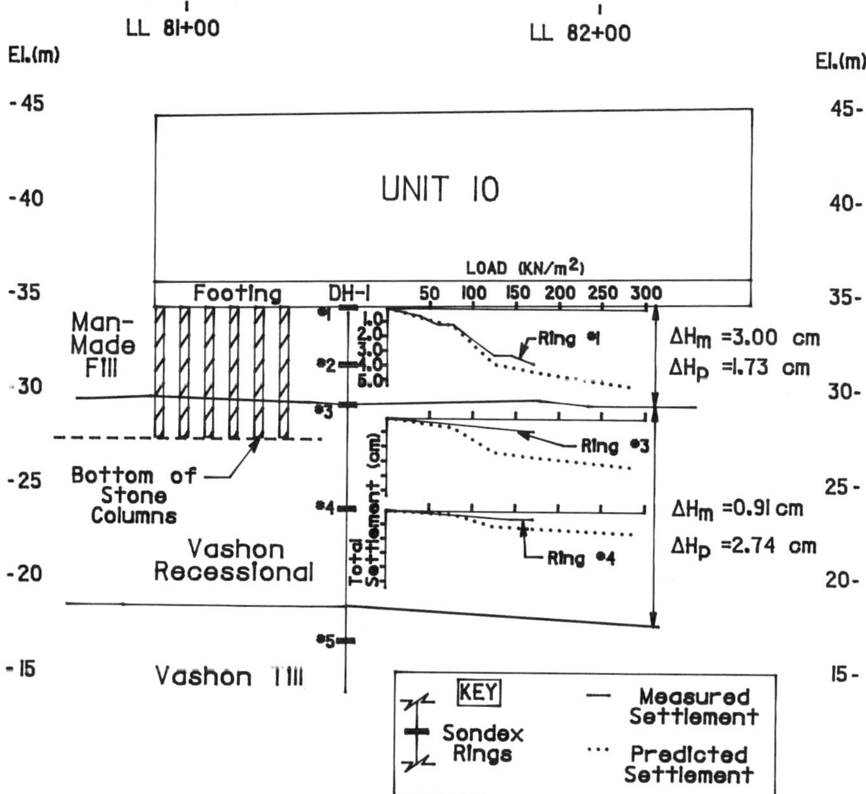

Figure 7: Measured and Predicted Footing Settlement At Unit 10

Predicted footing settlement based on the plate load test data, utilizing the Hough Method and Boussinesq pressure distributions, is shown in Figure 7. The total settlement predicted in this way is 4.47 cm (1.76 in.), which is approximately 12 percent greater than what was actually measured. The Hough Method is known to be conservative [3], which may partially account for this difference. When considering only the settlement occurring in the existing fill, however, the predicted settlement was considerably less than the measured settlement. Also, the predicted settlement was considerably greater than the measured settlement in the Vashon Recessional. One reason for the lack of agreement between the measured and predicted settlement in each soil unit may be related to the soil stress distribution. The overestimate of settlement in the Vashon Recessional may also be partially due to conservatism in the soil settlement parameters used for this soil unit.

If the settlement estimate based on the plate load test data is extrapolated to the maximum expected load for the footing, the total settlement will still be less than 5.1 cm (2.0 in.). Based on all of the available data, the stone column soil densification program was successful, reducing the footing settlement to within allowable limits.

CONCLUSIONS

The use of stone columns on this project was proven to be a cost effective method of ground improvement. It was shown that stone columns can improve the density of loose to medium dense, gravelly silty sand sufficiently to support major structure footings to loads of 290 kN/m^2 (6.0 ksf).

SPT results showed the stone columns effectively densified the site soils. This soil densification was obtained despite relatively high silt contents in the order 22 percent passing the No. 200 sieve.

Field measurements showed a footing settlement of 3.91 cm (1.54 in.) at an average footing load of 170 kN/m^2 (3.55 ksf). However, predicted settlement ranged from 4.47 cm (1.76 in.) from plate load test data to 7.6 cm (3.0 in.) using currently recommended semi-empirical methods. Improvements in the semi-empirical settlement prediction methods involving stone columns are needed, particularly for cases involving structural foundations.

Extrapolation of stone column load test data was found to reasonably predict footing settlement. The experiences from this project show that load test measurements should be used for ensuring acceptable stone column performance under critical structures.

ACKNOWLEDGEMENTS

The authors wish to thank Mr. Ron Erickson, WSDOT project engineer for his valuable data input into this paper and to the subcontractor for the stone column construction GKN Hayward Baker particularly Mr. Mark Koelling, who provided valuable background information during design. Civil Consulting Engineers Howard, Needles, Tamen & Bergendoff and geotechnical engineers Hart-Crowser and Associates provided support during redesign.

REFERENCES

[1]   Hough, B.K., "Compressibility as The Basis for Soil Bearing Value", JSMFD, ASCE, SM 4, August 1959, pp. 11-39.
[2]   Barksdale, R.D., and R.C. Bachus, Design and Construction of Stone Columns, Vol. 1, FHWA/RD-83/026, December 1983.
[3]   Gifford, D.G. et. al., Spread Footings for Highway Bridges, FHWA/RD-86/185, October 1987.

John R. Davie, Lloyd W. Young, Michael R. Lewis, and Francis J.
Swekosky

USE OF STONE COLUMNS TO IMPROVE THE STRUCTURAL PERFORMANCE OF COAL
WASTE DEPOSITS

REFERENCE: Davie, J. R., Young, L. W., Lewis, M. R., and
Swekosky, F. J., "Use of Stone Columns to Improve the Struc-
tural Performance of Coal Waste Deposits," _Deep Foundation
Improvements: Design, Construction, and Testing_, _ASTM STP
1089_, Melvin I. Esrig and Robert C. Bachus, Eds., American
Society for Testing and Materials, Philadelphia, 1991.

ABSTRACT: The Gilberton Power Project, located in central
Pennsylvania, makes use of waste anthracite coal, called culm,
for fuel. The culm processing area is underlain by this culm,
which is generally loose, and quite unsuitable to support the
heavily-loaded processing equipment. Vibro-replacement stone
columns were selected for ground modification. The stone columns
were installed to various depths, depending on the size and
loading of the structure. The paper describes the design and
installation of the vibro-replacement stone columns, difficulties
encountered during installation and how these were overcome, and
techniques used to confirm the effectiveness of the modified
ground, including soil borings, plate load testing, and performance monitoring.

KEYWORDS: stone columns, vibro-compaction, coal waste, plate load
test, settlement

Numerous cogeneration plants have been constructed recently in
the mining districts of east central Pennsylvania to burn anthracite
culm, a waste product of previous coal mining in the area. The
favorable economics of such plants depend to a large extent on the
distance the culm must be transported to the plant. The Gilberton
Cogeneration Plant, located in Frackville, Pennsylvania, has the
advantage of extensive reserves of culm near the plant. One drawback
to this large area of closely available culm is that, to minimize
conveyor lengths, the culm processing equipment must itself be
founded on the culm. This paper describes the steps taken and

Dr. Davie and Mr. Young are engineering supervisors for Bechtel
Corporation, 15740 Shady Grove Road, Gaithersburg, MD 20877;
Mr. Lewis is engineering supervisor for Bechtel Corporation, 3950 RCA
Boulevard, Palm Beach Gardens, FL 33410; Mr. Swekosky is Regional
Manager for GKN Hayward Baker, 800 E. Northwest Highway, Palatine, IL
60067.

problems encountered in providing an economical and successful
foundation for the processing equipment founded on over 25 ft (7.6 m)
of anthracite culm.

SUBSURFACE CONDITIONS BENEATH PROCESSING EQUIPMENT

Fig. 1 shows the layout of the culm processing equipment. The
dimensions and structural and foundation design data for the three
structures (culm silo, culm dryer and transfer tower) are given in
Table 1. All three structures have maximum bearing pressures of
6 ksf (287 kPa), with maximum allowable settlements of 2 in. (51 mm)
for the silo and dryer, and 1 in. (25 mm) for the transfer tower.

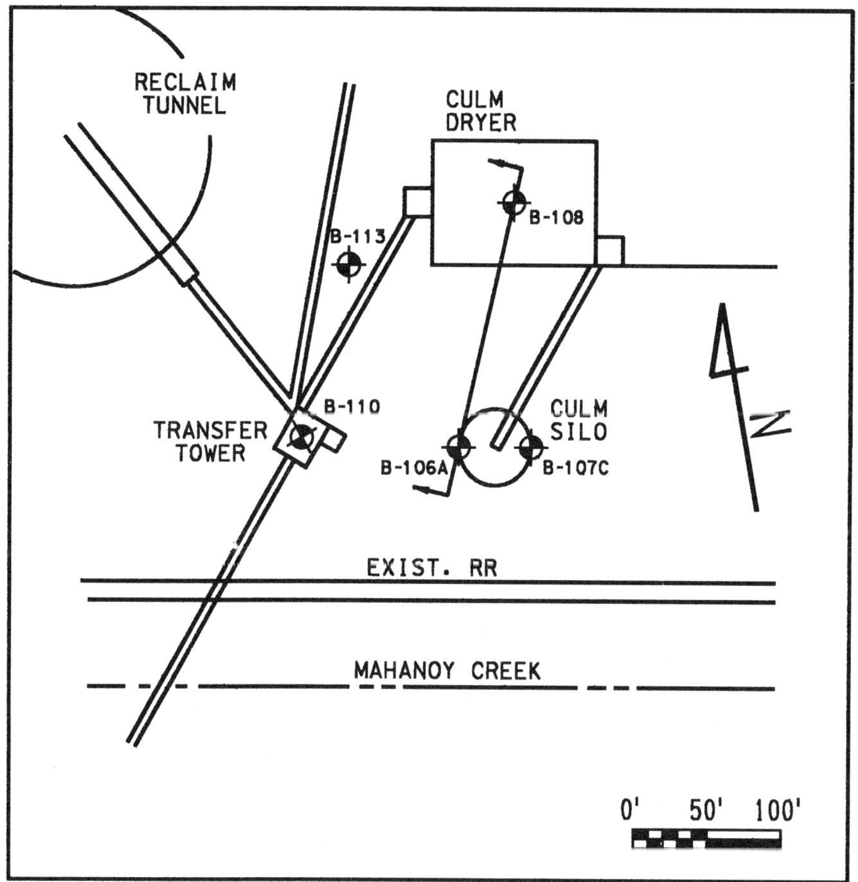

FIG. 1 -- Site and boring location plan

Five borings were drilled at the locations shown on Fig. 1. The
profile given in Fig. 2 is representative of the subsurface conditions throughout the culm processing area. The top 25 to 30 ft (7.6
to 9.1 m) are mainly anthracite culm fill resulting from past mining
operations (stripping, crushing, and sorting) beginning about
80 years ago.

TABLE 1 -- Foundation Design Data

	Culm Dryer	Culm Dryer	Transfer Tower
Plan Dimensions	50 ft Dia.	65 ft x 67 ft	24 ft x 36 ft
Maximum Height	150 ft	95 ft	80 ft
Foundation Type	Mat	Spread Footing	Spread Footing
Foundation Dimensions	66 ft Dia.	9 ft x 9 ft (typ)	7 ft x 7 ft (typ)
Dead Load, kips	4500	-	-
Live Load, kips	10870	-	-
Static Bearing Pressure, ksf	4.5	4.0	4.0
Maximum Bearing Pressure, ksf	6.1	6.0	6.0
Allowable Total Settlement, in.	2	2	1
Allowable Diff. Settlement, in.	1	1	0.5

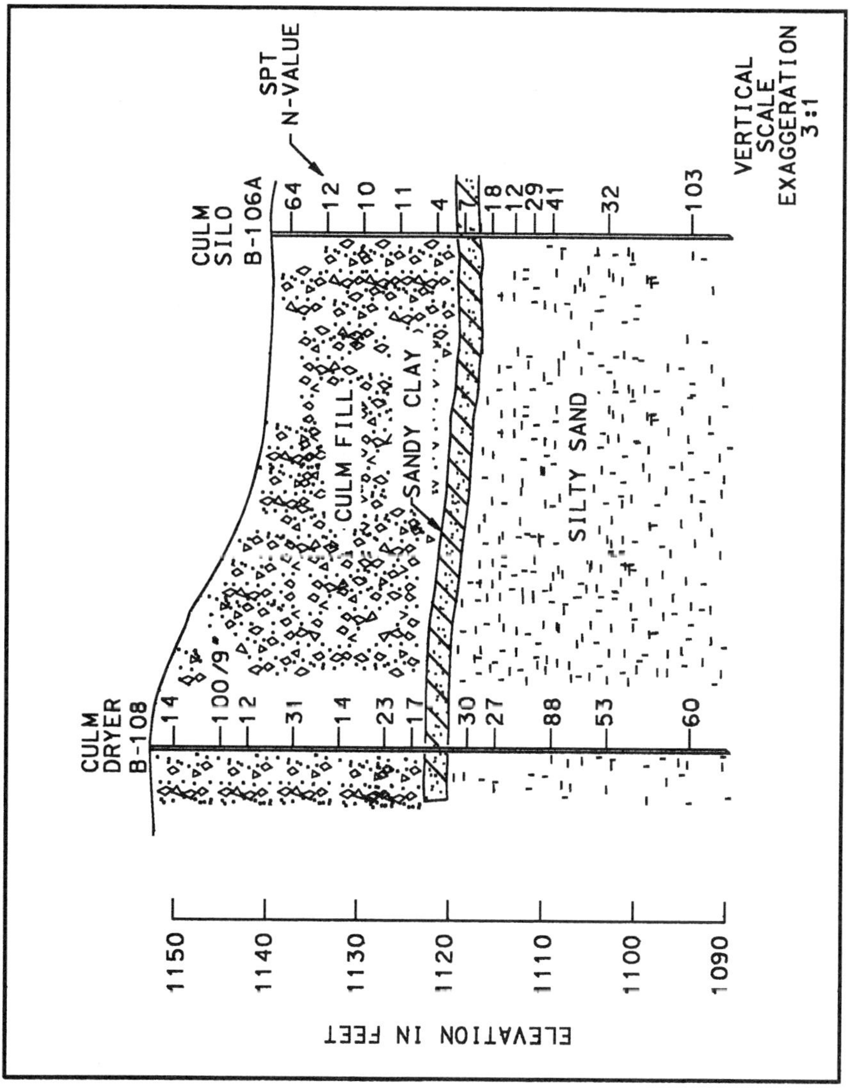

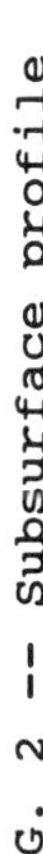

FIG. 2 -- Subsurface profile

The culm comprises chiefly sand or gravel size particles, with 5 to 12 percent finer than the # 200 (.075 mm) sieve, although isolated zones of coal silt and coal dust were encountered. Fig. 3 shows typical grain size curves. The heterogeneous nature of the culm was very apparent during inspection of exposed excavation faces. Standard Penetration Test (SPT) N-values ranged from 2 to over 50 blows per foot.

Groundwater is perched in the culm above the underlying clay. The water level in the observation well in B-106A ranged from about 15 to 24 ft (4.6 to 7.3 m) depth during the period of investigation and construction; the level appeared to vary with the water level in the adjacent Mahanoy Creek. A falling head test in B-106A indicated a permeability of around 1×10^{-4} cm/sec.

About 30 ft (9.1 m) of dense silty sand extends below the culm, separated from the culm by a thin layer of sandy and gravelly clay. The clay ranged from 0 to 7 ft (2.1 m) in thickness, averaging about 2.5 ft (0.8 m). Several thin-walled tube samples were attempted but were unsuccessful because of the gravelly nature of the clay. SPT N-values ranged from 2 to 20, with a median value of about 5. Liquid limit, plastic limit, and natural moisture content values averaged 29, 24, and 27 respectively.

FOUNDATION SELECTION

The SPT results in the culm indicated a variability of consistency both laterally and vertically that ruled out shallow foundations. It was recognized that the soft to firm clay layer beneath the culm would also affect foundation performance, particularly for the 66-ft (20.1-m) diameter culm silo mat foundation. As a result, driven piles were assumed in the conceptual foundation design.

Once detailed design was underway, several ground improvement options were considered in lieu of pile foundations. Vibro-compaction and deep dynamic compaction were discounted because of their inability to treat the clay layer beneath the culm. Also, deep dynamic compaction would have adversely affected the railroad tracks immediately adjacent to the culm silo. Cost factors ruled out compaction grouting, although this technique was used effectively for the reclaim tunnel excavation and ground support in culm just west of the processing area.

The system of ground improvement selected for support of the culm processing facility was vibro-replacement stone columns (vrsc). This system satisfied the design requirements, with several advantages:

o   The vibro-compaction could increase the relative density of the culm to the extent necessary to successfully support the foundations within the allowable settlement tolerances.

o   The stone columns could penetrate the soft clay layer, thereby transferring load to the underlying dense sand.

o   Cost savings of $250,000 could be realized by using vrsc instead of piles.

o   The schedule for foundation construction could be cut in half.

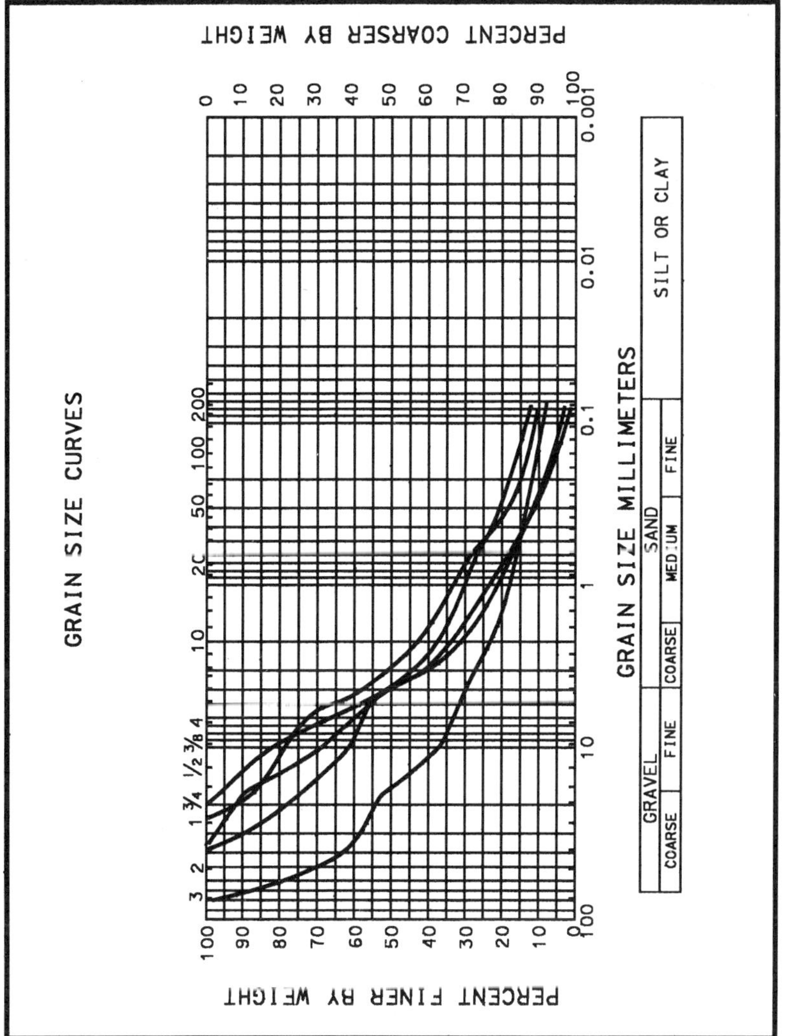

FIG. 3 -- Typical grain size curves

STONE COLUMN ANALYSIS AND DESIGN

The analysis, design, and installation of the vrsc system was
performed by GKN-Hayward Baker (GKN-HB) of Odenton, Maryland. The
design criteria, especially for the culm silo, presented a particular
challenge. Typically, stone columns supporting such major structures
have been limited to 2 to 4 ksf (96 to 192 kPa) contact pressure,
rather than the 6 ksf (287 kPa) required here.

The analysis was based on the assumption that the culm would
behave as a granular material, and that any pockets or isolated zones
of fine-grained material would be bridged if not adequately
densified. The granular behavior meant that bearing capacity
requirements would be satisfied as long as settlement limitations
were met. The strain influence method developed by Schmertman [1 and
2] was used to compute settlement. Since this method considers
settlement to a depth of two times the foundation width below the
foundation, the large culm silo mat presented the critical case.
Stone columns at the culm silo were assumed to extend through the
culm and soft clay and into the dense underlying silty sand,
requiring column lengths of up to 30 ft (9.1 m).

Computed settlement in the unimproved condition assuming 6-ksf
(287-kPa) loading on the silo mat was 5 in. (127 mm), made up of
3 in. (76 mm) in the culm, 1 in. (25 mm) in the soft clay beneath the
culm, and 1 in. (25 mm) in the underlying soils. This analysis was
repeated assuming the culm had been densified by vibro-compaction,
but neglecting the contribution to ground improvement provided by the
stone columns. The densification assumed the median SPT N-value in
the culm to be increased to at least 20. The N-values were converted
to equivalent cone penetrometer resistance values for the analysis.
In addition, the ratio of elastic modulus to cone resistance of the
improved culm was increased to adjust for its reduced compressibility
[3]. The resulting estimated settlement values were just over
0.5 in. (13 mm) in the culm, with the other values remaining as
before, i.e. 1 in. (25 mm) in the clay and 1 in. (25 mm) in the
underlying soil.

The contribution of the stress concentration on the stone columns
in the soft clay was then analyzed using improvement factors based on
work by Priebe [4]. The stone columns were assumed to be 3.5 ft
(1.1 m) in diameter spaced on a 7-ft-(2.1-m-) square grid. The
coarse gravel in the stone columns was assigned an internal friction
angle of 40 degrees. The computed settlement in the soft clay layer
was reduced to just under 0.5 in. (13 mm), giving an overall predicted maximum settlement in the improved culm and clay of 1 in.
(25 mm), with a further 1 in. (25 mm) in the underlying soil. This
latter deep-seated settlement would be almost certainly uniform
under the silo area, and thus the differential settlement limit of
1 in. (25 mm) would be satisfied. The design for the culm silo
called for 132 stone columns. The columns were placed to a distance
of 10 ft (3 m) outside the foundation perimeter to allow for load
spreading from the mat. Fig. 4 shows the stone column locations in
the culm silo area.

The design proposed that the culm dryer and transfer tower be
placed on stone-column-supported footings, typically 9 x 9 ft (2.7 x
2.7 m) in plan dimension. Although the contact pressure on the
footings was similar to the silo mat, the depth of influence was much

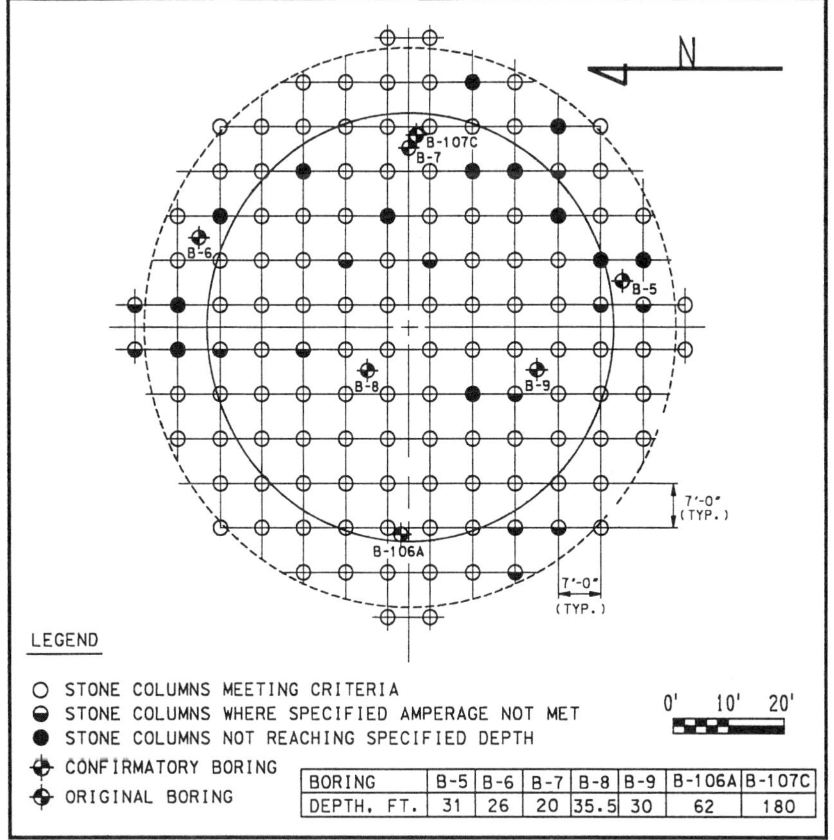

BORING	B-5	B-6	B-7	B-8	B-9	B-106A	B-107C
DEPTH, FT.	31	26	20	35.5	30	62	180

FIG. 4 -- Stone columns and boring locations at culm silo

less because of the far smaller footing area, certainly not extending
to the bottom of the culm or to the clay below. The columns chosen
for the dryer and transfer tower were 3.5 ft (1.1 m) in diameter and
either 10 ft (3 m) or 18 ft (5.5 m) deep, based on a simplified
settlement analysis. The stone column spacing reflected the size of
the footings (e.g., columns were spaced on 6-ft (1.8-m) centers under
9 x 9-ft (2.7 x 2.7-m) footings). Figs. 5 and 6 show the stone
column locations for the transfer tower and culm dryer, respectively.

INSTALLING THE STONE COLUMNS

The electrically operated vibroflot used for installing the vrscs
was a GKN Keller S-type device, with a 33-ft-(10.1-m-) long follower,
connected to a 500-gpm (1.9-cu m/min) pump and a 120 kW diesel
generator. The vibroflot and diesel generator were mounted on an
American 7250-C crane. Fig. 7 shows the vibroflot and selected
technical data. Installation of the stone columns followed these
steps:

1.   Insert the vibroflot to the required stone column depth, using
     the self-weight of the vibroflot and follower, vibration, and
     jetting from twin nozzles at the vibroflot base.

2. Withdraw the vibroflot a few feet, push crushed stone down the
    annulus between the vibroflot and the culm, drop the vibroflot
    assembly onto the stone and vibrate and jet until a target power
    reading is achieved.

3. Repeat step 2 in 1- or 2-ft (0.3- or 0.6-m) lifts until the stone
    column is within about 4 ft (1.2 m) of the ground surface, where
    there is generally insufficient overburden to achieve the desired
    densification. (This top 4 ft (1.2 m) is compacted by tradi-
    tional methods).

4. Record total depth, amperage, and amount of stone used at each
    lift for each stone column.

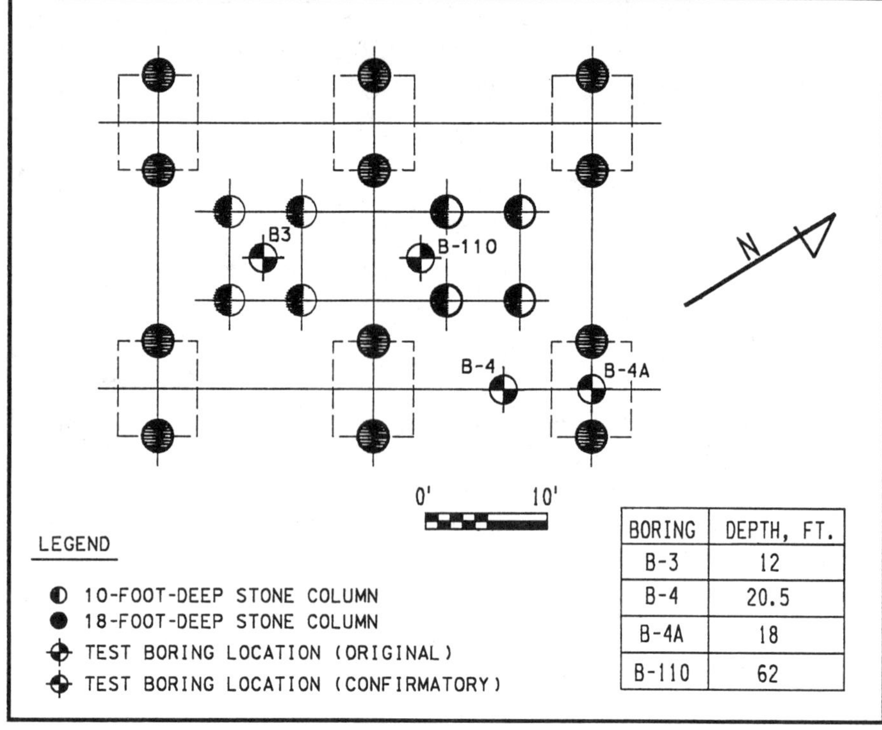

BORING	DEPTH, FT.
B-3	12
B-4	20.5
B-4A	18
B-110	62

FIG. 5 -- Stone columns and boring locations at
transfer tower

The jetting rate ranged from 200 to 400 gpm (0.76 to
1.52 cu m/min). Jetting not only accelerated the operation, but also
helped to provide a larger diameter stone column. The jetting
process caused an immediate increase in porewater pressure in the
vicinity of the installation; however, drainage through the stone
column itself quickly reduced this excess pressure.

The material used in the initial columns was 4-in. (102-mm)
maximum-sized crushed stone. However, 2-in. (51-mm) stone was used
in the remaining columns since it fell more easily down the annulus.
As might be expected, the quantity of stone used per hole varied,
depending on the initial density of the culm. Average amounts used

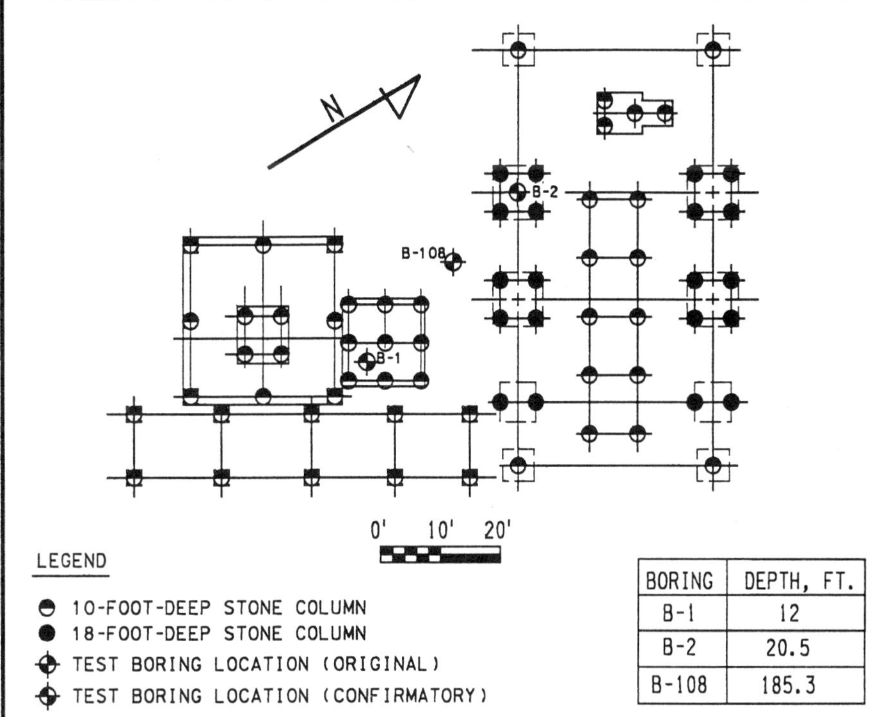

BORING	DEPTH, FT.
B-1	12
B-2	20.5
B-108	185.3

FIG. 6 -- Stone columns and boring locations at culm dryer

were 3.5 cu yd (2.68 cu m) for 10-ft (3-m) columns, 6 cu yd (4.6 cu m) for 18-ft (5.5-m) columns, and 10 cu yd (7.65 cu m) for 30-ft (9.1-m) columns.

To achieve the desired level of compaction in the stone columns, the vibroflot operator used the amperage reading on the diesel generator as a guide to the vibratory effort being applied. In loose culm, amperage readings were typically in the 120- to 160-amp range. Since the vibroflot had a constant amplitude of vibration and constant voltage, increased resistance to movement (and hence, increased compaction) required increased power which resulted in a higher amperage reading. GKN-HB considered a reading of 200 amps or more as an indication of sufficient compaction. It frequently took several minutes and repeated lifting and lowering of the vibroflot to achieve the 200-amp reading, particularly near the bottom of the deeper columns.

INSTALLATION DELAYS

It took 20 working days to install the 220 stone columns. This time period included some unanticipated delays. The first delay occurred when installation began in the culm silo area. Three out of the first five columns attempted could only penetrate 4 or 5 ft (1.2 or 1.5 m) before refusal. A bulldozer was brought in and soon exposed a layer of large boulders extending across the silo area. (Neither of the soil borings made in the silo area during the

exploration program (Fig. 1) had indicated these boulders). The
dozer stripped off the top 6 to 8 ft (1.8 to 2.4 m); silt and sand
sized culm was used as a replacement. Even then, some of the columns
stopped well short of their intended depth, presumably on top of
large boulders. Other columns achieved their design depth but were
unable to reach the 200-amp criterion.

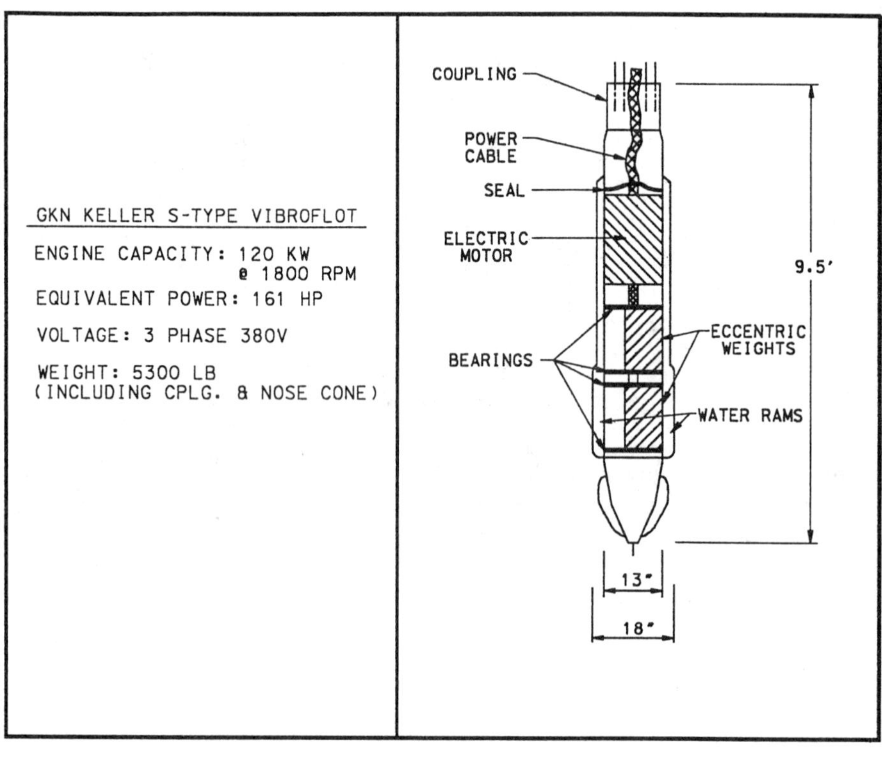

FIG. 7 -- GKN Keller S-type vibroflot

    Fig. 4 shows the locations of the short and the "below 200 amps"
columns. For the isolated "out of specification" columns, the
overall design was sufficiently conservative to assume bridging to
adjacent columns would occur if the column and surrounding culm did
not have adequate bearing capacity. Although Fig. 4 shows no large
groupings of "out of specification" columns, there are several small
groups. Confirmatory borings were made between selected columns, as
discussed in the PERFORMANCE ASSESSMENT section.

    The second major delay occurred when the operator of an adjacent
culm processing plant complained that a footing on his conveyor had
settled, possibly due to vibration from the vrsc installation.
Installation was halted, and Vibrotech Engineering, Inc. of Hazelton,
Pennsylvania was brought in to measure vibration levels caused by the
vrsc installation. The footing in question was about 300 ft (91 m)
from the vibroflot. Vibration measurements were taken at 5 locations
between the footing and the vibroflot, at distances ranging from 125
to 280 ft (38 to 85 m) from the vibrator. The maximum peak particle
velocity recorded was 0.024 in./sec (0.6 mm/sec), only a small

fraction of the 2 in./sec (51 mm/sec) minimum peak particle velocity generally assumed to cause damage. Therefore, it was concluded that settlement of the footing was not caused by the vibroflot vibration. A reading of 0.8 in./sec (20 mm/sec) was later recorded about 7 ft (2.1 m) from the vibroflot. This level of vibration might be expected from a relatively small pile driving (impact) hammer 7 ft (2.1 m) from the pile.

PERFORMANCE ASSESSMENT

In addition to the previously noted onsite installation inspection, where depth, amperage, and amount of stone used in each column were recorded, three independent means of confirming that the vrscs had achieved the desired results were adopted. Soil borings and plate load tests provided an estimate of the increased density and elastic modulus of the culm, while settlement monitoring gave a direct indication of whether the design criteria had been met.

Nine confirmatory borings were drilled -- two at the culm dryer, two at the transfer tower, and five at the culm silo, as shown in Figs. 4, 5, and 6. The borings were located in the center of the stone column group, to test the culm rather than the stone columns themselves. The vrsc design was based on achieving a minimum N-value of 20 in the improved culm, and a median N-value of 25. Fig. 8 shows that the vrscs achieved a substantial increase in N-values in practically every case. The target minimum N of 20 was reached at some elevations, although a minimum N of 15 appears to be a more realistic value. The target median N-value of 25 was achieved at almost every depth.

In confirmatory boring B-2 at the culm dryer, the N-values between 10- and 20-ft (3- and 6-m) depth were low, ranging from 4 to 11, with a median of 8 (Fig. 8). This boring was drilled in the center of four 18-ft-(5.5-m-) deep stone columns as shown on Fig. 6. The installation records for the columns were examined and showed no anomalies. Two actions were therefore taken. First, a plate load test was performed at the B-2 location. Later, settlement of the B-2 footing and adjacent footings was monitored, both during and after construction of the dryer.

The plate load test at B-2 was performed using a 5-ft-(1.5-m-) diameter stiffened steel plate, 2 in. (51 mm) thick, positioned symmetrically between the columns as shown on Fig. 9. The plate was jacked against a weighted steel frame to a maximum load of 90 tons (801 kN), which was 50 percent more than the footing design load. As shown on Fig. 9, a maximum settlement of just over 0.5 in. (13 mm) was recorded for sustained loading, indicating a very satisfactory performance. However, with only about 10 percent of the surface pressure being transferred to the soils below 10 ft (3 m), it could not be considered a real test of the looser soils below 10 ft (3 m) indicated in Boring B-2.

The 9 x 9-ft (2.7 x 2.7-m) footing at boring B-2 and three adjacent footings were monitored for settlement during and after construction. The maximum settlement recorded was about 0.25 in. (6 mm) under a loading of 4 ksf (192 kPa). Since each of the footings monitored was supported on four 18-ft-(5.5-m-) deep stone columns, these readings document clearly the performance of the

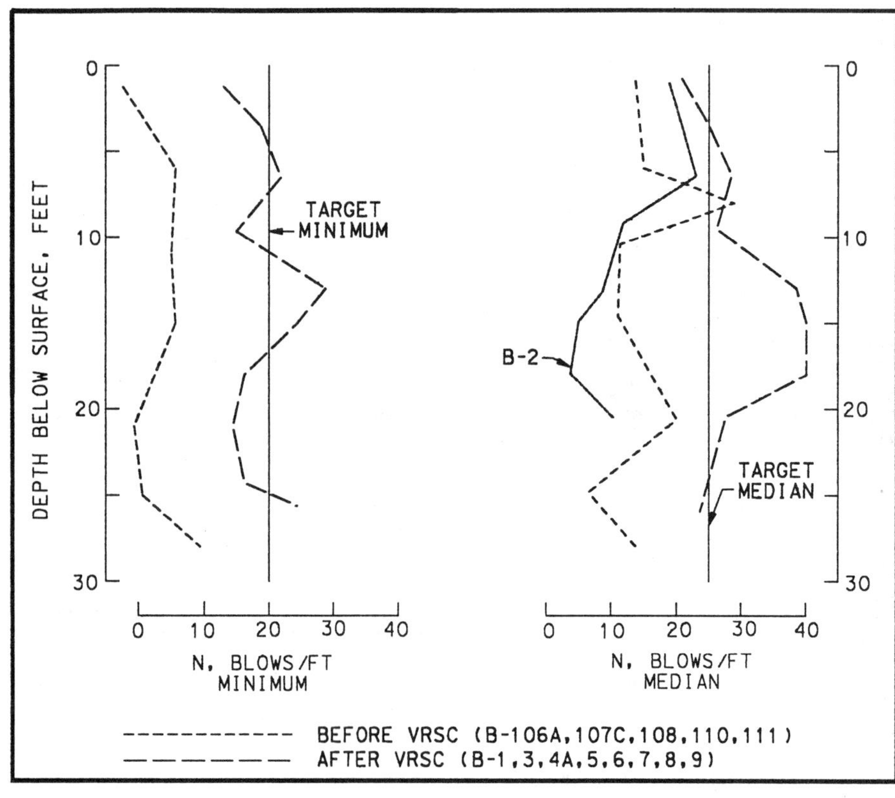

FIG. 8 -- SPT values before and after stone
column installation

columns. If relatively unimproved culm had been present between 10-
and 20-ft (3- and 6-m) depth, it apparently did not affect the
footing performance.

Confirmatory boring B-4 at the transfer tower was replaced by
B-4A, since B-4 was not located between adjacent columns (Fig. 5).
B-4 showed a median N-value of 12 between depths of 6 and 20 ft (1.8
and 6 m), while B-4A, located between adjacent columns at 7-ft
(2.1-m) centers, showed a corresponding median N-value of 30. The
error in locating boring B-4 thus provided a good demonstration of
the densification resulting from the vrscs, particularly towards the
bottom of the column.

Fig. 4 shows the location of the five confirmatory borings at the
culm silo. B-5 was drilled at the center of four columns that had
only been able to penetrate to between 23- and 25-ft (7- and 7.6-m)
depth, rather than the specified 30 ft (9.1 m). The boring indicated
N-values greater than 20 between 25- and 31-ft (7.6- and 9.4-m)
depth, i.e., there was no loose material below the boulders at 25 ft
(7.6 m).

The culm silo was monitored for settlement during construction.
When the silo was filled with a static load of about 4.5 ksf
(216 kPa), maximum total settlement recorded was about 1.25 in.

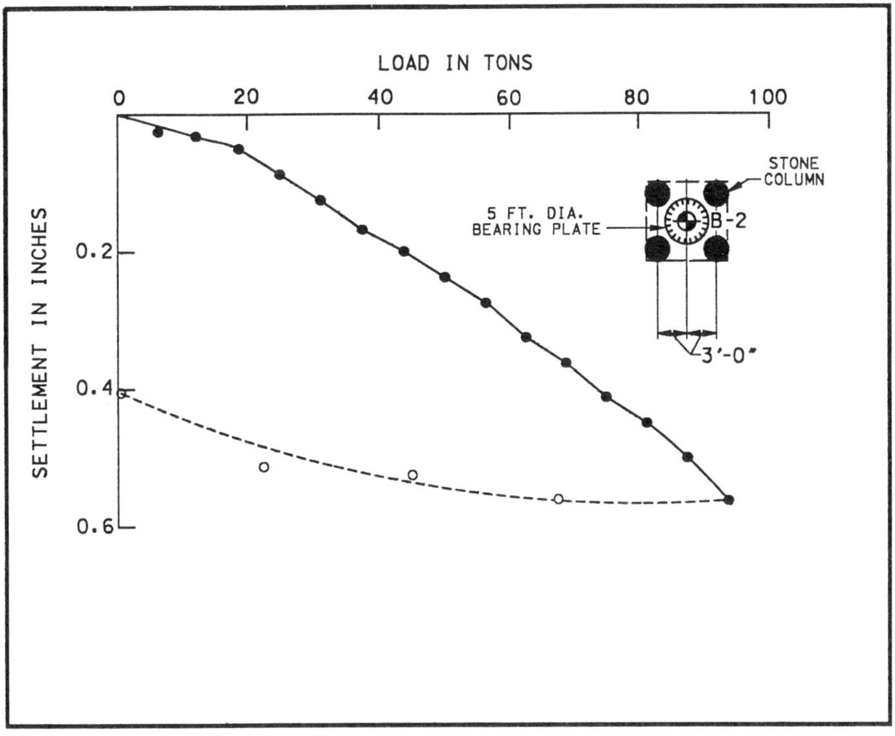

FIG. 9 -- Plate bearing test

(32 mm), with an average of about 1 in. (25 mm) and a maximum differential settlement of less than 0.25 in. (6 mm).

CONCLUSIONS AND LESSONS LEARNED

The vibro-replacement stone columns provided a satisfactory
foundation that not only met the structural design criteria, but
demonstrated distinct benefits from a cost and schedule standpoint.

The heterogeneous nature of the culm was continually apparent
during the compaction operation. No single series of compaction
centers exhibited the same response as monitored by amperage and
backfill consumption. However, as with pile driving, vibro-replacement stone columns are somewhat self-compensating. Instead of
blowcounts, there is a combination of depth, amperage, and backfill
consumption that provides a target installation criterion.

Despite its heterogeneous behavior, the culm behaved overall as a
granular material, as anticipated in the design. This was demonstrated by the satisfactory level of densification achieved between
the columns on 7-ft (2.1-m) centers. The isolated instances where
the desired amperage could not be attained probably reflected pockets
or zones of fine grained material.

The median target N-value of 25 in the confirmatory borings was
achieved, although the minimum target value of 20 was not always met.

Setting such a minimum target value was probably unrealistic, given the heterogeneous nature of the soil.

Although the stone columns were completed in less than 4 weeks, rather than the 10 weeks envisaged for installing and testing piles, further time could have been cut from the schedule if the layer of near-surface boulders at the culm silo had been recognized before starting installation. A series of test pits would have almost certainly located the presence of boulders, whereas isolated test borings did not.

REFERENCES

1. Schmertman, J. H., "Static Cone to Compute Static Settlement Over Sand," _ASCE Journal of the Soil Mechanics and Foundation Division_, Vol. 96, No. SM 3, May 1970.
2. Schmertman, J. H., Hartman, J. P., and Brown, P. R., "Improved Strain Influence Factor Diagrams," _ASCE Journal of the Geotechnical Engineering Division_, Vol. 104, No. GT 8, August 1978.
3. Vesic, A. S., "A Study of Bearing Capacity of Deep Foundations," Final Report, Project B-189, Georgia Institute of Technology, March 1967.
4. Priebe, H., "Estimating the Shear Resistance of a Soil Improved by Vibro-Replacement," in German, _Die Bautechnik_, Vol. 55, 1978.

R. Robert Goughnour, J. Teh Sung, and John S. Ramsey

SLIDE CORRECTION BY STONE COLUMNS

REFERENCE: Goughnour, R. R., Sung, J. T., and Ramsey, J. S., "Slide Correction by Stone Columns," <u>Deep Foundation Improvements: Design, Construction, and Testing</u>, <u>ASTM STP 1089</u>, Melvin I. Esrig and Robert C. Bachus, Eds., American Society for Testing and Materials, Philadelphia, 1991.

ABSTRACT: A review of design procedures for slope stabilization by stone columns is given. Stone column design and construction is reviewed for three projects where natural slopes were stabilized by stone columns. Stone column construction was achieved by the wet vibro-replacement method, a dry preaugering technique, and the dry, bottom-feed method. Instrumentation is described and performance of the stone columns on these projects is reviewed. Slope movement has stopped completely in all cases.

KEYWORDS: stone columns, slope stabilization, slide correction, ground improvement, vibro-replacement, vibro-displacement

Stone columns have been an accepted technique for soil reinforcement since the modernization of the installation process in the late 1950's [1]. Applications have included soil stabilization to limit settlements and to increase foundation strength under reinforced earth walls, tank farms, dam and highway embankments, bridge abutments, and buildings. Stone columns also function as efficient gravel drains in providing a path for relief of pore water pressures, thus reducing potential for liquefaction during an earthquake.

A less known application is the stabilization and prevention of landslides. For slope stability purposes, this ground improvement technique increases the average shear resistance of the soil along a potential slip surface by replacing or displacing the in-situ soil with a series of closely-spaced, large-diameter columns of compacted stone.

Dr. Goughnour is vice president of Geotechnics America, Inc., P.O. Box 2324, Peachtree City, GA 30269; Mr. Sung is a senior soils Engineer at New York State DOT, Soil Mech. Bureau, Bldg 7A, 1220 Washington Ave., Albany, NY 12232; Mr. Ramsey is a graduate research assistant at the Univ. of Del., Dept of Ocean Engr., Newark, DE 19716.

A review of current design procedures is first presented, followed by case histories of three projects where active slides were stabilized through the use of stone columns. The first case, involving an unstable soil mass that had damaged one of the main piers of the Steel Bayou bridge on Highway 465 in Mississippi, utilized the conventional vibroflotation (vibro-replacement) wet process.

The second case, located at the Nemadji River bridge in Superior, Wisconsin, involved a large clay mass slipping toward the Nemadji River and carrying the west bridge abutment and roadway with it. The soil at this location was stiff enough that augered holes stayed open down to the failure plane. Stone columns were installed by placing stone directly into preaugered holes in lifts of about 1 to 2 ft (0.3 to 0.6 m), each lift being compacted by a vibroflot.

The most recent case involved a natural slope along New York Route 22, near Wadhams, in soils too weak to use the augering technique. The area is environmentally sensitive. The necessary treatment and disposal of large amounts of silt-laden effluent would have made the wet installation method very difficult and expensive. Installation was accomplished by a totally dry, displacement technique, which utilized bottom-feed equipment.

CURRENT DESIGN PROCEDURES

Figure 1 illustrates typical stone column treatment patterns and definitions of the treated areas. The area ratio, A_r, defined also on Fig. 1, is the area of the stone (horizontal projection) divided by the total horizontal treated area, i.e. A_s/A.

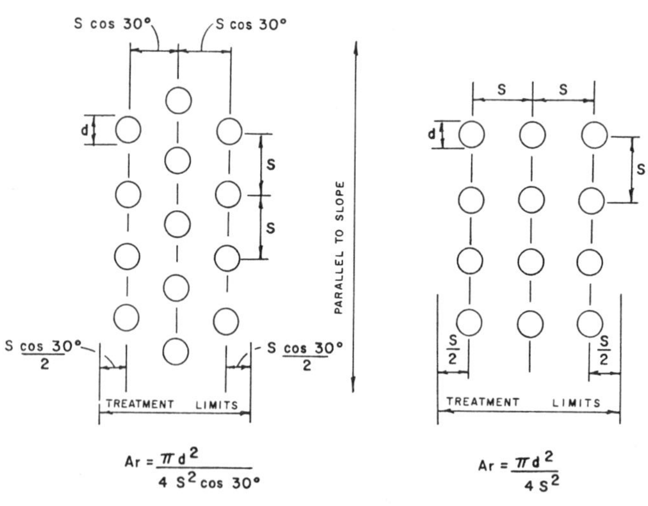

$$Ar = \frac{\pi d^2}{4 S^2 \cos 30°}$$

$$Ar = \frac{\pi d^2}{4 S^2}$$

a. Triangular array   b. Square array

FIG. 1 -- Definition of stone column treatment limits.

The shear strength of stone column treated soil depends on the shear strength of the untreated soil, the transverse shear strength of the columns, and the area ratio. All current design methods seek to define some average shear strength that can be applied to the stone column treated soil. Stability calculations are then carried out as usual by means of Bishop's, Janbu's, or other slope stability method.

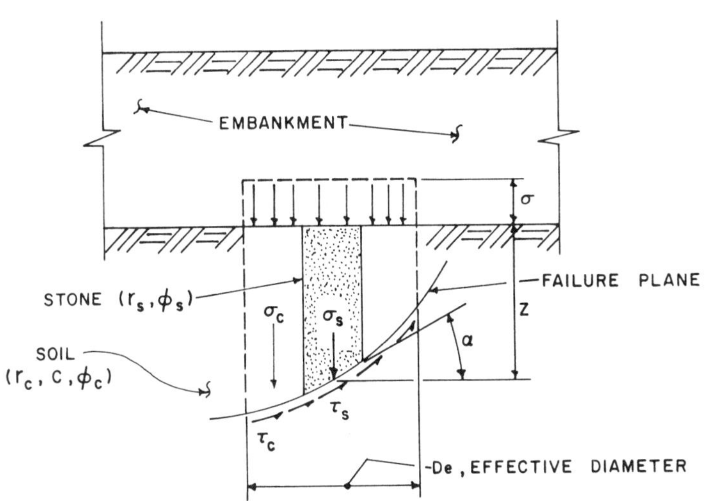

FIG. 2 -- Definition sketch for Japanese method.

Japanese Method

Although this method is applied to sand columns in Japan [2], the theoretical basis applies equally well to stone columns. Within a selected column at the depth where the failure surface intersects the centerline of the column (Fig. 2), the effective vertical stress due to the weight of the column and applied loading, σ_s, can be expressed as

$$\sigma_s = \Gamma_s z + \sigma\mu_s \tag{1}$$

where Γ_s is the unit weight of the column material (buoyant if below the ground water table), z is the depth below the ground surface, σ is the stress due to any embankment loading (zero in the case of a natural slope), and μ_s is the stress concentration factor for the column given by

$$\mu_s = S_{rv}/[1+(S_{rv}-1)A_r] \tag{2}$$

where S_{rv} is the stress ratio or the vertical stress in the column divided by the vertical stress in the in-situ soil, σ_s/σ_c. The shear strength of the column, τ_s, can then be expressed as

$$\tau_s = (\sigma_s \cos\alpha) \tan\phi_s \tag{3}$$

where α is the angle of inclination of the shear surface with respect to the horizontal, and ϕ_s is the angle of internal friction of the column material. The average weighted shear strength, τ_{av}, within the area tributary to the column then becomes

$$\tau_{av} = (1-A_r)\,\tau_c + A_r\,\tau_s\,\cos\alpha \tag{4}$$

where τ_c is the shear strength of the in-situ soil, and the other terms have been previously defined.

Average Strength Parameters Method

Average soil strength parameters within the treated soil are given by the following [3]:

$$c_{av} = c_c(1-A_r) \tag{5}$$

where c_{av} is the average cohesion to be used for the treated soil, and c_c is the cohesion of the in-situ soil, and

$$\tan\phi_{av} = \frac{(1-A_r)\tan\phi_c + S_r A_r \tan\phi_s}{1 + A_r(S_r-1)} \tag{6}$$

where ϕ_{av} is the average internal friction angle to be used for the treated soil, ϕ_s is the internal friction angle of the stone, ϕ_c is the internal friction angle of the in-situ soil, and S_r is the stress ratio appropriate to the orientation of the failure surface at that location, given by

$$S_r = 1+(S_{rv}-1)\cos\alpha \tag{7}$$

where α is the inclination of the failure surface from the horizontal. The average unit weight to be used in the treated zone, Γ_{av}, is

$$\Gamma_{av} = (1-A_r)\Gamma_c + A_r\Gamma_s \tag{8}$$

where Γ_c is the unit weight of the in-situ soil, and Γ_s is the unit weight of the stone.

Stability analyses may be performed using a total stress approach by assigning $\phi_c = 0$ for end-of-construction conditions, or using an effective stress approach by assigning $c_c = 0$ for long-term conditions.

Design Parameters

Parameters commonly used in Japan for sand columns are $\phi_s = 30$ degrees, $S_{rv} = 3$ to 5, and factor of safety = 1.2 to 1.3 [2]. Note that the Japanese method applies the stress ratio only to added loads, i.e., an embankment. Thus, if there is no added load, use of this method is equivalent to using a stress ratio approximating one with the Average Parameters Method (or more precisely a stress ratio value equal to the ratio of the unit weight of the stone to the unit weight of the in-situ soil - refer to eq 1).

When load is added over stone columns, stress ratios for the Average Parameters Method commonly used in the United States range from 2 to 3 [3,4]. If no load is added a stress ratio of 1 is common.

Values of the internal friction angle of the stone measured by shear tests on stone columns, large scale triaxial compression tests and large scale shear box tests range from 50 to 55 degrees for crushed, sound, well-compacted stone [4]. Since the friction angle of stone decreases with increased confining stress, these values should be corrected for depth and confining stress on the column [4].

For stabilization of natural slopes by use of the Average Parameters Method the authors recommend the following parameters: friction angle of the stone equal to 42 degrees when sound crushed stone is used, stress ratio of unity, and a factor of safety of 1.2 to 1.3, depending on the accuracy of the soil strength parameters. The same parameters apply when using the Japanese method with sound, crushed stone.

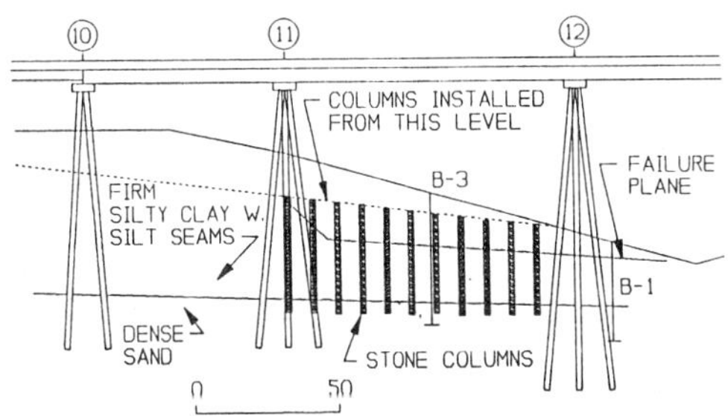

FIG. 3 -- Cross section of stone column area, Steel Bayou Bridge (section A-A of Fig.4).

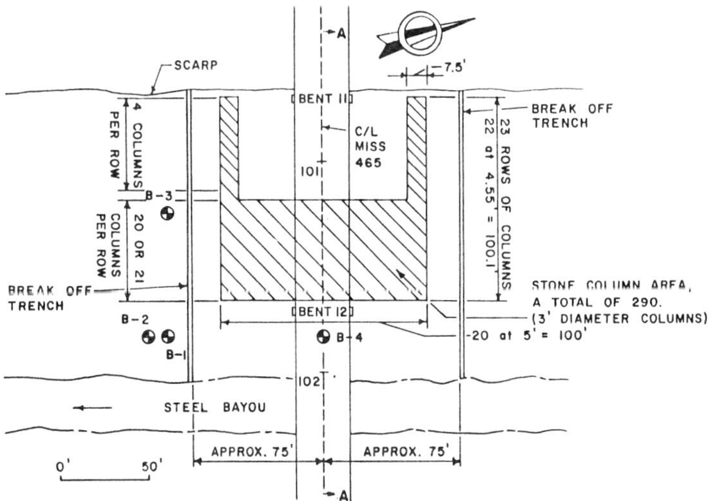

FIG. 4 -- Stone column layout at Steel Bayou Bridge.

CASE I - STEEL BAYOU BRIDGE

This project involved emergency repair of damage caused by a landslide which displaced a main river bent (bent #12, Fig. 3) of the Steel Bayou bridge on Mississippi Route 465. The slide, first reported in February, 1977, caused vertical ground displacement of about 4 ft (1.2 m) and displaced the main river bent horizontally at the ground line approximately 6 ft (1.8 m). Vertical displacement of the bridge deck was 0.93 ft (0.28 m). Soil underlying the site consisted of approximately 25 ft (7.6 m) of soft, gray, silty clay with silt seams, overlying gray sand. The spacing of the silt seams increased with depth.

The failure scarp, centered at the roadway, extended about 300 ft (91 m) in each direction parallel to the shoreline, see Fig. 4. Stability analyses were performed for initial conditions using inclinometer data and the failure scarp to define the failure surface. A sliding block failure along a silt seam with an average depth of 12 ft (3.6 m) was found to be the most critical, see Fig. 3.

Solutions considered included a rock buttress, a sand shear key, and stone columns. The stone column alternative was chosen as the most cost effective.

Since a rounded, natural, local stone material was used for column construction, two test columns were exposed and field shear tests were performed to determine column diameter and internal friction angle. Based on these tests, conservative values for effective friction angle, 35 degrees, and column diameter, 3 ft (0.91 m), were used for design. (The authors now believe that, because of local bearing failure of the stone columns against the in-situ soil, the reaction ring type shear test used in this case produced friction angles lower than actually existed [4]).

The shear strength contribution of each column along the failure plane was calculated in a manner similar to the Japanese method. To achieve a factor of safety 1.5, required 12 rows of columns at a 5 ft (1.5 m) triangular spacing placed between bent 11 and bent 12, extending 50 ft (15 m) from the centerline in each direction, see Fig. 4. In addition, 4 rows of columns extended on both sides of the roadway to add extra protection against further movement of bent 11.

Stone columns were installed by the standard vibroflotation process using water as the jetting medium. Although the original failure surface only extended to a depth of 12 ft (3.6 m), columns were placed through the silty clay and keyed into the underlying sand layer to prevent the possibility of a future failure in the deeper clay. No movement has been observed since completion of the project.

CASE II - NEMADJI RIVER BRIDGE

Seven years after major repairs had been performed at both abutments of a single span steel bridge across the Nemadji River in the city of Superior, the Wisconsin Department of Transportation discovered significant damage to the west abutment and approach slab. Inspections in 1982 indicated settlement of the approach slab and movement of the abutment toward the river. Inclinometers located at the site indicated more than 1 in. (2.5 cm) of movement in a southeasterly direction over a 4 month period in that same year.

Soil stratification at the site included 2 to 8 ft (0.6 to 2.4 m) of clayey and silty sand fill, underlain by 70 to 80 ft (21 to 24 m) of clay. Below the clay was a glacial till. The clay was composed of 2 layers; a reddish-brown clay, predominantly found above the water table, and a mottled reddish-brown and gray clay below the water table. Leaching of iron and manganese out of the clay caused the gray coloration. Some layers of the reddish-brown clay found above the water table indicated zones of saturation or trapped water. The failure probably developed because of an increase in hydrostatic pressure in these areas of trapped water.

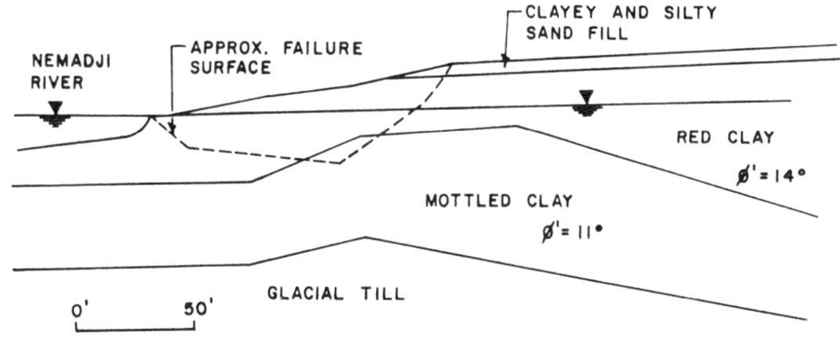

FIG. 5 -- Assumed failure surface at Nemadji River Bridge
(Section A-A of Fig. 6).

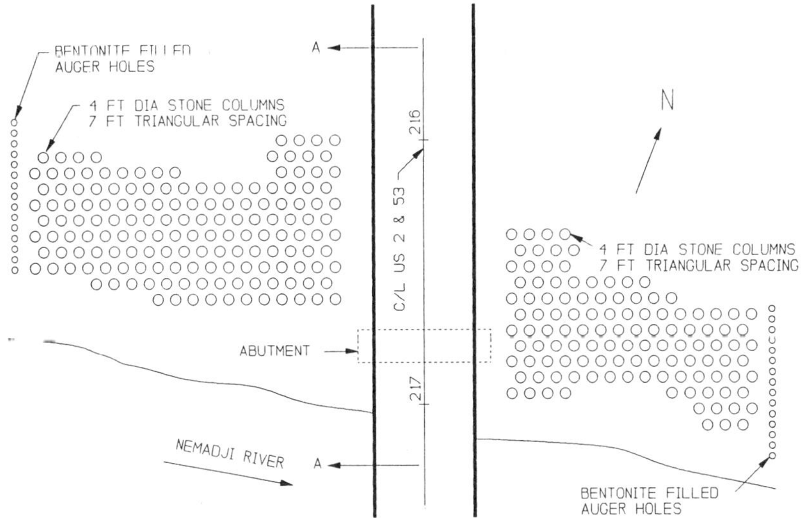

FIG. 6 -- Stone Column Layout at the Nemadji River Bridge.

The failure surface was generally defined by the location of cracks in the approach pavement and inclinometer data, but the shape of the failure surface was unknown. Both wedge and circular failure surfaces were modeled for stability of existing conditions, using the location of the cracks and the movement at a 30 ft (9 m) depth below the abutment as reference points. The assumed form of the failure surface is shown in Fig. 5.

Two correction measures were evaluated; (1) removal of a large wedge of soil at the top of the slope to be replaced with a structure supported roadway, and (2) stone columns. The stone column alternative was chosen for its economy, speed of installation, and minimal disturbance to the existing structure. By placing columns on either side of the abutment the slope was stabilized while traffic across the bridge was maintained. A series of closely spaced auger holes filled with bentonite were placed adjacent to both outer extremes of the stone column area to minimize disturbance from vibration outside the construction area.

The final design, using the Average Strength Parameters Method, required that 7 rows of columns be placed at a 6 ft (1.8 m) spacing perpendicular to the river and a 7 ft (2.1 m) spacing parallel to the river. Additional columns were placed adjacent to both sides of the approach to compensate for columns missing from this pattern under the approach (see fig. 6). This design produced a safety factor of 1.25 using an effective friction angle of 42 degrees and a diameter of 3.5 ft (1.07 m) for the stone columns.

Soil conditions at this site allowed preaugered holes to stay open to their full depth of 45 ft (13.7 m). Stone columns were installed using a 4 ft (1.2 m) diameter auger which made the use of water as a jetting medium unnecessary. Thus, the need for treatment and disposal of large quantities of silt-laden effluent was avoided. In addition, there was no need for environmental concerns over the possible intrusion of effluent into the river caused by the installation process. Stone size used on the project consisted of minus 3 in. (75 mm) crushed stone with less than 10 percent fines.

Inclinometer readings indicated a small amount of movement following construction in 1983. No movement has been measured at the site since 1985. The movement before this time may be attributed to the displacement required to mobilize the column shear strength.

CASE III - NEW YORK ROUTE 22

Construction of 156 stone columns by a bottom-feed, vibro-replacement, dry method stabilized a 220 ft (67 m) long natural slope on New York State Route 22 in the town of Wadhams. Seven borings in the failure area indicated a 10 ft (3 m) thick layer of grayish, brown, silty clay, overlying a 10 to 20 ft (3 to 6 m) thick layer of soft, gray, silty clay. Beneath this was a layer of silty gravel, with an artesian head of up to 5 ft (1.5 m). Liquidity index and activity of the soft clay were 1.0 and 0.5 respectively. Distribution of the moisture content, Atterberg limits, undrained shear strength, and sensitivity prior to construction are shown on Fig. 7.

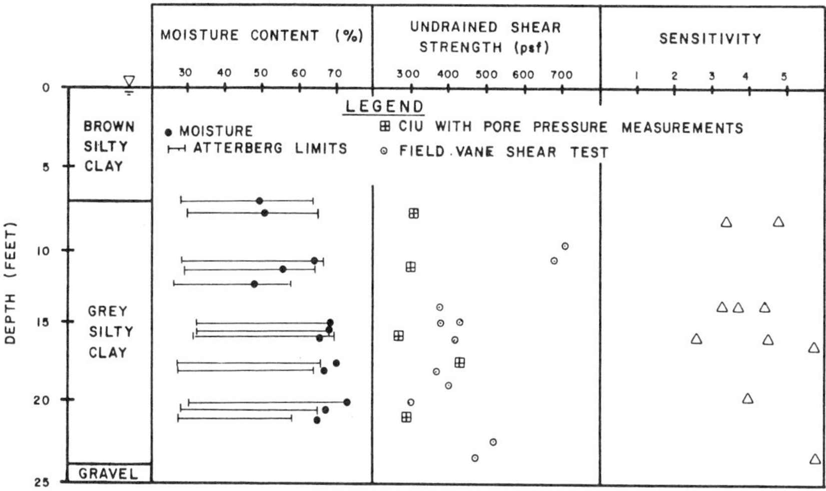

FIG. 7 -- Soil profile and engineering properties.

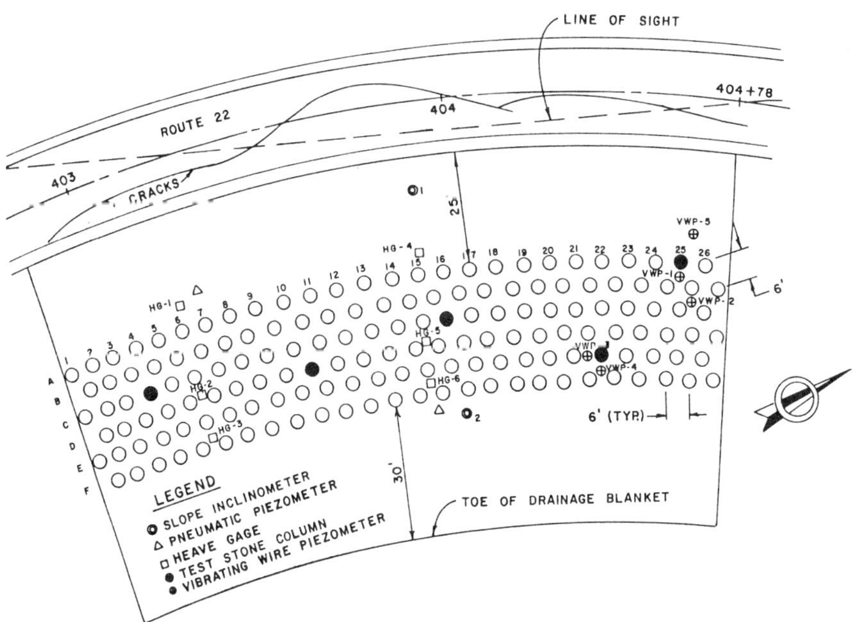

FIG. 8 -- Stone column and instrumentation plan.

Treatment Alternatives and Design

Three types of treatment were analyzed; stone columns, stabilizing berm, and shear key. The stone column alternative was selected because of environmental considerations, feasibility of construction, and economy.

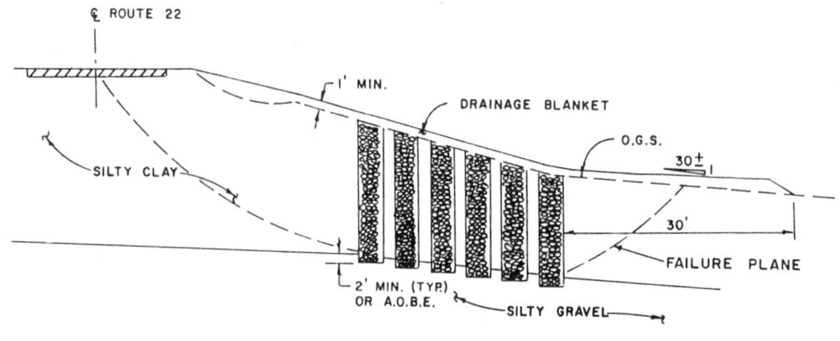

FIG. 9 -- Cross section with stone columns.

The design procedure followed the Japanese Method for selection of column spacing, number of rows and column diameter. A total stress analysis for end-of-construction conditions used an average undrained shear strength for the clay estimated to be 270 psf (12.9 KN/M^2). This was based on results of field vane shear tests, considering a remolding zone surrounding the column. For long-term analysis a single effective angle of shearing resistance assigned to the entire clay deposit was 21 degrees. This was based on a back-figure analysis of existing slope conditions and results of consolidated undrained, triaxial compression tests with pore pressure measurements. Stability analyses using the simplified Janbu method for both end-of-construction and long-term conditions were performed for various locations of the stone column zone to find the optimum position on the slope for the installation. Based on the friction angle of the stone, taken as 40 degrees, the calculated safety factor was 1.20.

As shown on Fig. 8, a total of 156, 3.5 ft (1.07 m) diameter columns, arranged in 6 rows of 26 columns each, were installed in a triangular spacing of 6 ft (1.8 m) on center. This yields a replacement ratio, A_r, of 0.31. A drainage blanket of 1 ft (0.3 m) minimum thickness was placed over the area as shown on Fig. 9.

Equipment

Specialty equipment consisted of a stone delivery system, follower tubes, and a vibroflot. Other support equipment included a 50-ton crawler crane, a front end loader, an air compressor, and an electric generator. Only the stone delivery system is unique to the bottom-feed installation method, the remaining specialty equipment being identical to that used for the vibro-replacement (wet) method.

The stone delivery system is composed of a ground-based hopper containing a blow tank unit connected by a pneumatic stone delivery line to a stone receiving tank at the top of the follower tubes. Two separate chambers, the stone receiver and a pressure tank, are mounted at the top of the follower tubes. The outlet at the bottom of the pressure tank is connected to a 6 in. (15 cm) stone delivery pipe located alongside the follower tube and the vibrating probe. This stone delivery pipe terminates at the probe tip, thus assuring positive delivery of a known amount of stone to a known depth in the ground. The equipment arrangement and specifications are detailed elsewhere [5].

Stone Column Material

NYSDOT No. 2 stone initially used for column construction, was found to contain several particles in excess of 4 in. (100 mm) in the long dimension, and caused clogging problems within the stone delivery system. A smaller stone, NYSDOT No. 1, was selected to replace the No. 2 stone and assure smooth operation of the system. The gradations and the compacted dry-rodded densities, determined in accordance with ASTM C29, for these two types of stone are shown in Table 1.

TABLE 1 -- Gradation and Densities for No. 1 and No. 2 Stone

Sieve Size	Percent Passing	
	No. 2 Stone	No. 1 Stone
1-1/2"	100.0	100.0
1"	93.2	98.9
1/2"	3.9	74.8
1/4"	0.0	8.2
Compacted Density (pcf)	99	95
Loose Density (pcf)	90	88

1 in. - 2.54 cm
1 pcf = 0.15771 KN/M^3

Installation Procedure

Stone column installation began by allowing the vibroflot to penetrate by its own weight aided by 25 to 40 psi (1.7 to 2.7 atm) of air pressure through the in-situ soil to the required depth. The vibroflot was then lifted 3 to 4 ft (0.9 to 1.2 m) while depositing stone with the aid of air pressure into the void left by the vibroflot, and then repenetrated into this freshly placed stone, compacting it and forcing it radially into the in-situ soil. This process was repeated with repenetration each time to the original depth, until the acceptance criterion described below was satisfied. Once this acceptance criterion was met for a particular depth, repenetration was made to a depth equal to 1 ft (0.3 m) less than the previous repenetration depth, repeating this process as necessary to reach the ground surface.

The acceptance criterion was based on information obtained during the installation of 5 test columns. In general the top 10 ft (3 m) of clay was stiff and could provide a larger lateral resistance than the underlying soft clay. The acceptance criterion in this upper region was based on achieving a peak motor amperage of 140, rather than achieving a column diameter of 3.5 ft (1.07 m). In the softer material, below a depth of about 10 ft (3 m), the requirement for a 3.5 ft (1.07 m) diameter column was maintained. This was assured by requiring that at least one batch of stone was exhausted for each 1 ft (0.3 m) of column. A batch of stone was determined by weight and proportioned to be that amount of stone necessary to form a column 1 ft (0.3 m) in height and 3.5 ft (1.07 m) in diameter at a density corresponding to 95 percent relative density as determined in accordance with ASTM C29.

Observed Behavior

During initial penetration the vibroflot motor current varied from 100 to 140 amperes depending on the stiffness of the clay surrounding the probe. The lowest amperage readings were observed in the very soft clay at the clay-gravel interface. Attempting to penetrate the gravel layer caused an abrupt increase in amperage to over 140. After the probe reached the gravel layer the contractor was allowed to begin stone backfill and compaction.

The very soft clays, found immediately above the gravel layer, were remolded locally around the vibroflot during installation. The contractor first attempted to achieve an amperage of 140 for all depths regardless of the amount of stone deposited. However, the very soft clays, after remolding, could only provide peak power consumption of about 130 amperes. This usually occurred after two batches of stone had been deposited. To prevent excessive remolding of the soil fabric, the probe was allowed to repenetrate a stone charge no more than twice, and never permitted to rest within the very soft clay layer.

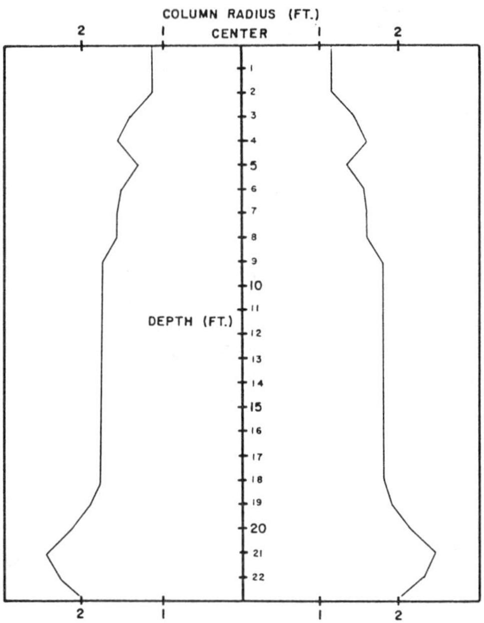

FIG. 10 -- Typical calculated stone column diameter.

Column Diameter

The in-place diameter of the columns was estimated using the method of batch distribution developed by the senior writer, outlined in detail in reference [6]. It was assumed that each batch of stone was deposited equally over a specified length of column, 4 ft (1.2 m) in this case. The average column diameter for the top 10 ft (3 m) was 2.5 ft (0.76 m). The average overall in-place column diameter ranged from 2.9 to 3.4 ft (0.88 to 1.04 m). The smaller top diameter reflects the effect of the stiff surface layer. A typical distribution of column size with depth is shown in Fig. 10.

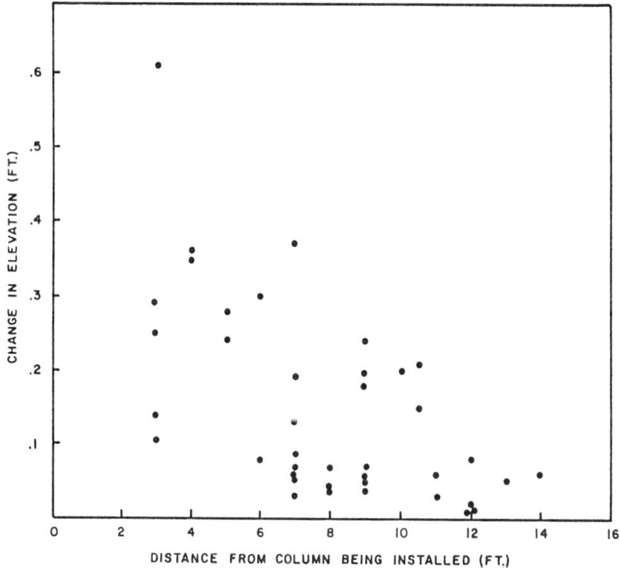

FIG. 11 -- Change in surface elevation versus distance from column.

Heave

Heave was monitored for columns within 14 ft (4.3 m) of 6 Borros type heave gages anchored about 3 ft (0.9 m) below the surface. See Fig. 8 for location of instrumentation. Although soft clay extruding up around the probe made valid calculation of actual displacement difficult, heave gage readings indicated that initial penetration only produced 0.5 to 2 in. (1.3 to 5.1 cm) of surface heave. Compaction of the columns resulted in approximately 1 to 2 ft (0.3 to 0.6 m) of vertical lift, heave and extruded material combined, over an estimated influence area of 3 ft (0.9 m) from the center of the column. Although the surface heave records are scattered as shown in Fig. 11, the amount of heave generally decreased with increasing distance from the column. The maximum heave was about 0.6 ft (0.18 m) when the column was installed 3 ft (0.9 m) from the gage. The radius of the heaved area extended to a distance of about 12 ft (3.7 m).

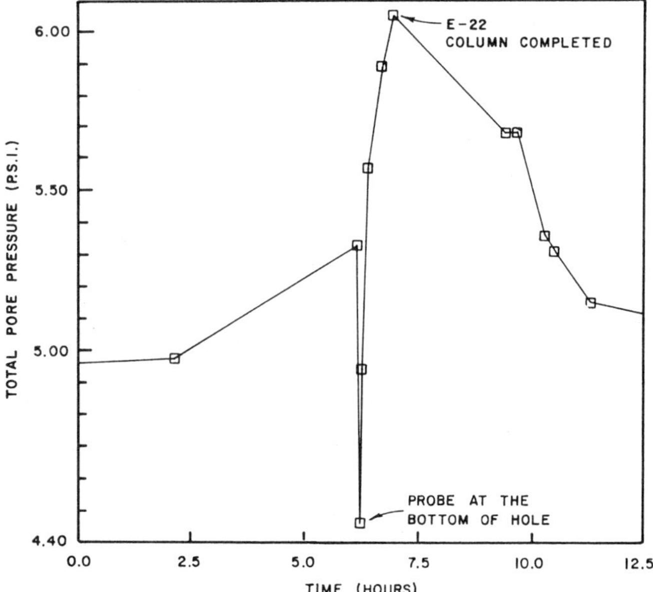

FIG. 12 -- Piezometer VMP-4 reading versus time.

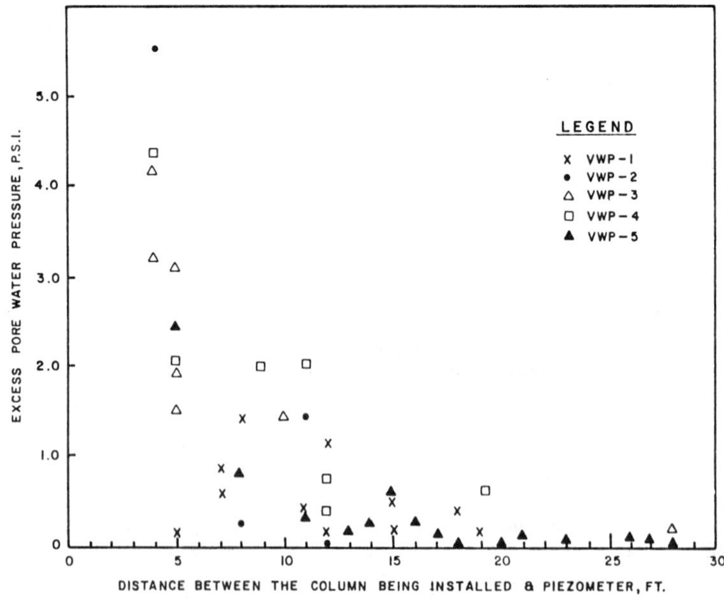

FIG. 13 -- Pore pressure versus distance from column.

Excess Pore Pressure

Five vibrating wire piezometers were installed midway between selected columns in the soft clay at depths of either 12 or 18 ft (3.7 or 5.5 m). Typical pore pressure readings versus time are shown for VWP-4 in Fig. 12. Most of the piezometer readings indicated that negative pore pressures were developed during initial penetration of the probe. This could be due to a combination of soil remolding, reduction in lateral pressure, and pore water being forced away from the piezometer by the 25-40 psi (1.7 to 2.7 atm) air pressure. Once stone placement and compaction started, the pore pressure rose sharply. Dissipation of the resulting excess pore pressures started immediately after compaction as shown in Fig. 12. Complete dissipation generally took between 5 hours and 2 days, which indicates rapid reconsolidation of the soil between columns and that any strength loss due to remolding during installation was temporary. At a distance greater than 15 ft (4.6 m), the maximum excess pore pressure was negligible as shown on Fig. 13.

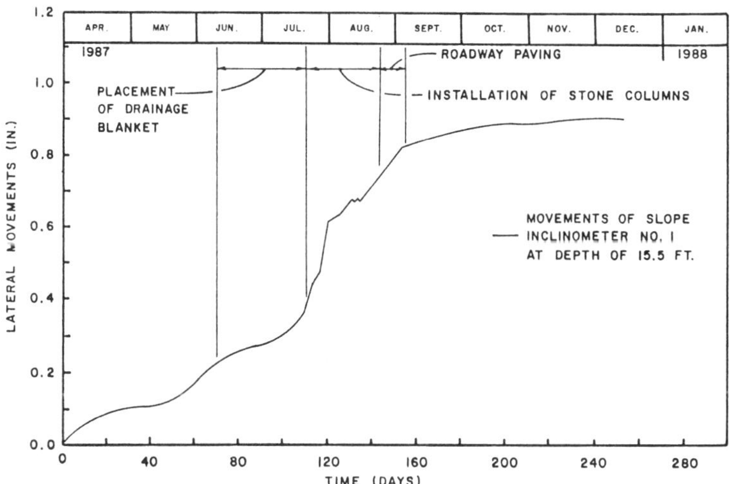

FIG. 14 -- Lateral movement versus time.

Lateral Displacement

Inclinometer readings confirmed that the slip plane was at a depth of 15.5 to 16 ft (4.7 to 4.9 m), and that virtually all of the slip occurred at that depth. As shown in Fig. 14, movement prior to construction amounted to 0.003 in. (0.076 mm) per day. However, during construction of the bottom two rows, the displacement rate increased up to 0.03 in. (0.76 mm) per day; for a total additional movement of 0.3 in. (7.6 mm). After completion of the bottom two rows, the rate of movement decreased and additional movement of only 0.13 in. (3.3 mm) took place during installation of the remaining columns. To date inclinometers are being monitored at a frequency of once per six months, and no additional movement has taken place beyond that shown on Fig. 14.

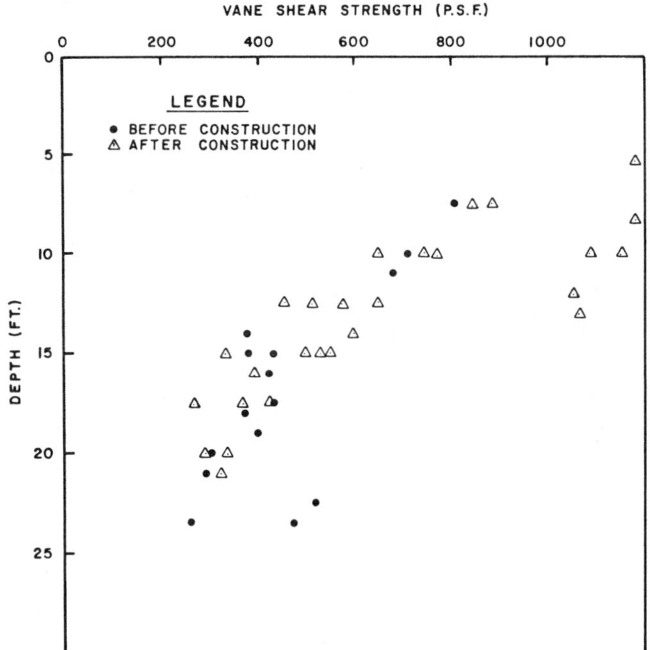

FIG. 15 -- Undrained shear strengths before and after construction.

Strength Change of In-Situ Soil

Referring to Fig. 15, a significant increase of vane shear strength occurred in the upper 10 ft (3 m) of soil with a smaller increase to about 15 ft (4.6 m). Essentially no improvement of the undrained strength of the in-situ soft soil resulted from the placement of columns below a depth of 15 ft (4.6 m). All of the vane shear data were taken within 2 weeks of the completion of the columns. The upper 10 ft (3 m) of soil was stiffened with column installation as evidenced by slower rates of probe penetration as the project progressed. This stiffening took place in soils which were reconsolidated over their initial stress condition by the lateral stress imparted by the column installation. In the deeper, softer clay it appears that after remolding, reconsolidation to approximately the stress condition before installation took place.

SUMMARY

Stone column design and construction has been reviewed for three slope stabilization projects. Instrumentation has been described and performance of the stone columns on these three projects has been presented. A review of two methods of calculating shear strength improvement by stone columns, along with recommended design parameters for use in each has been given.

Appropriate construction methods chosen for each project were based on individual site conditions and project requirements. In all three cases the stone column method has provided a positive, cost effective solution to the slope stability problem. Slope movement has stopped completely in all cases.

ACKNOWLEDGEMENT

Data and information were furnished by the Mississippi State Highway Department, the Wisconsin Department of Transportation, and the State of New York Department of Transportation. Their cooperation is gratefully acknowledged.

REFERENCES

[1] Mitchell, J.K., "Soil Improvement, State-of-the-Art-Report," Proc. of Tenth Int. Conf. on Soil Mechanics and Found. Engr., Stockholm, 1981, Vol. 4, pp 509-565.

[2] Aboshi, H., Ichimoto, E., Enoki, M., and Harada, K., "The 'Composer'- A Method to Improve Characteristics of Soft Clays by Inclusion of Large Diameter Sand Columns," Intern. Conf. on Soil Reinforcement, March, 1979, Paris, pp 211-216.

[3] Goughnour, R.R. and Jones, J.S., "Design and Construction of Stone Column Reinforced Cofferdams," Found. Engr. Congress, Evanston, Ill., June, 1989, pp 231-243.

[4] Barksdale, R.D. and Bachus, R.C., (1983). Design and Construction of Stone Columns, Report SCEGIT-83-10 submitted to the Federal Highway Administration, School of Civil Engineering, Georgia Institute of Technology, Atlanta Georgia.

[5] Sung, J.T. and Ramsey, J.S., "Slope Stabilization by Stone Columns at Wadhams, New York," NYSDOT, Soil Mechanics Bureau, February, 1988.

[6] "Stone Column Inspection Guide for a Bottom Feed Vibro-Displacement Method," NYSDOT, Soil Mechanics Bureau, April, 1987.

David A. Greenwood

LOAD TESTS ON STONE COLUMNS

REFERENCE: Greenwood, D. A., "Load Tests on Stone Columns,"
_Deep Foundation Improvements: Design, Construction, and
Testing, ASTM STP 1089_, Melvin I. Esrig and Robert C. Bachus,
Eds., American Society for Testing and Materials, Philadelphia,
1991.

ABSTRACT: Stone column foundations are often incorrectly regarded
as piles and tested accordingly. For single columns load tests
are appropriate. However full scale field tests in which
stresses on and between columns were measured clearly demonstrate
how the performance of columns under widespread loads depends on
loading circumstances which cannot be simulated by small scale
tests on single columns. Such tests are misleading for contract
purposes. A more cost effective approach is to concentrate on
high grade site investigation and control during design and
construction; supplemented by instrumentation to monitor response
to initial loading of the prototype.

KEYWORDS: stone columns, load testing, column/soil stress ratios,
construction monitoring.

Many engineers regard stone columns constructed to strengthen
cohesive soils as piles. It seems logical therefore to test them as
piles by direct loading at ground surface. The fundamental mechanisms
of stress transfer from the load to ground are the same for columns
and piles. However interaction of columns with the soil is so extreme
as to make them effectively a form of soil reinforcement rather than
of load transfer to lower ground horizons. Concrete piles are a part
of the structure by-passing weak ground; but stone columns form a
composite with weak ground to stiffen it and reduce its deformation.
There is no virtue in stone columns for load bearing unless they are
cheaper to construct than piles or more convenient and appropriate to
the site circumstances.

The paper outlines full scale field experiments which demonstrate
the reinforcing behaviour of groups of stone columns. The changing
ratio of stresses on columns and intervening soil as the load is
applied confirms load sharing between columns and soil: the variation
in relative stiffness of column and soil determines whether response
is mainly of reinforcing or load transfer to lower strata.

Dr. Greenwood is a Director of Cementation Piling & Foundations
Ltd, Maple Cross House, Maple Cross, Rickmansworth, Herts WD3 2SW, U.K

In general, reinforcement of the soil is the major role. Because of this difference from pile behaviour, tests which do not simulate full size loaded areas are misleading, as with small plate tests for spread footings. The cost of such large scale tests tends to prohibit their use, especially for small scale works. Small projects may even be eliminated since the cost of testing dominates project costs.

It is better instead to concentrate during construction on the critical column details to ensure that appropriate methods are used to assure achievement of the designed properties of the foundation: also to monitor its behaviour during first loading. Attention so directed is probably more worthwhile than direct loading tests and of considerably greater cost benefit.

BASIC STABILITY OF STONE COLUMNS AND PILES

Fig. 1 illustrates stresses generated in the soil surrounding a single column subject to load, and round a group of columns under widespread loads. These are common to both piles and stone columns or any similar columnar structure of stiffness which differs from that of the soil. The differing behaviour of piles and stone columns arises from the hugely different ratio of their stiffnesses to that of the soil. Piles are relatively very stiff - about 10,000 times more than the soil - but stone columns are only about 2 to 20 times stiffer. Unlike piles their relative stiffness can change significantly as load is applied. The stiffness and resistance to bursting, of timber, concrete and steel piles is such that their cross-section barely changes during loading and their radial enlargement has no significance for the soil. Thus radial soil stress σ_R remains virtually unchanged after pile construction and during loading. The weaker stone column, with no tensile strength, compresses and bulges in response to load and the contact stress σ_R increases bringing into play the resistance of the soil, analagous to a pressure meter.

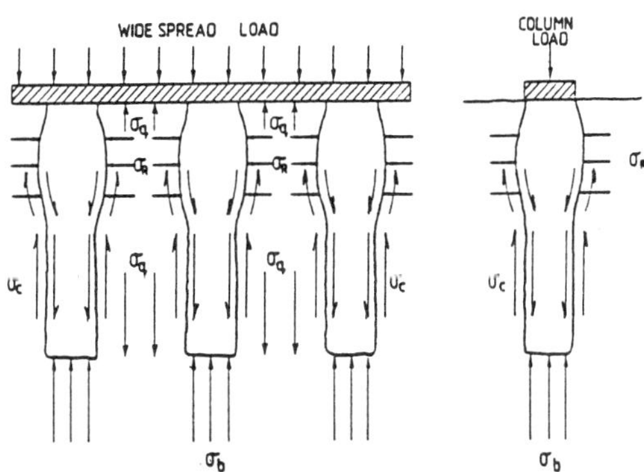

FIG. 1 Stress constraints on soils surrounding stone columns or piles

DEFORMATION OF STONE COLUMNS

Under widespread load a group of columns deforms much more than a group of rigid piles assuming both have solid toe bearing. The contained soil is subject to much greater compression and consolidation than in a pile group and is thereby strengthened: the same action between piles is insignificant in strengthening the soil. This in turn increases the resistance of the stone column, so the stresses are interactive. The interaction between columns and soil is marked. Moreover, the strength of both columns and soil is continually changing as load is applied, unlike that of piles and contained soil, which remains substantially constant during load application.

Too rapid an application of load can induce high pore pressures in the soil which negates the strengthening effect. At each load increment the radial constraining stress on bulging is approximately: $4c_u + \sigma_{r_0} - u$ according to Hughes and Withers [1]. The undrained cohesion of the soil (c_u) and passive stress (σ_{r_0}) are reduced by excess pore pressure (u). σ_{r_0} can be taken as K_s ($\gamma z + yp$). K_s is the ratio of vertical to horizontal stresses in the soil prior to loading. γ is the effective unit weight at depth z. p is the surface load stress on the soil and y a stress distribution factor such as Boussinesq or Westergaard. For ground reinforcement, columns in groups are usually closely spaced: also they are usually of free draining grading relative to the soil. In most situations drainage of the columns is almost instantaneous, but that of intervening soil will depend on its fabric. Design is usually based on neutral soil pore pressures and undrained strength. If the soil is fully drained its effective stress parameters should be used. It is important to be assured that these design assumptions apply, especially during initial loading. An example of failure is given later where this was not done.

The range of drained modular ratio for stone columns to soil is similar to that of steel to concrete and the analogy with compression reinforcement of concrete is obvious. Elastic theories have been developed as for concrete structures [2] for estimation of deformation of stone columns. Elasto-plastic models have offered further refinement [3]. However it is difficult for such theories to cope with the constantly changing strength of the granular materials in the column and soil.

One of the most recent approaches [4] takes account of interactive behaviour of soil and column material as pore pressures dissipate summing results of calculations of deformations for discrete intervals of soil/column depth in a manner similar to consolidation settlement calculations. The method appears to give results of the correct order [5].

The behaviour of the composite soil/stone column foundation is extremely complex with simultaneous and inter-dependent changes of stress ratios, pore pressures and resulting stiffnesses in both soil and column. Current theoretical models for deformation estimates are either oversimplified, or with increasing complexity, require input parameters which are difficult if not impracticable to determine by soil investigation. Hence the need for testing at full scale for important structures. However, only testing which accurately simulates the constraints on the soil is valid in such a situation.

FULL SCALE LOAD TESTING

Confirmation of this general behaviour has been obtained from field tests. In the late 1960's and early 1970's in Britain an intermittent series of full scale field tests was undertaken concomitant with development of understanding of stone column behaviour. Usually these tests exploited contract works opportunistically and so were often undertaken with limited preparation time and with site investigation data obtained for the contract purposes rather than specifically for the tests. Relevant soil properties were not always available in adequate detail for test analysis therefore.

Nevertheless direct information on the magnitude of load sharing was obtained as a useful sidelight on theoretical development. An outline of some of these tests is given below with their outcome. They illustrate why single column load tests are irrelevant for stone columns under widespread loads.

Load shedding along stone columns

Hughes and Withers [1] pointed out that with stress transfer through skin friction to the soil the direct vertical stress in a column would rapidly diminish so that a single column would be unlikely to bulge except near the top. This is both because the direct stresses are highest at this level, and the containing radial

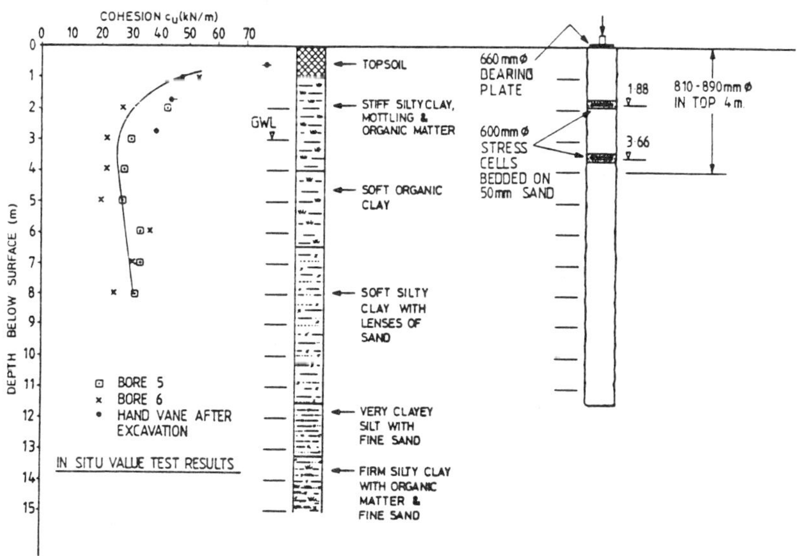

FIG. 2 Uskmouth - load test circumstances.

stress is likely to be a minimum because there is little overburden weight and the strength of normally consolidated clay just below any surface crust is also low. Thus a single column almost invariably fails by bulging just below the crust with little stress transferred downwards beyond 4 or 5 column diameters. The critical length, at which end bearing and skin friction resistances equate, is typically short.

Uskmouth: An isolated column was constructed for plate load testing on to a firm toe bearing stratum at 11m depth. Two stress gauges were incorporated within it at depths of 1.83 and 3.66 metres (Fig. 2). Thus stresses in the column were known at three levels (including the surface). The column was formed in a conventional manner by wet technique vibroflotation. However, at each gauge elevation the vibroflot was removed and a 50mm layer of 20mm gravel poured in followed by a similar thickness of sand which was levelled before the 600 mm diameter gauge was lowered onto it. The vibroflot then gently tapped the cell to seat it and it was covered by similar layers of sand and gravel in reverse order before constructing the column normally to the next cell horizon.

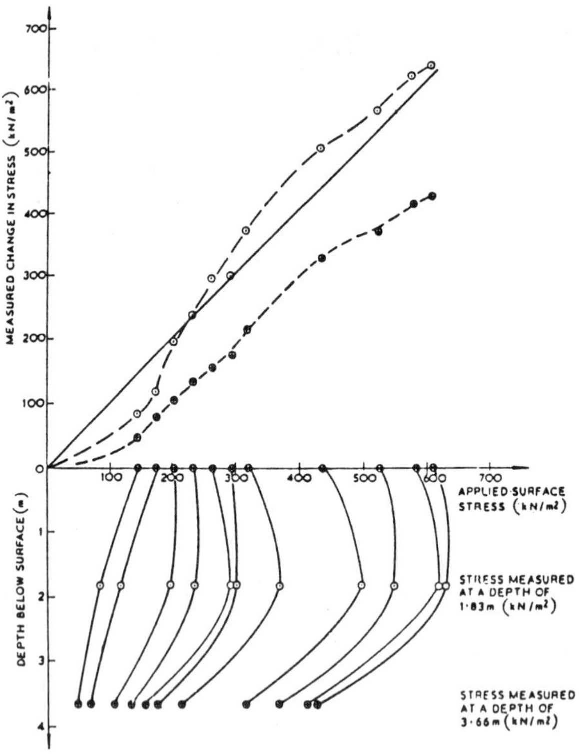

FIG. 3 Uskmouth - measured stresses in stone column.

For the test a load was applied to the surface of the column through a 660mm diameter steel plate by 30 tonne hydraulic jack. Immediately after the test the upper 4 metres of column was carefully excavated and its diameter measured at between 810 and 890mm with a mean of 850mm and the maximum at 2m depth below the soil crust. At the same time hand vane tests were made in the clay surrounding the column, which in this upper zone at least, showed enhanced values of cohesion by a factor of approximately 1.5.

The test results are summarised in Fig. 3 for maximum applied stress of 630 kN/m^2. Settlements were measured only to the extent of 25mm which occurred at a surface bearing stress of 330kN/m^2, settlement and stress readings were allowed to become sensibly constant at each increment. A Chin plot [6] of the recorded settlement suggests an ultimate strength of 704 kN/m^2. Bulging resistance was calculated at 630 kN/m^2 on the basis of initial soil strength: this would be enhanced if treatment resulted in a real soil strength gain as implied by the hand vanes.

From Fig. 3 it will be seen that at the higher loadings the upper cell registered higher stress levels than those applied at surface which prima facie seems impossible. A possible explanation is stress redistribution due to deformation of the soft clay below the crust causing the crust to transfer its weight to the column by skin friction.

Despite these reservations the form of the result is valid. In the early stages of loading little stress was transferred deep into the column because skin friction against the strong soil crust sustained stress distributed from the plate. With an average crustal cohesion of 45 kN/m^2 over a two metre depth the plate stress would have to reach about 420kN/m^2 before this was fully mobilised as reflected in the changing gradient of the plots. Projection of the stress/depth curves suggests that load stress effect would be eliminated at a depth of about 5.5 metres or 6½ diameters. If account is taken of the effect of the crust this is quite reasonable.

The test confirms the Hughes/Withers hypothesis but demonstrates the practical influence of a stiff crust over soft material.

East Brent (Somerset) By contrast a loading test on a group of columns constructed under a widespread embankment load, reported by McKenna et al [7] shows how the effect of containment on soil between columns enhanced their bulging resistance, so that they transmitted stress to full depth, acting as piles. The field circumstances for this trial are reproduced from their paper in Fig. 4.

It is important to note that the distinction between the grey silty sand and the soft grey silty clay is by no means clear. Estuarine sediments of the River Severn have been continuously deposited from Pleistocine to recent times, such that the horizons below 11 metres depth in contrast to those above, tend to be more sandy and silty, than clayey and silty, but there is no sharp

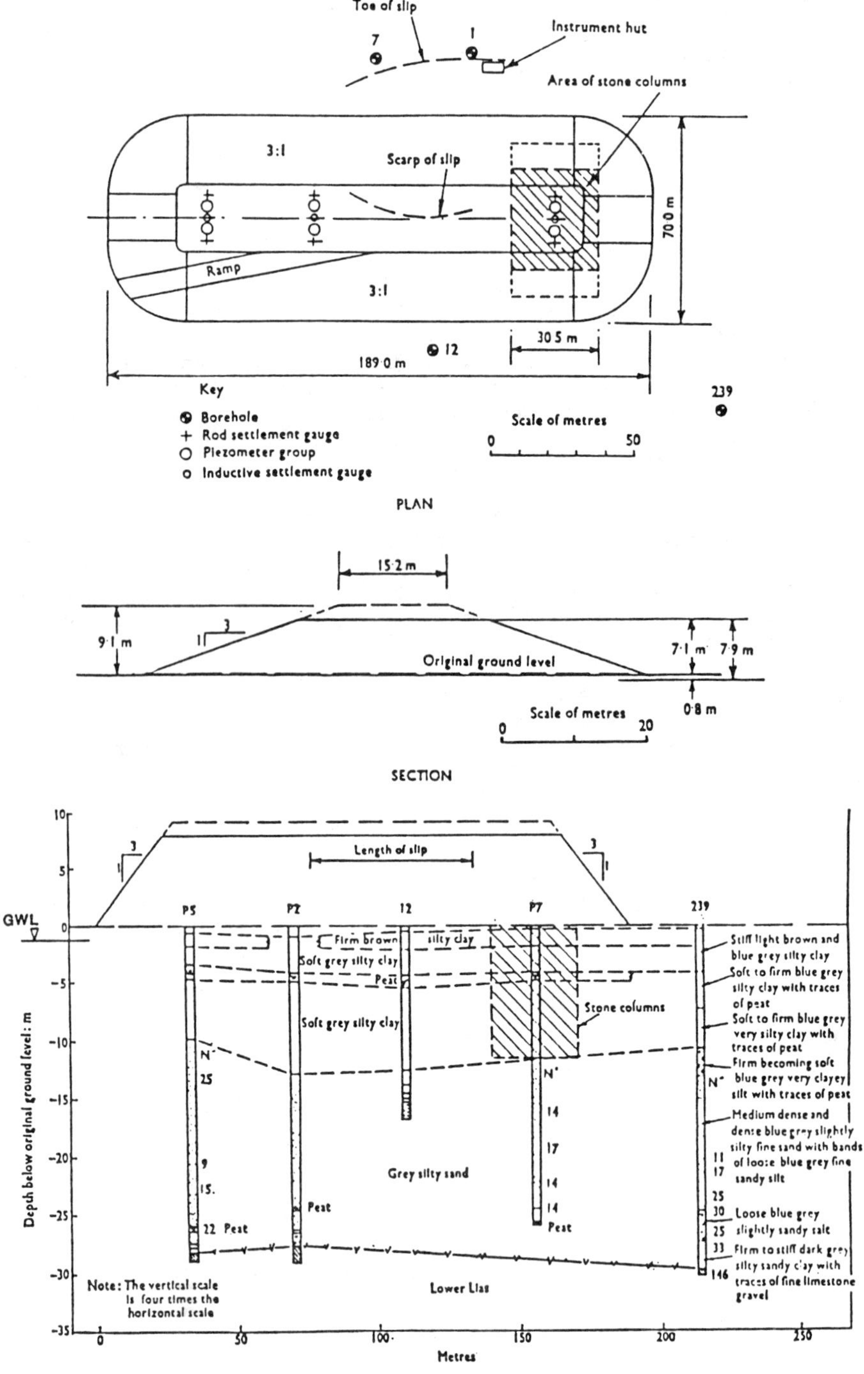

FIG. 4   East Brent – embankment load test circumstances.

divide. The lower layers are inter-bedded with clay laminae and intercalated with peat lenses, especially near the base. Undisturbed samples can frequently be taken within the "sand".

It will be noted that stone columns were constructed in a small area under the highest part of one end of the trial embankment. They were spaced at 2.45m on a triangular pattern. Average diameters inferred from stone consumption were 0.9m.

The authors show (Fig. 5) that the untreated end of the embankment settled considerably less than that with stone columns. The untreated central section, which slid after 90 days of loading when the embankment was 7.1 metres high, had settled almost identically with the stone column end immediately prior to the slide. As a result, the Engineers concluded that the columns were unsatisfactory for the support of the embankment and they were not used in the main project. The reasons postulated for this behaviour by McKenna et al were: no drainage to the columns due to remoulding during the construction process; and the loss of clay volume by its interpenetration in the coarse stone column under the load stresses.

These explanations do not stand examination. Greenwood [8] demonstrated how piezometric measurements during construction of columns had clearly shown free drainage. By reference to shear resistance required for soil to penetrate soil pore spaces [9] the strength of clay, even softened by remoulding, would inhibit interpenetration. In any event columns constructed by wet vibroflotation as in this case, have gravel void spaces filled with sand from the parent soil, the fines having been removed by water.

A more plausible explanation supported by the original authors' own piezometric data (Fig. 6) is that columns performed as friction piles. For the average value of c_u = 26 kN/m^2 the critical length of column is about 10 metres which is approximately the same as the length of the columns constructed at 11.3 metres.   The bulging resistance by Hughes and Withers method [1],

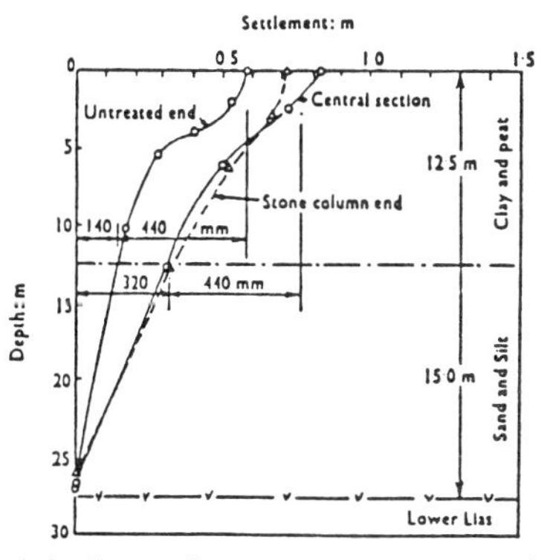

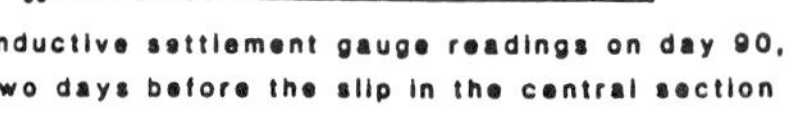
Inductive settlement gauge readings on day 90, two days before the slip in the central section

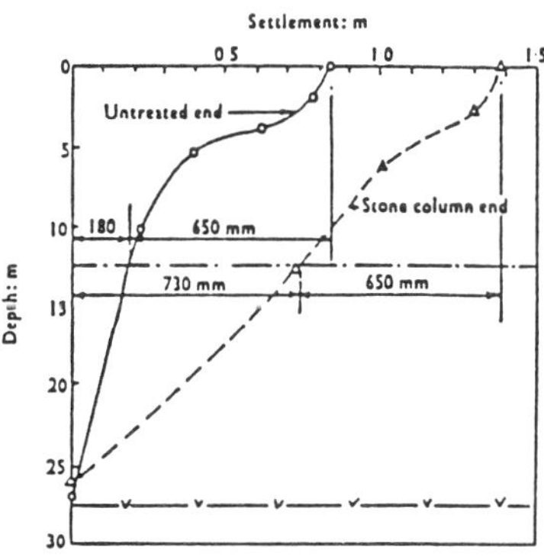

FIG. 5 East Brent - settlements recorded.

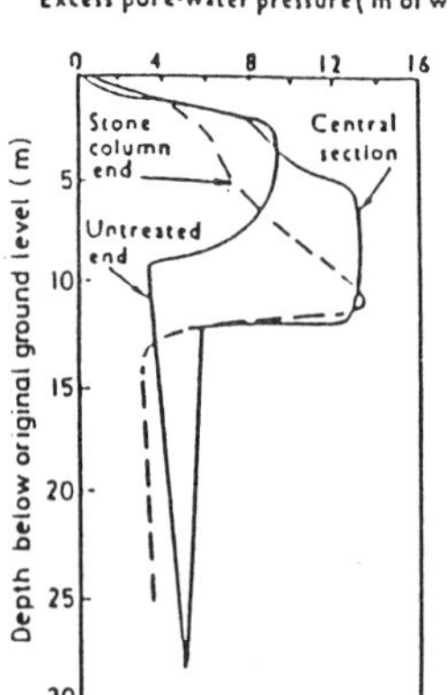

FIG. 6   East Brent – peizometric records.

taking account both of the minimum cohesion of $20kN/m^2$ at about 5 metres depth, and of the contribution of the load itself, was at least equal to the stress due to the weight of the embankment, so that bulging probably did not occur. The columns should therefore have just sustained the load as rigid piles.

However, the clays were slightly sensitive and when fully remoulded had strengths of about 8 to 9 kN/m^2. The effect of this would be to reduce resistance of the column to punching penetration whilst leaving the bulging resistance unimpaired. This is because during wet stone column construction any silty clay sheared and softened by the lateral gyratory impacts of the vibroflot is immediately removed by the upflowing water velocity in the annulus between the machine and the soil. The space is made good by falling coarse stone (50mm size) and the impacts of the machine act on this pad of stone, through which water can flow freely. Thus as the process progresses, the gravel between the machine and the soil gradually thickens until the amplitude of shears transmitted to the soil is insufficient to cause shear degradation. The column then builds upwards and the process is repeated at the next level. With this hypothesis, backed by observations of excavated columns in clays, it is unlikely that bulging resistance, which depends on the bulk properties of the soil remote from the column, was influenced. It is possible that skin friction in a thin layer adjacent to the column could have been diminished marginally, but assuming a rough contact it is unlikely to have regressed to its limit.

The pore pressure measurements recorded before failure of the central section (which also wrecked the recording station) show that in the stone column zone pore pressures increased more or less proportionally with depth to the base of the columns. This behaviour is consistent with increasing relative movement with depth between the column and soil, suggesting punching. The load of the rapidly constructed embankment was transferred almost fully to the level of the toe of the columns.

By contrast piezometric measurements in the central and remote end zones without columns showed high pore pressures at the elevation of the peat layer between 4 to 5 metres depth. The presence of the peat is barely reflected at the stone column end where pore water pressure dissipation appears to have taken place. The central section which slipped was found to have failed on and just below the peat. Fig. 5 shows that the untreated end settled – presumably by shearing displacement – above the peat layer, whereas the stone column end showed uniform settlement throughout the depth of the sediments.

This seems to be convincing evidence of the columns having acted as rigid piles. The conclusion must be that the widespread load contained the intervening clay, preventing column bulging at any depth, so allowing stress transfer down the columns. They did not control the settlement because they were of inadequate depth to do so.

The behaviour of these two examples of a single loaded column and a group of columns under widespread load demonstrates the crucial influence of containment of the soil on column behaviour. It is clear from this that test loading of a single column, although satisfactory for isolated columns, cannot represent the performance of columns under widespread loading. A single column fails primarily because of bulging at shallow depths, but under widespread load columns may fail by bulging or can be like a rigid pile according to circumstances of constraint. Since relative column and soil stiffnesses are not known precisely, and bulging is sensitive to their ratio, the prototype circumstances of constraint must be reflected in the test for the result to be reliable.

Measurement of load sharing between column and soil

Field tests are described for three different cases: a simple strip footing on a drained fine grained granular fill; a widespread flexible load on columns in soft clay; and a widespread relatively rigid load on stone columns in soft clay, each with measurements of direct vertical stresses on the stone columns and on the soil between columns. Differing behaviour was recorded.

St. Helens The arrangements were as illustrated in Fig. 7. A dummy footing was fitted with 150mm diameter stress cells in its base to locate on and between columns conventionally constructed by wet vibroflotation. The soil was unusual: it comprised siliceous particles with a content of jewellers rouge. These had been used for glass grinding and hence had developed almost spherical particles of inert material of silt size. This fill material was present in great depths and undergoing wet vibro compaction for construction works. Ground water table was low and the material affecting the test was unsaturated and behaved as drained during compaction and testing.

The load was applied by hydraulic jacks on a spreader beam to the dummy footing to a magnitude approximately half the ultimate bearing capacity of the untreated ground so that there was no likelihood of shear failure.   In fact as the cycled test results showed the material behaved elastically at all the values of stress adopted.

At low load stresses the ratio of column stress to soil stress was approximately 3.5 declining with increasing load to about 2.5. It

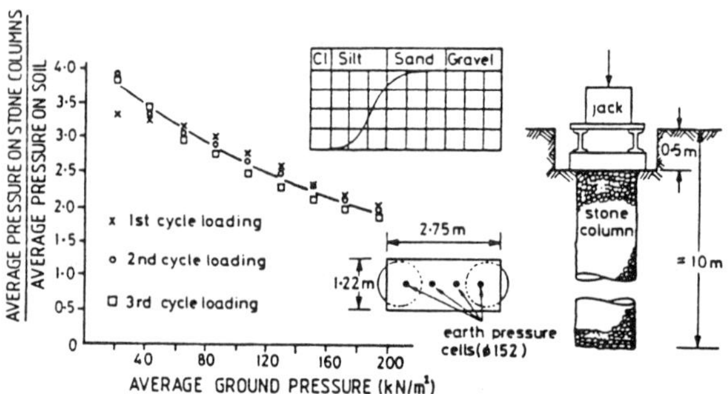

FIG. 7 St. Helens - loading a strip footing on stone columns.

is interesting to note that a ratio of 2.05 is derived for fully mobilised $\emptyset'$ (ultimate failure) in both the column and soil from Mohr stress plots, assuming values of $\emptyset' = 42°$ for the column and $\emptyset' = 30°$ for the silty fill. This suggests that with elastic loading of drained material the stress ratio reflects simply the value of mobilised friction angle contributing to the stiffness of each component of the ground.

<u>Canvey Island</u> Column/soil stress ratio at ground surface was measured below a 36m diameter, by 12m high steel oil storage tank with conical roof. This was constructed on a conventional free draining rolled gravel, asphalt topped pad foundation resting on original ground surface. Stone columns were formed to support the pad from original surface by wet vibroflotation to a depth of 10 metres on a triangular spacing of 1.52m covering the area of the tank plus an annular strip 6 metres wide. Sample measurements of column diameter near surface showed an average of 750mm. The soil conditions are shown in Fig. 8. The silty clays were recent estuarine deposits of the River Thames.

Pressure cells 600mm diameter were placed on ground surface prior to construction of the granular pad and were bedded in sand as at Uskmouth.

Measurements of stress were made during initial water test loading of the tank which took place slowly over 100 days. This was followed by rapid unloading and equally rapid filling with oil during which the stresses were also measured.

Loading rate and column spacing should have ensured drained loading conditions although no piezometric measurements were made. The stepped shape of the settlement curve suggests that drainage was occurring under incremental loading. Also the silty clays were laminated which would enhance radial drainage to the columns.

The pressure cells were placed in an area close to the centre of the tank and would not be affected by edge shears in the soil.

Measured stresses against settlement, loading and time are shown in Fig. 9. An apparent small zero error in the measured stresses was registered. This was probably induced by rolling the sand/gravel pad over the cells. Rebound of either the cell strain diaphragms or of the soil would almost certainly register some effect. If $20kN/m^2$ adjustment of datum is made the deduced loadings fit the known surcharge reasonably well. However the ratio of stresses on soil and columns is not altered much by a datum change and is shown in Fig. 9 against surcharge stress. At low load stresses the ratio was very high at 25 (Fig. 10).

Corresponding settlements were small at about 50mm measured at the edge of the tank. Calculations of bulging resistance show this to be very high in relation to the applied stresses both at the edge and centre of the tank. Under the centre where surcharge would contribute to stiffness bulging would be unlikely at column stresses less than $1250 kN/m^2$ and approximately half that near the edge of the tank. The columns thus would act as relatively rigid piles. The initial high ratios of stress are confirmation of this expectation having regard to the close spacing of columns. After surcharge of about $50 kN/m^2$ corresponding to a column load of 155 kN/m^2 some plastic yielding occurred but there is no obvious explanation for the onset. Stress ratio then rapidly fell at a decreasing rate until at full loading of $130 kN/m^2$ it reached a ratio of about 5. Settlement also advanced in concert.

Generally the experiment showed that the field loading was reflected fairly accurately under the flexible base of the tank. It is clear that as settlement increased the soil accepted a progressively larger proportion of the applied stress. The ratio

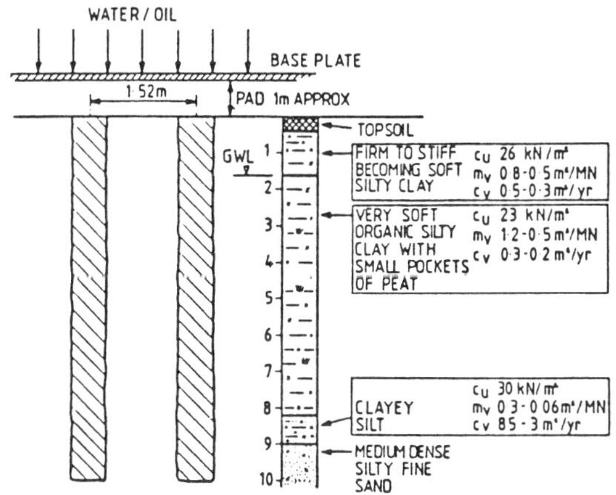

FIG. 8 Canvey Island - soil properties

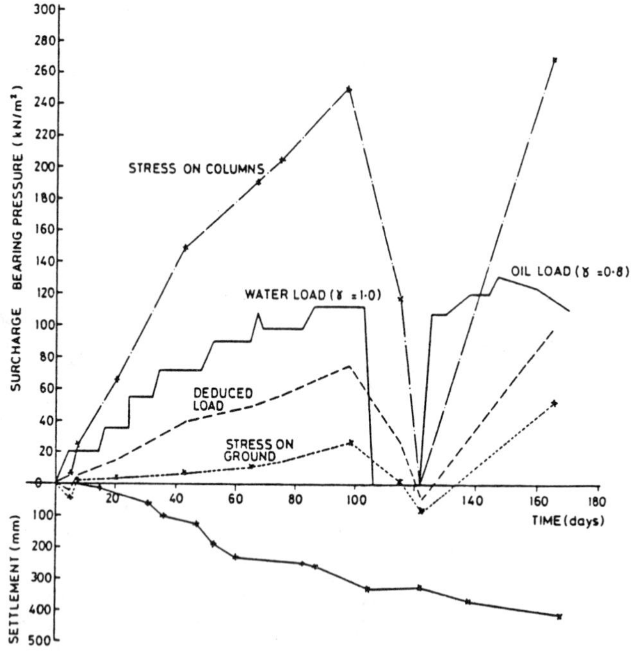

FIG. 9   Canvey Island - measured stresses and settlement.

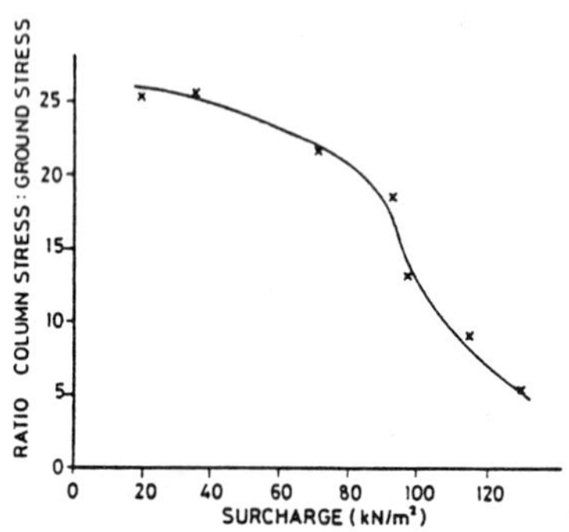

FIG. 10   Canvey Island - measured stress ratios.

of 5 is much higher than that for the St. Helens test because of
the relative weakness of the soil at Canvey. The final ratio of 5
suggests an isotropic stress on the clay since for $\emptyset'$ assumed
fully mobilised and no plastic bulging in the columns, the ratio
of principal stresses in the column would also be about 5. Thus
stresses on the soil both vertically and radially would be
approximately equal, and there would be little shear stress in the
soil at this stage. This accords with the relatively small total
settlements measured after 100 days.

<u>Humber Bridge South Approach</u> As a precursor to construction of the southern approach to the Humber Bridge a trial of stone columns for limiting settlements was carried out under a rolled chalk fill embankment. The embankment was founded on soft organic silty clays (Fig. 11) which had been stiffened with stone columns. Those under the highest part of the fill were at 2.25m triangular spacing. The columns were constructed from ground surface by wet vibroflotation and were 9 metres long to rest in stiff boulder clay.

Roughly one metre depth overburden was excavated both to allow measurement of stone column diameters (average 775mm) and to place the pressure cells. The cells were bedded on about 150mm of fine sand both on and at mid points between columns. Further sand was placed on top and the area backfilled to original ground level with crushed chalk. The embankment was then raised in the stages indicated in Fig. 11 by means of conventional heavy construction plant, placing and rolling chalk insitu. Density determinations of the fill gave an average 2.08 tonnes/m^3: this is much denser than natural local chalk and suggests collapse of the chalk structure under rolling stresses.

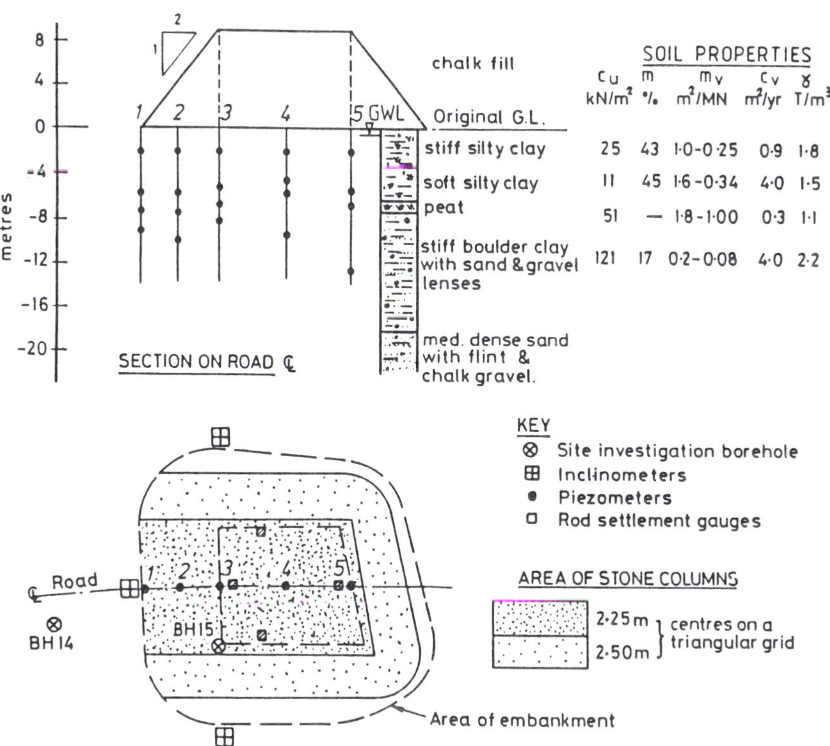

FIG. 11 Humber Bridge - embankment load test circumstances.

The site was instrumented with induction settlement gauges at three depths and with rod settlement gauges encased in independent tubes through the fill; with piezometers and inclinometers. Settlements under the centre of the bank are shown in Fig. 12.

Relative to estimates of untreated settlement under slow loading final settlements were 1.3 times less which corresponds almost exactly with predictions both empirical and elastic for a modular ratio of 5, and for the area ratio (A/Ac) of 9.3 [10]. Furthermore there is a clear step in settlement in the softest stratum on reaching the highest load, implying plastic bulging in this stratum.

Using the Hughes and Withers hypothesis [1] and taking account both of the overburden constraint due to applied load, and of measured pore pressures (Fig. 12), it can be shown that bulging should not occur at any stage of loading. However, if the constraint due to the rolled fill is not taken into account it can be shown that although bulging should not have occurred during the first plateau of loading, the second would have induced it. This is significant in the light of the measurements of stress on the columns and soil shown in Fig. 13.

The stresses recorded on the soil were initially high at about 80 kN/m^2 and remained almost constant throughout the loading showing a tendency to rise only in the final stages when settlement was approaching its maximum of almost a metre. The experimenters were concerned about zero errors on the cells in contact with granular material giving variable distribution of stress across the cell. However, strain gauges were sited on the diaphragms both near centre and edge and readings were averaged to give the recorded stress. There is no reason to suspect errors in readings as several cells corroborated each other. A plausible alternative explanation is that the effect of the heavy compaction of the chalk was to create intense local direct stress which was partially maintained by capillary suction to leave a residual stress akin to that in an overconsolidated crust. This appears to be about twice overburden weight above the cells. When properly compacted at correct moisture contents the fine grained chalk structure collapses and becomes exceedingly strong. It seems in this case to have formed a raft sufficiently rigid to span the columns and adhering through soil suctions to the partly saturated clay crust below. The measured stress on the soil remained fairly constant until overburden weight exceeded the prestress imposed by the rollers so only the columns would reflect the increase in weight beforehand.

There is some further evidence of this rafting from the fall in column stress as pore pressures dissipated after the first plateau of loading. If both soil and columns had been yielding uniformly to load, pore pressure dissipation should have increased resistance to bulging allowing the column stress to increase or at least to be maintained; and corresponding settlement would continue instead of levelling off. If however, the fill was rafting to some extent over the soil, the effect of pore pressure dissipation would be to allow radial consolidation as bulging

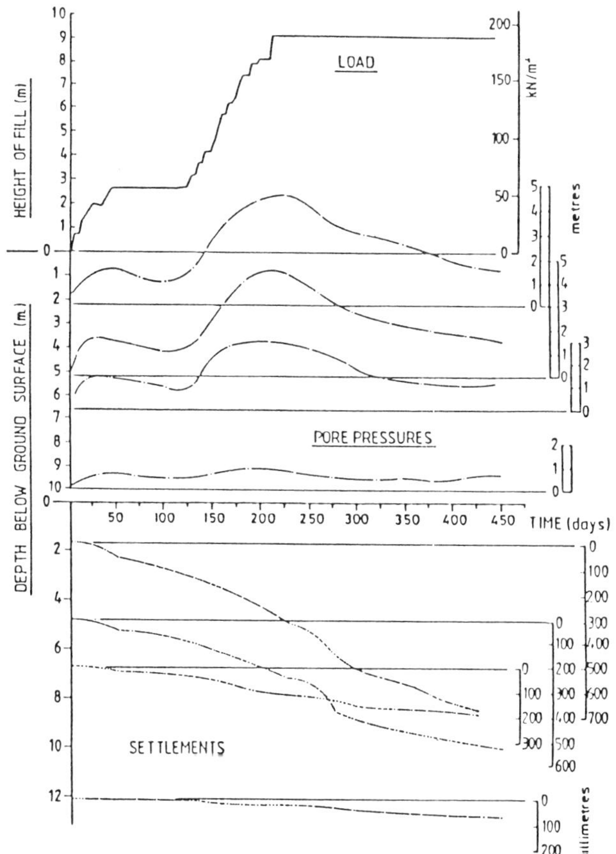

FIG. 12   Humber Bridge - measured pore pressures and settlements.

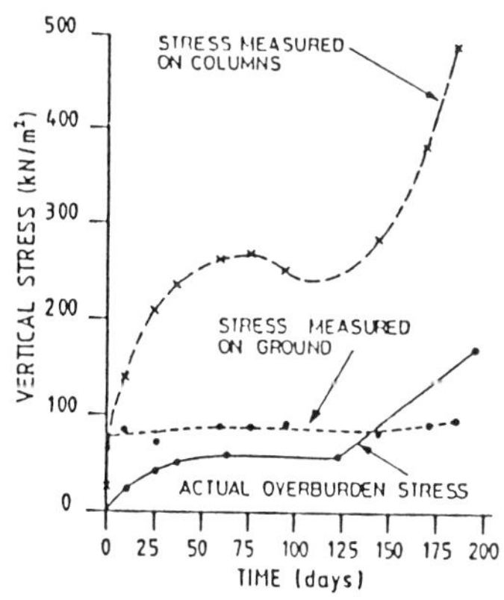

FIG. 13   Humber Bridge - measured stresses.

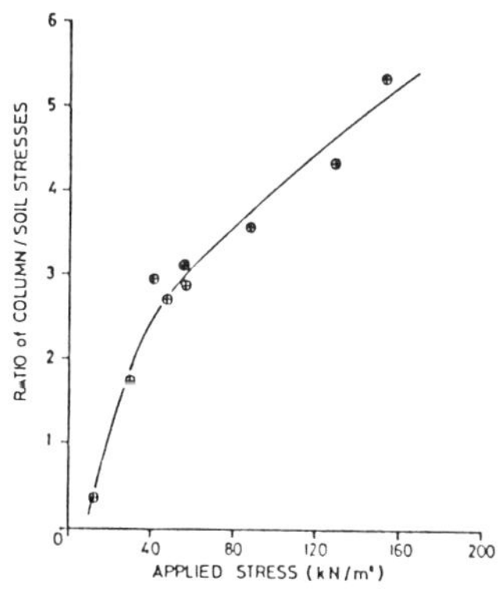

FIG. 14   Humber Bridge - measured stress ratios.

occurred with consequent reduction of stress on the column as recorded. With the increase of load to the second plateau, the bulging would be exacerbated and the rafting effect partially overcome throwing more load on to the soil.

Unfortunately construction plant destroyed the connections to the cells after 185 days whilst raising the fill between 7 and 7.5 metres so the ultimate development of stress ratios is unknown.

Calculations taking account of the measured pore pressure suggest that when the load cell connections were damaged the stress ratio should have been about 13 instead of 5 as measured, if no column bulging occurred, and load was shared with the soil. This is further evidence of earlier plastic bulging, probably due to rafting action. At this final stage the surcharge exceeded the pre-stress and the soil pressure cells were proably recording a genuine ratio of stresses due to applied load.

The conclusion from these examples is that loading conditions dominate the performance of stone columns. It is apparent from Fig. 10 representing a wide flexible foundation, that at low loadings the columns, being stiffer than the soil, accept most of the load: mobilised column friction rises towards the peak value ($\emptyset'$). Little extra stress is applied to the soil to increase radial constraint. Column spacing is virtually irrelevant in this context, except for promoting drainage. With further load the column begins to bulge (as $\emptyset'$ tends towards $\emptyset$cv). A greater proportion is carried by the soil: radial constraint increases to keep mobilised $\emptyset$ close to $\emptyset'$. This reflects a falling stress ratio between column and soil. There is a dynamic re-adjustment as load is applied, with the majority falling on the soil at higher loadings: constraint increases and mobilised column friction remains just past peak until the balance is overcome and $\emptyset$cv reached. Only during translation from $\emptyset'$ to $\emptyset$cv does column spacing have any significant influence on residual bulging resistance by affecting the general soil stress level.

The significance of Fig. 14 is that due to pre-stressing the soil stress barely changed with applied load. Extra load went into the columns, and stress ratios increased, thus reversing the expected behaviour.

Both the "flexible" and "pre-stressed" foundations represented by Canvey and Humber Bridge reduced to a stress ratio of just over 5 after some bulging of columns.

This is consistent with a column principal stress ratio ($1 + \sin \emptyset/1 - \sin \emptyset$) close to 5 with a soil Ks value of 1. The column stress ratio implies a value of $\emptyset$ close to 42°, whilst $K_S = 1$ is appropriate to soft clay passively loaded. Thus in neither example is it likely that columns had reached $\emptyset$cv with gross bulging.

MIS-USE OF LOADING TESTS

At a recent project near Bombay, India, four 18 metre diameter liquid natural gas spheres were being constructed on concrete pedestal foundations which effectively provided a rigid surface raft (Fig. 15). This was founded on stone columns placed on a square grid at 1.2 metres apart and nominally of 0.9 metres in diameter estimated from stone consumption. The columns extended to bedrock at a depth of between 10 and 12 metres under a 14.4 metre diameter raft. The soil was an amorphous soft marine clay of liquid limit 110, plasticity index 65, and moisture content 70–80%.

Recognising load sharing between stone columns and soil, the Superintending Engineer ordered load tests on concrete footings on single columns, and bridging two columns. Results of the three load tests conducted are given (Fig. 16). Load was applied quickly as settlement stabilised at each increment and each test took only a few days. Recorded settlements for stresses exceeding the intended design load (265 kN/m^2) for the structure foundation were deemed satisfactory and were about 15 to 50mm. On this apparently satisfactory basis the construction proceeded and the first steel sphere was water tested. Its capacity was approximately 3,000 tonnes. The water was pumped into the sphere and allowed to stand at a number of incremental levels. Within 110 hours a total of 1,700 tonnes of water had been added to the dead weight of 1,300 tonnes and foundation tilt had reached 91mm with an average settlement of 300mm. The tilt then slowly exaggerated resulting during a few minutes in total failure (Fig. 17). This was accompanied by ground heave and cracking of the surface crust (Fig. 18) over a distance of about 3 metres. Heave slowly continued over a few days before stabilising. The plot of the water test is shown in Fig. 19.

Back calculations suggest that pore pressure dissipation between columns was at most 15% despite the close spacing of columns, with resulting reduction in radial constraint and loss of strength of the column by a factor of 2.4 times. Just prior to failure, the ratio of stresses on the columns and soil was calculated as almost 10 because of loss of soil strength.

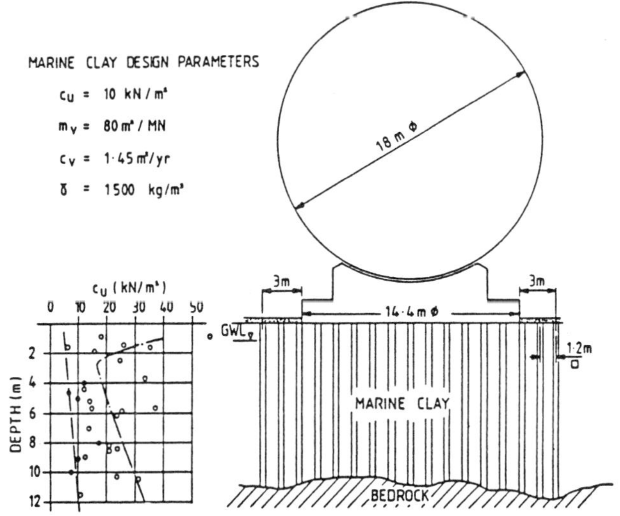

FIG. 15   India LNG spheres – circumstances.

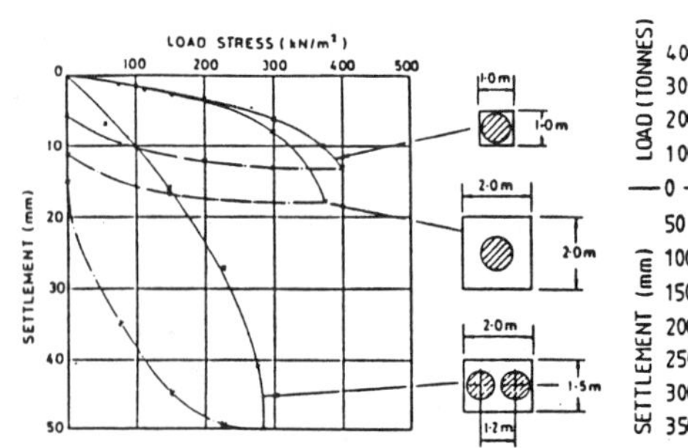

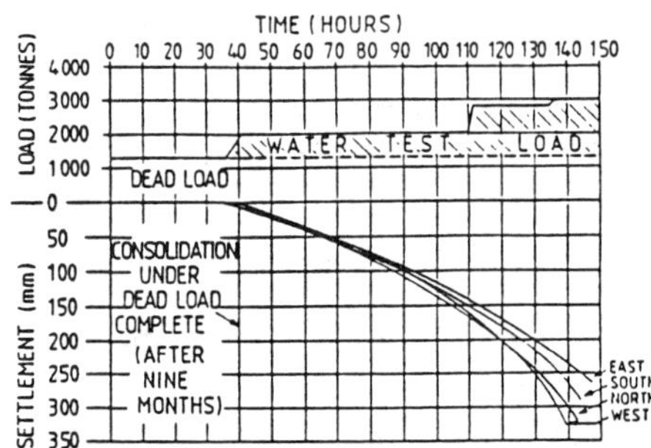

FIG. 16   India LNG spheres
- small scale load test results.

FIG. 19   India LNG spheres
- full scale initial water
test loading record.

Clearly small scale tests were inappropriate in this instance!

ALTERNATIVE TO SMALL SCALE LOAD TESTS TO VERIFY PERFORMANCE

It is apparent from the above experiences that anything other than full scale load testing is potentially misleading when assessing a foundation on soft clays strengthened by stone columns. The stiffness of the composite system is dependent to a very large degree on the interaction between the columns and clay: mobilised strengths in each are continually changing with applied load as each deforms to accommodate and equate stresses.

Small scale tests on single columns are appropriate only if they simulate prototype loading in every respect. Otherwise their worth is only for general study of stone column behaviour, and not for design and performance verification.

Furthermore field load testing is expensive. Often the cost is substantial in relation to the value of stone column contracts and might negate use of the method if tests were to be implemented.

Within the limitations of soil mechanics practice methods exist for predicting total settlements and load bearing capacities of columns with reasonable accuracy for the conditions in which such a foundation would be employed.

Having regard to the limitations of load testing of stone columns it is perhaps more cost effective to concentrate on quality assurance during design and construction of stone columns. It is then important to know what can be controlled and to which parameters most attention must be given.

Hughes and Withers equation for column strengths is as follows:-

$$Q_c = A_c \frac{(1 + \sin \phi')}{(1 - \sin \phi')} (4 c_u + \sigma_{r_o} - u)$$

and

$$\sigma_R = K_s (\gamma_s z + y p)$$

FIG. 17   India LNG spheres - foundation failure.

Considering parameters in the equation and attributing to each a variation factor which is the ratio of assessed maximum to minimum value of the parameter, the following values may be assigned:

Parameter		Factor	Remarks
Principal stress ratio in soil $(\sigma_h/\sigma_v)^s$	K_s	2	Depends on soil type and stress conditions.
Load stress distribution	y	1	e.g., Boussinesq/or Westergaard. No variation.
Load stress	p	∞	i.e., 0 to full value of p. Pre-determined.
Soil density	γ_s	1.05	Estimation of density is fairly accurate.
Depth	z	1.1	Estimate for depth to σ_{ro} minimum.
Undrained soil strength	c_u	1.5	Scatter in soil test results.
Column maximum friction angle	$\emptyset'$	1.05	Typical value for packed stone 42°.
Column stress ratio	K_c	1.1	Follows from $(1+ \sin \emptyset')/ (1- \sin \emptyset')$.
Column area	A_c	3	Sensitive to diameter.
Pore pressure in soil	u	f(p)	Function of permeability of soil and time for which p is applied. Can be measured insitu, but usually low therefore not much variation.

Little can be done to influence the accuracy of y, γ_s, z; and u must be accepted within accuracy of measurement or theory.

It is perhaps paradoxical that the factor which has infinite range, (p) is one which also contributes to stability thus neutralising its effect in this context. In most cases p is ultimately large in relation to the soil overburden weight by a factor often about 10. Typically this produces a variation in   of about 6 but all in concert with the level of application of load. The variation thus simply shifts the datum, increasing bulging resistance subject to low excess pore pressure.

With good quality site investigation the value of the c_u can be obtained with reasonable precision.

The initial value of K_s is best be determined insitu by self boring pressure meter. In soft clays appropriate to stone column construction it is reasonable to assume a value of K_s approaching 1. Certainly the response of such clays to relatively rapid loading is

often without significant volume change and the direction of change of
K_s is likely to be increasing rather than decreasing in the composite
column/soil system. The range is likely to fall within about 0.6 to 1
for these soft materials. This parameter thus may not vary greatly
but it is difficult to measure directly. It may not be practical to
measure its value for all loading stages and a value has otherwise to
be inferred from indirect information.

The value of $\emptyset'$ for the column material does not vary much for
the coarse gravels usually employed for constructing columns by
vibroflotation or ramming. All techniques of construction aim to
leave the gravel well compacted insitu and the potential range of
friction values for such materials is unlikely to be more than 2° in
40° [11]. The resulting variation in the derived column stress ratio
term $(1 + \sin \emptyset'/1 - \sin \emptyset')$ is then about 10%.

The prediction of pore pressures in the ground in a dynamic state
of loading is unreliable to varying degrees. Moreover, the prediction
of dissipation of pore pressure is often substantially in error.
Direct measurements with piezometers located at carefully chosen zones
of anticipated high shear stress are probably the most reliable guide
to rising pressures. However, sometimes the dissipation of pore
pressure needs to be judged from settlement records rather than from
piezometers which can be misleading [12].

The cross-section area of columns A_c is fortunately one of the
most easy parameters to control and it is one of the most sensitive
influences on column capacity. For columns constructed by
vibroflotation techniques it would be advantageous to instrument the
machines for depth measurement; and in conjunction to use weighing
shovels on tractors to discharge the stone. Shovels should be shaped
to funnel all stone weighed into the hole so that none is spilled on
the ground. A telemetry link from the tractor enables all data to be
logged and processed on the base crane, to give real time guidance to
the operator and a printed record of each column. The volume of stone
for each unit depth of column can be obtained, and coupled with
compaction energy or preferably directly measured amplitude records
[13]. Thus the consistency of column diameter and compaction with
depth can be assured. Such instrumentation is actively being
developed by Cementation in U.K. An alternative is the Keller vibrocat
which introduces a volume of stone pre-determined by being contained
in a hopper strapped to the vibroflot which discharges through the
nose cone. Coupled with a depth gauge this can also give reasonable
assurance of volume of stone in the ground. Columns formed by ramming
from a tubular casing also have a pre-determined minimum diameter.
However all displacement systems can be unreliable in sensitive clays
and wet vibroflotation is preferable for column formation in any soil
which loses strength when sheared.

Finally when first loading a wide foundation on ground
strengthened by a stone column array it is prudent to provide adequate
instrumentation for pore pressure and settlement monitoring.
Appropriately trained staff with authority to control rate of loading,
should be present to ensure that unsafe conditions do not arise at
this stage. Re-loading without restriction of loading rate is usually
safe thereafter.

CONCLUSIONS

1.   Stone columns differ from rigid piles in the degree to which bulging causes interaction with the soil. Columns usually behave as a soil reinforcement but occasionally may behave comparably to end bearing piles.

2.   Direct loading tests can be useful for understanding stone column behaviour.

3.   Isolated columns loaded by plates verify the general behaviour predicted by the Hughes and Withers hypothesis.

4.   Columns in soils with neutral pore pressures reflect performance which is closely elastic. The ultimate column/soil stress ratios relate to the effective stress friction properties of each.

5.   Under widespread vertical loads, ground strengthened by arrays of columns behaves in complex ways, depending on the magnitude and rate of loading. If excess pore pressures in the soil are low during loading, the effect of loading is to make the columns progressively stronger. If soil pore pressures are allowed to lessen soil strength, the columns become weaker.   There is no unique factor of safety.

6.   For widespread loading on columns in soil in which excess pore pressures are insignificant, the column/soil stress ratio progressively reduces. Most columns work just past peak $\emptyset'$ in soft clays. Typically this yields a column/soil stress ratio of about 4 to 6 at working load, similar to the principal stress ratio in the column, implying a vertical to horizontal stress ratio in the soft clay close to 1.

7.   There is some indication from field tests that rolling of granular pad foundations, or earth fills, may change the interdependent stresses in columns and soil.

8.   Small scale load tests which do not simulate the circumstances of the prototype accurately are irrelevant for control of construction, and can mislead with regard to performance of the structure. Such small scale tests are appropriate for verification of construction technique only for isolated or strip arrangements of columns.

9. For widespread loads, it is cost effective to abandon small scale load tests in favour of monitoring construction techniques, and settlements during initial loading. Instrumentation on plant can ensure that the key parameter of column diameter, consistent with design for each unit depth, is achieved. Monitoring of excess soil pore pressure by well sited piezometers, and of structure settlements ensures a safe rate of loading. Erratic behaviour can then be allowed to settle down, or can be corrected by wedging and jacking the structure, if necessary. Usually there is no problem providing the design data are correct, and rate of loading precludes local shear distortions.

ACKNOWLEDGEMENTS

The Author whilst expressing personal opinions in this paper, is grateful to Cementation Research Ltd (now incorporated in Trafalgar House Technology Ltd) and to Cementation Ground Engineering Ltd (now amalgamated in Cementation Piling & Foundations Ltd) for use of data from field works initiated by the Author.

REFERENCES

[1] Hughes, J.M.O. and Withers, N.J., "Reinforcing of Soft Cohesive Soils with Stone Columns", Ground Engineering, May 1974, pp 42-49.

[2] Priebe, H., "Abschatzung des Setzungverhaltens eines durch Stopverdichtung Verbesserten Baugrundes", Die Bautechnik, H.5. 1976.

[3] Balaam, N.P., Poulos, H.G., and Brown, P.T., "Settlement analysis of soft clays reinforced with granular piles", University of Sydney, Civil Engineering Laboratories, Research Report 305, July 1977.

[4] Goughnour, R.R. and Bayuk, A.A., (a) "Analysis of stone column and soil matrix interaction under vertical load". Colloquium Institute Renforcement des Sols ENPC, Paris 1970, pp 271-277.

[5] Goughnour, R.R. and Bayuk, A.A. "A field study of long term settlements of loads supported by stone columns in soft ground", Colloquium Institute Renforcement des Sols ENPC, Paris 1979, pp 279-285.

[6] Chin Fung Kee, "Estimation of the ultimate load of piles from tests not carried to failure", Proceedings 2nd S.E. Asian Conference SM&FE, 1970, pp 81-90.

[7] McKenna, J.M., Eyre, W.A. and Wolstenholme D.R., "Performance of an embankment supported by stone columns on soft ground", Geotechnique, March 1975, pp 46-51.

[8] Greenwood, D.A., Discussion, Ground treatment by deep compaction, Institution of Civil Engineers, London, U.K. 1976, pp 123-125.

[9] Raffle, J.F. and Greenwood, D.A., "The relation between the rheological characteristics of grouts and their capacity to permeate soil", 5th International Conference Soil Mech. & Foundation Engineering, Paris, Vol II, 1961, pp 789-793.

[10] Greenwood, D.A. and Kirsch, K., "Specialist ground treatment by vibratory and dynamic methods - State of Art", Piling & Ground Treatment, International Conference, Inst. Civil Eng. London U.K. 1984, pp 17-45.

[11] Leslie, D.D., "Large scale triaxial tests on gravelly soils", 2nd Pan American Conference on Soil Mechanics & Foundation Engineering, Sao Paulo, 1963, pp 181-202.

[12] Hansbo, S., Jamiolkowski, M. and Kok, L., "Consolidation by vertical drains", Geotechnique, 33, 1, March 1981, pp 45-66.

[13] Morgan, J.G.D. and Thomson, G.H., "Instrumentation methods for control of ground density in deep vibrocompaction", 8th European Conference Soil Mechanics & Foundation Engineering, Helsinki, Vol. 1, 1983, pp 59-64.

Robert F. Hayden, P.E. and Christine M. Welch, P.E.

DESIGN AND INSTALLATION OF STONE COLUMNS AT NAVAL AIR STATION

REFERENCE: Hayden, R. F. and Welch, C. M., "Design and Installation of Stone Columns at Naval Air Station," <u>Deep Foundation Improvements: Design, Construction, and Testing</u>, <u>ASTM STP 1089</u>, Melvin I. Esrig and Robert C. Bachus, Eds., American Society for Testing and Materials, Philadelphia, 1991.

ABSTRACT: Stone columns were used to increase bearing capacity, reduce settlement, and mitigate liquefaction potential for one- to four-story structures on a housing project at Fallon Naval Air Station, Fallon, Nevada. Eighteen field plate load tests were used to measure load versus settlement characteristics of the stone columns. Standard penetration test borings and cone penetrometer measurements in the improved soils at various distances from the columns were also used to evaluate load tests and to predict building foundation performance.

KEYWORDS: Stone columns, bearing capacity, settlement, liquefaction mitigation, load testing, standard penetration testing, cone penetrometer testing, soil improvement.

PROJECT DESCRIPTION

The Unaccompanied Enlisted Personnel Housing (U.E.P.H.) project involved the construction of three one- to four-story masonry structures covering approximately 65,000 square feet (6042 square meters) at the Fallon Naval Air Station. Loose, saturated sands underlying the project site were determined to be highly liquefiable when subjected to the seismic acceleration values recommended by the Navy. Even under static design conditions, total and differential settlement concerns would have dictated the use of a low design bearing value for loads transferred to the native soils via conventional spread footings. In order to mitigate liquefaction potential and to reduce settlement, the following alternatives were considered:

• Densification and partial replacement of the liquefiable soils using vibroreplacement stone columns. Construction of a gravel blanket "cap" vented to the atmosphere also would aid in dissipating excess pore water

Mr. Hayden is Senior Project Manager with GKN Hayward Baker, Inc., P.O Box 7690, Ventura, CA, 93006; Ms. Welch is Project Engineer with Kleinfelder, Inc., 3189 Mill Street, Reno, NV 89502

pressures induced by seismic loading.
- Construction of closely spaced gravel drains to prevent build up of pore pressure during earthquake events. Partial replacement of the loose sands by the drain rock would also reduce settlements somewhat, although not as much as the stone column option.
- Supporting the structures on a grade beam and friction pile system which would be designed to transfer loads through the liquefiable soils to underlying stiff clays.

The above options were evaluated for their ability to reduce liquefaction potential, increase bearing capacities and decrease settlements, and for their overall economy, based on estimated construction costs. The vibroreplacement stone column and pile options were chosen over the gravel drains option as the alternatives most likely to provide the protection sought against liquefaction while also improving bearing values and decreasing anticipated settlements under static loading conditions.   The stone column option was chosen over pile foundations when the economic analysis indicated substantial cost savings for the ground improvement alternative.

SITE CHARACTERISTICS

Geologic and Seismic Conditions

The Naval Air Station is located within the western portion of the Basin and Range physiographic province, a region characterized by north-south trending mountain ranges separated by sediment-filled valleys. The project site is located within the Lahontan Valley and is underlain by a thick sequence of Quaternary age lake sediments.

The site is located just west of an active seismic area referred to as the 118° meridian seismic zone. This seismic zone is responsible for most of the large earthquakes recorded in the State of Nevada. Since no active faults were known to cross the site, surface rupture was not considered a major hazard for the project. A larger concern was for groundshaking during earthquake events along nearby faults. Several active faults are located within 35 miles (56 kilometers) of the site. At least seven earthquakes of Richter Magnitude 6.3 or higher have originated along these fault zones during historic time, with a magnitude 7.3 earthquake registered in 1954.

The current Naval Facilities Engineering Command (NAVFAC) Earthquake Safety Investigation [1] criteria specifies using a ground surface acceleration value with an 80 percent probability of not being exceeded in 50 years, for seismic safety evaluation. This event is characterized by a sustained acceleration of approximately 0.32g for the project site. For foundation design and liquefaction analyses, this full design event was used.

Geotechnical Conditions

The subsurface exploration program for the geotechnical investigation was performed by Kleinfelder Inc. of Reno, Nevada [2] and, included drilling six borings and performing four piezoelectric cone penetrometer (CPT) tests. Borings were advanced to approximate depths of 13 to 76 (4 to 23 m) feet below the existing ground surface using a hollow stem auger drill rig.   Standard penetration tests were performed at average intervals of 5 feet (1.5 m). Continuous logging of soil conditions was provided at the CPT locations, which extended to depths up to approximately 80 feet (24 m).

Surface soils throughout the site generally consisted of medium dense, very silty fine grained sand which extended to a depth of 2 to 5 feet (0.6 to 1.5 m) below the existing ground surface.   In limited areas of the site, near-surface lenses of medium stiff silt or clay soils were present.   Underlying soils extending to depths of approximately 20 feet (6.1 m), generally consisted of loose to medium dense clean to silty sand with occasional interbeds of medium stiff sandy silt and clay.   In general, soils between approximately 13 and 20 feet (4.0 and 6.1 m) contained more frequent interbeds of the silty and clayey materials than were present above.   A zone of "heaving" or flowing sand was encountered at an average depth of 10 to 15 feet (3.0 to 4.6 m) in the borings.   Between approximately 20 feet (6.1 m) and 45 feet (13.7 m), medium stiff, highly plastic clay was present.   Below 45 feet (13.7 m) interbedded black, loose or soft silts, sands and clays were encountered.   This frequent interfingering and lateral variation of site soils are characteristic of lake-deposited sediments.   A typical soil profile of the upper 30 feet (9 m) is presented on Figure 1.

Soils below a depth of 5 feet were very wet to saturated, and static ground water levels ranged from 5 to 6 feet below the ground surface.   Boreholes generally caved at or slightly below the ground water table.

Laboratory testing of selected samples was performed to evaluate gradation, plasticity, in-place moisture content and dry density, and shear strength.   Selected laboratory test results are shown on Figure 1.

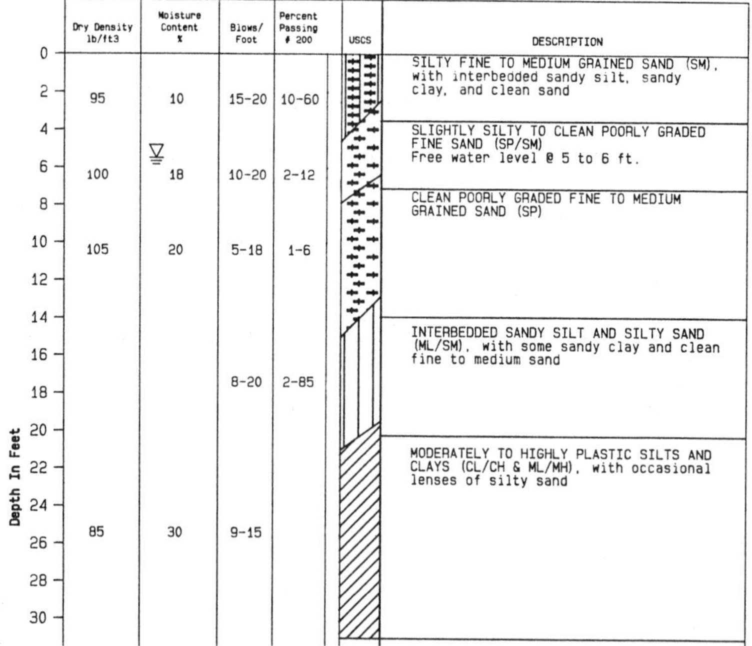

FIG. 1 -- Generalized soil profile.

DESIGN FEATURES

The building foundation design included densification of loose subsurface sands by the construction of vibroreplacement stone columns on typical 8-foot square grid spacings. This spacing was designed to allow an increase of the bearing pressure from 3000 to 4500 psf (215 kn/m^2), while limiting post-construction foundation settlements to approximately 3/4 inch (20 mm) and also mitigating the potential for liquefaction-induced structural damage.

The specified stone column depths were to have a minimum 1 foot (0.3m) penetration into the underlying stiff clay. Prior to installation, a series of 30 CPT tests were performed over the site to depths of 25 feet (7.6 m) to provide a basis for establishing the final penetration depth which was found to average 21 feet (6.4 m). This cone penetration testing was performed on the average rate of one test per 2500 square feet (232 square meters) of treated ground surface area.

The maximum 150 ton structural column loads imposed by the 4-story buildings resulted in footing dimensions of up to 8 foot (2.4 m) square. Continuous wall footing widths up to 5 feet (1.5 m) were used for the maximum wall loading of 10 tons/lineal foot (3 ton/ln m). Footings rested on 24 inches (0.6 m) of compacted fill and a 12-inch (0.3 m) thick gravel drainage blanket which was placed above the stone columns over the building footprint areas. The gravel blanket material consisted of clean, durable minus 2-inch (5 cm) crushed stone having less than 5 percent passing the No. 200 sieve.

The dry, bottom-feed vibro process was used for stone column installation. With this method, the 165 horsepower vibrator penetrated to the final depth under its own weight. Stone was then introduced at the lower tip with air pressure assistance. Stone used to form the 3-foot (0.9 m) diameter columns consisted of well graded, slightly rounded gravel ranging in size from 3/8 to 1-1/2 inches (9 to 38 mm).

STONE COLUMN LOAD TESTING PROGRAM

Test Program Overview

Field load tests were performed on individual production stone columns in 15 areas of the site. The purpose of the testing was to confirm the design and installation techniques. Maximum stone column test loads were 24 tons which equates to a contact pressure of 6800 psf (325 kn/m^2) on the 3-foot (0.9 m) diameter plates. This was approximately 150 percent of the 4500 psf (215 kn/m^2) design bearing pressure. Test locations were spaced over the site to provide coverage of the variable soil conditions. Load test evaluation criteria limited the maximum acceptable deflection to 0.5 inches (13 mm). Test procedures called for increments and durations as described below.

Load Test Procedure

Test loadings were applied with a 50-ton capacity hydraulic ram jacking against a 34-ton kentledge weight. An electronic load cell with a strain readout box was used to measure load magnitudes. A 36-inch diameter (0.9 m) steel base plate was placed concentrically over the individual stone column prior to placing the jack system. Uniform fill sand was used to level the ground prior to placing the base plate, and steel shims were used to further remove any excess slack.

Two dial gauges accurate to 0.001 inch (0.025 mm) were mounted on steel reference beams to measure vertical column deflections. The test setup is shown in Figure 2. After zeroing out the dial gauges and registering an initial strain to seat the jack, loads were applied to the columns in 4-ton increments.

The load test procedure was modeled after that described by Barksdale and Backus [3]. Initial deflection readings were taken at each load level immediately after application. Deflection readings were then taken every 5 minutes until the rate of settlement was less than 0.01 inches per hour (0.25 mm/hr). Deflection readings were made at the maximum load and at the zero load rebound condition until the rate of settlement was less than 0.005 inches/hr (0.12 mm/hr).

Load Test Results

Load tests were conducted on a total of 18 production stone columns with results as presented in Table 1. Three of the tests originally exhibited deflections in excess of the specified allowable 0.5 inches (13 mm). Vibrations from construction equipment and excessive amounts of fill sand placed on top of two columns were the apparent causes of the excessive deflections Tests 7A, 8A, and 11A were performed as retests for these three stone columns.

TABLE 1 -- Summary of Stone Column Load Test Results

Test No.	Settlement At Design Load (Inches)[a]	Settlement At 1.5 Times Design Load (Inches)[a]	Stone Column Spacing (Feet)	Near-Surface Soil Type	Stone Column Location In Group
1	0.10	0.14	8	Sand	Inside
2	0.07	0.10	8	Sand	Inside
3	0.16	0.23	8	Silt	Inside
4	0.14	0.24	6	Sand	Corner
5	0.12	0.15	8	Silt/Clay	Inside
6	0.17	0.24	8	Silt	Inside
7	0.51	0.74[b]	8	Sand	Outside
7A	0.19	0.24	8	Sand	Corner
8	0.47	0.55[c]	8	Clay	Outside
8A	0.19	0.49	8	Clay	Outside
9	0.19	0.24	8	Silt	Inside
10	0.20	0.48	8	Clay	Inside
11	0.62	1.28[c]	6	Clay	Inside
11A	0.22	0.44	6	Clay	Outside
12	0.31	0.49	8	Clay	Inside
13	0.08	0.13	8	Silt	Inside
14	0.25	0.45	8	Clay	Inside
15	0.22	0.42	8	Clay	Inside

[a] Metric Conversion: 1 inch = 25.4 mm
[b] Excessive settlement attributed to vibrating equipment
[c] Excessive settlement attributed to loose sand fill over column

The predominant factors in the test result variations appear to be soil type

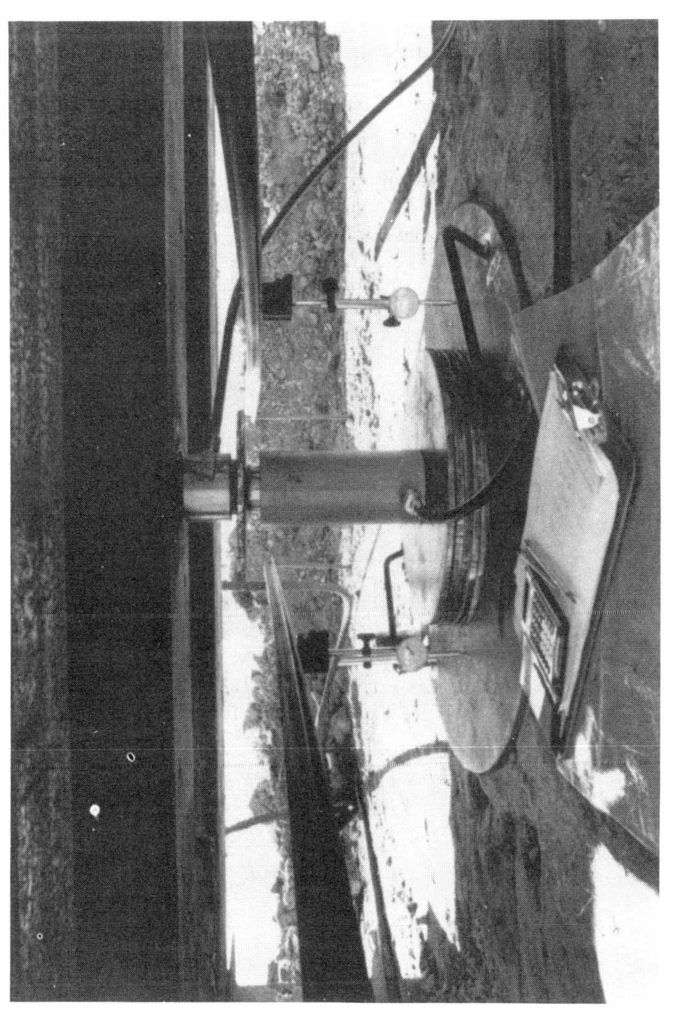

FIG. 2 -- Load test setup.

and strength within the upper 6 to 8 feet (1.8 to 2.4 m). As shown in Figure 3, less deflection was generally measured where stone columns were constructed in areas of cohensionless near-surface soils. This is attributed to the increased lateral confinement provided by the densified sandy soils compared to silty or clayey soils. These fine grained soils would not be expected to respond to the densification process to the same extent as the clean sandy soils. Spacing of columns and location of the column in the group appeared to have no observable effect on the results of the individual column tests.

The short-term load tests were effective in demonstrating that columns had an adequate factor of safety against shear failure and that proper construction techniques had been followed. As shown by the load-deflection curves, the column responses were still essentially in the elastic range at 150 percent of the design load. Ultimate loads were calculated to be between 125 and 160 tons for the range of clayey to sandy near-surface soils conditions for the 3-foot (0.9 m) diameter plate load tests. Because the columns were installed primarily in cohensionless soil and because the footing dimensions of 8 feet (2.4 m) maximum were not large enough to extend stresses deep into clayey soils, the short-term plate load testing was judged to be satisfactory for use in verifying the design and installation of the stone columns.

QUALITY ASSURANCE EVALUATIONS

<u>Design Settlement Verification</u>

Based on the above reported load test results, the stone columns were accepted to provide adequate settlement control of the structures. As shown in Figure 3, the maximum allowable test deflection of 0.5 inch (13 mm) at maximum

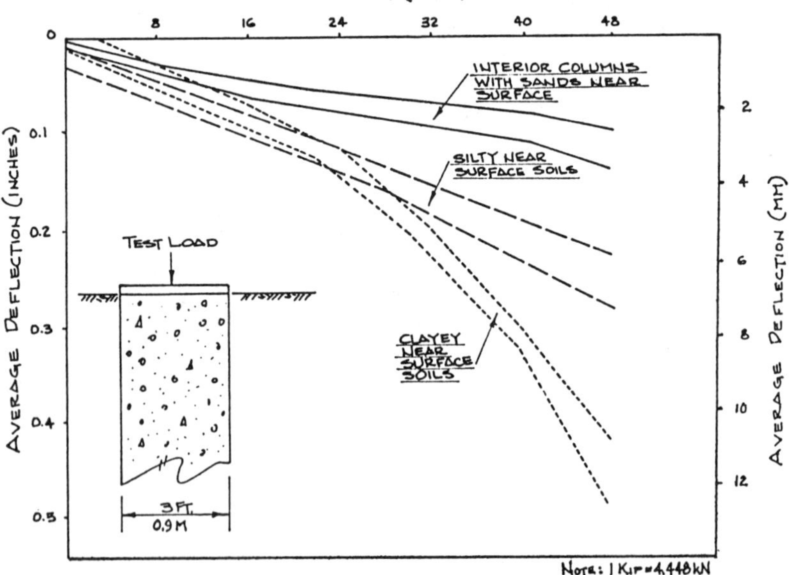

FIG. 3 -- Typical load-deflection curves.

test load (1.5 x design load) indicated less than 0.25 inch (6.4 mm) deflection at the design load for the non-linear load settlement response of the clayey surface soils. The coarser soils deflected less at this load level.This magnitude of deflection in the load test was intuitively considered to be conservatively within the range needed to limit total long-term foundation settlements to the 3/4 inch (19 mm) maximum recommended by the geotechnical engineer and adopted in the structural design.

<u>In-Situ Soil Testing</u>

Following installation of the stone columns, a series of 10 CPT (cone penetrometer) tests and 11 DMT (dilatometer) tests were performed to measure the degree of densification achieved. Tests were located at varying distances from stone columns, interior and exterior to groups, as shown in Figure 4. CPT tip resistances, measured at varying distances from the edge of the stone columns, are shown in Figure 5. The data shows improvement of 1.5 to 3 times the CPT tip resistance in the sandy strata at distances up to 5 feet (1.5 m) from the column face. Densification in the lower portions of the stone columns was less significant, since the conditions became progressively more silty and clayey with depth as represented by CPT friction ratios of 1 to 2. DMT data, as presented in Figure 6, showed consistent increases of 2 to 3 times the initial one dimensional constrained modulus within the upper 13 foot zone where sandy soils were generally located. Below that depth, modulus values in the silty and clayey soils were fairly constant regardless of distance from the face of columns.

<u>Individual Load Test Analyses</u>

The premise used in evaluating individual load test results was that the extensive post-construction in-situ measurements at distances from 0.5 to 6 feet (0.2 to 1.8 m) away from columns would be beneficial in evaluating the load test results.

Results of individual load tests were evaluated to determine if load-deflection relationships could be predicted for future tests on individual stone columns.   First, the approach advanced by Priebe [4] was used to calculate predicted load test deflections in the predominately sandy soils above depths of approximately 20 feet. This method involves the following assumptions:

- The individual stone column is in a state of plastic equilibrium under a triaxial stress state;
- Soils surrounding the column and stone are idealized as linearly elastic materials; and
- The vertical stress distribution in the column is according to the Boussinesq theory.

Using this method, the radial expansion of the stone column into the surrounding soil was calculated for a depth equal to two column diameters. To accomplish this, the active horizontal pressure was used to compute the lateral stress from the column as the vertical test load was applied. Elastic moduli for the sandy soil were developed based on correlations proposed by Schmertmann [5] using 2.5 times the CPT tip resistance.

Elastic moduli were based on CPT tests which were taken within 2 feet (0.6 m) of the edge of stone columns and the radial deformations of the column into

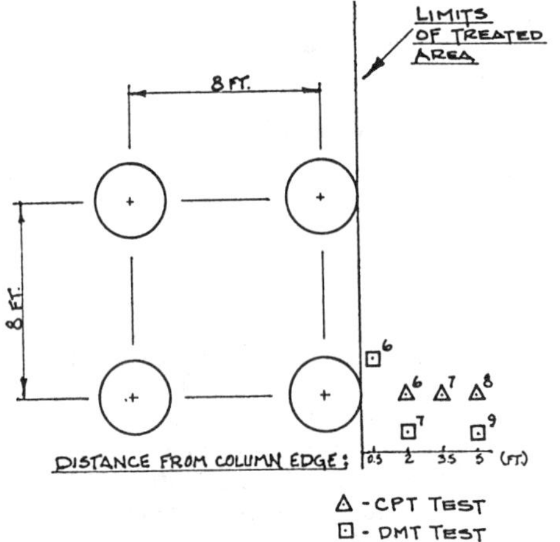

FIG. 4 -- Locations of post-construction cone penetrometer and dilatometer testing.

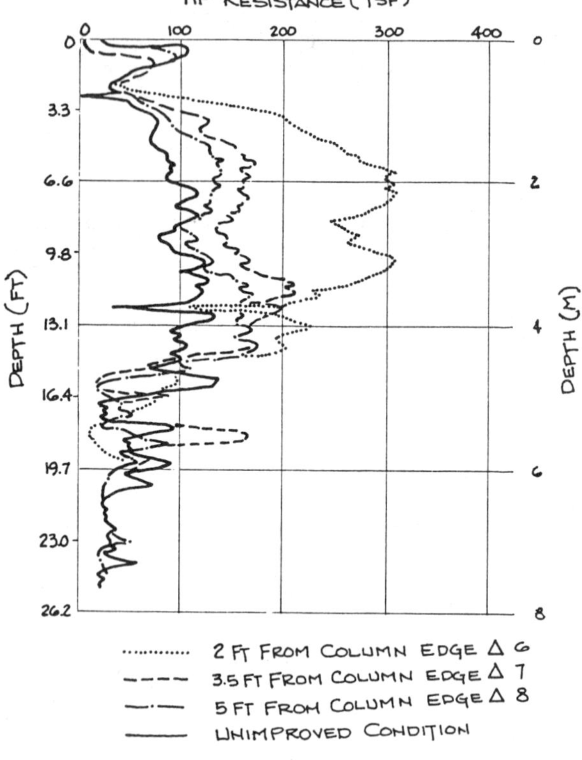

FIG. 5 -- Cone penetrometer tip resistance vs. depth at various distances from edge of stone columns.

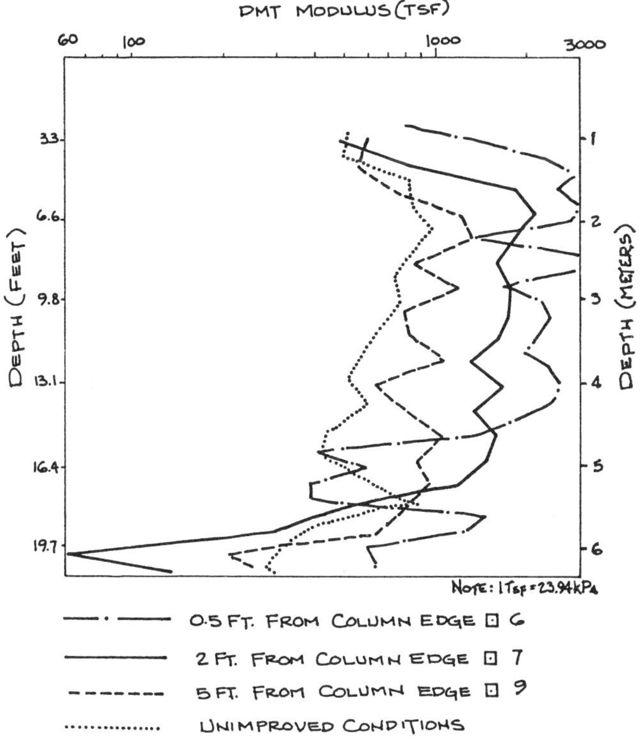

FIG. 6 -- Dilatometer modulus vs. depth at various distances from edge of stone columns.

the surrounding soil were based on compression of a 1-foot (0.3 m) thick ring of soil at the column perimeter. This assumption was made upon inspection of the lateral extent of the Boussinesq pressure bulb beneath a loaded circular plate. Radial expansion of the column in the top 2D column depth was then equated to vertical shortening.

Elastic vertical compression was added to the above described compression due to the radial expansion, resulting in a computed vertical test deformation of 0.14 inch (3.5 mm) at the 24 ton maximum load. This compares favorably with the 0.18 inch (4.6 mm) average measured deflection of columns in the sandy near-surface soil condition.

The same elastic approach was used for the clayey near-surface soil condition. The computed column deflection was 0.47 inch (11.9 mm). Again, this is very close to the measured average deflection of 0.46 inch (11.7 mm).

The close agreement of the predicted deflections obtained from the elastic approach with the field measured deflections may be attributed to one or both of the following:

- Site soils obtained their strength from both cohesion and phi angle;
- The applied loads being well below the yield point of the soils.

In very soft clay conditions, or where soil response is not essentially within the elastic range, more rigorous analysis methods may be necessary.

A finite element analysis using the axisymmetric computer program SAP IV [6] was also performed to evaluate the single load test results in the sandy near-surface soil condition.   The same linear elastic soil moduli obtained from correlation with the CPT testing were used in the finite element analysis. Load deflections were computed to be 0.3 inch (7.6 mm), somewhat higher than the measured average of 0.18 inch (4.6 mm).   The reason for the difference is unknown.

Prediction of Long-Term Structure Performance

The Fricke and Finite element methods were also used to develop predictions of settlement for the typical loaded footing shown in Figure 7. While actual post-construction building settlements are unknown, the computed settlements obtained from the above elastic method generally agree with those predicted by the unit cell improvement factors published by Barksdale & Bachus [3] lending further credence to the approach. Support is also given to selecting the original 0.5 inch (13 mm) load test deflection criteria to limit the total allowable footing settlement to 3/4 inch (19 mm).

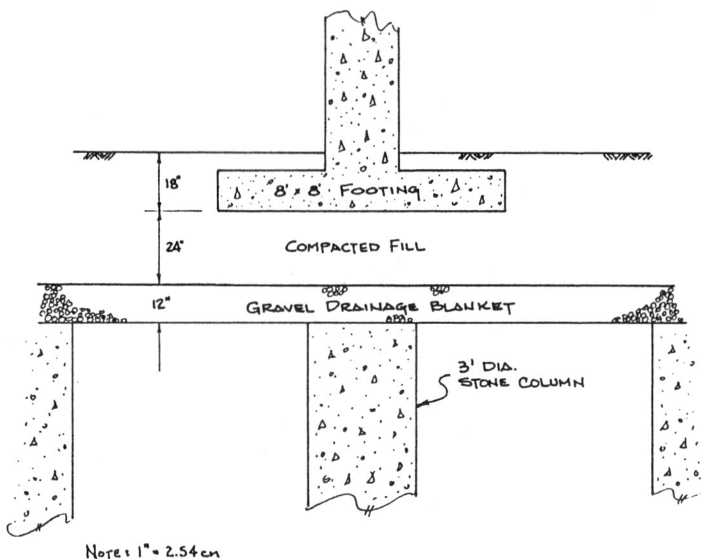

FIG. 7 -- Typical footing geometry.

Computing the distribution of vertical stress beneath the typical 8 by 8-foot (2.4 by 2.4 m) footing as before, and using active lateral forces to evaluate radial expansion of the column into the surrounding soil, a footing settlement of 0.79 inch (20.1 mm) was calculated for the clayey near-surface soil condition.

To further evaluate the anticipated long-term settlement of the footings, the unit cell improvement method as described by Barksdale & Bachus [3] was used.

Based upon an angle of internal friction of 40 degrees for the river run gravel in the column, a load concentration factor of 2:1 (i.e., vertical column stress versus stress in the surrounding soil) in the column and an area ratio of total cell size to column area of 9.0, an improvement factor of 1.65 was developed. Starting with a footing settlement of 1.2 inches (30.5 mm) for the unimproved clayey near-surface soil condition, a settlement of 0.80 inch (20.3 mm) was computed for the improved condition. Compression of the fill beneath the footing and underlying densified sand was also included. This agrees well with the above method and is in the range of allowed footing settlements.

Results of the finite element analysis of the long-term footing settlement in the predominant sandy soil conditions were 0.20 inch (5.1 mm). This compares favorably with 0.24 inch (6.1 mm) computed with the above elastic method. The finite element analysis was of secondary importance, given the excellent agreement found with the above methods of Priebe [4] and Barksdale & Bachus [3]. The use of sophisticated finite element analyses and long term load testing is considered more beneficial in soft, colusive deposits.

Experienced-based correlations of 8 to 10 times the settlement for foundations versus those measured in single column tests as reported by Mitchell [7] do not apply to spread footings of relatively small size. Further, it appears that load-deflection response of stone columns tested with small diameter plates may not predict long-term structure performance unless the sizes of the footings are specifically taken into account in the analysis. Based on the foregoing analyses, proven elastic methods exist for making settlement predictions based upon penetration testing of the treated sandy soils. The use of long-term load testing and sophisticated finite element analyses may be more appropriate in soft, cohesive deposits.

CONCLUSIONS

The following conclusions can be made regarding vibro-replacement (stone columns) use and the implications for future construction:

- Stone columns provided an attractive alternative to pile foundations based on economic analyses.
- The ground improvement allowed an increase in bearing capacity for shallow foundations to 4500 psf (215 kn/m^2), while reducing total settlement and mitigating the liquefaction hazard.
- Load testing was successfully used to evaluate the ability of the stone column system to meet the design requirements.
- Footing settlement calculations (based on the stone column load test results) appear to substantiate the empirical 0.5 inch (13 mm) stone column load test criteria for the given soil conditions, individual footing sizes, and maximum allowable footing settlements.
- The deflection response of stone columns in load tests using small diameter plates was insensitive to the spacing of adjacent columns and to the location of the test column in the group.
- The soil type and amount of lateral confinement provided the stone column in the top 2D to 3D distance is the primary factor in settlement response in plate load tests.
- A detailed analysis, as opposed to empirical relationships, is useful to predict foundation settlements based on plate load tests on individual

stone columns.
- CPT in-situ soil testing before and after ground improvement with stone columns can be used to predict bearing capacity and settlement response of a foundation using fairly simple calculations.
- The finite element method for analysis of the load test results provided insufficient improvement in accuracy to justify the cost for this routine project in mostly granular soil.

ACKNOWLEDGEMENTS

The U.S. Navy graciously allowed the release of information on the project. All conclusions drawn, however, are made by the authors. We are indebted to our associates who reviewed the draft and provided support.

REFERENCES

[1]   Glick, D.C. and Farris, G.R., "Earthquake Safety Investigation Geotechnical Report for Phase I, Naval Air Station, Fallon, Nevada," Report No. 8BE, September 1983.

[2]   Welch, C.M. and Salontai, G.J., "Geotechnical Investigation for the Proposed UEPH Project at the Fallon Naval Air Station, Fallon, Nevada," Kleinfelder Report R-1536-1, September 1984.

[3]   Barksdale, R.D., and Backus, R.C., "Design and Construction of Stone Columns, Volume I," U.S. Department of Transportation, Federal Highway Administration Report No. FHWA/RD-83/026, 1983, pp. 37, 61, 124-126.

[4]   Priebe, H.J., "Design Criteria for Ground Improvement by Stone Columns," Die Bautechnik 55,448, 1978, pp 281-284.

[5]   Schmertmann, J.H., "Guidelines for Cone Penetration Test, Performance and Design," Federal Highway Administration, Report No. FHWA/-TS-209, Washington DC, 1978.

[6]   Bath, K.J., Wilson, E.L., Peterson, F.E., "SAP IV - A Structural Analysis Program for Static and Dynamic Response of Linear Systems," Report to National Science Foundation, Report No. EERC 73-11, June 1973.

[7]   Mitchell, J.K., and Huber, T.R. "Performance of a Stone Column Foundation," Journal of Geotechnical Engineering, ASCE, Vol. III, No. 2, 1985.

James D. Hussin and Juan I. Baez

ANALYSIS OF QUICK LOAD TESTS ON STONE COLUMNS: CASE HISTORIES

REFERENCE: Hussin, J. D. and Baez, J. I., "Analysis
of Quick Load Tests on Stone Columns: Case Histories,"
Deep Foundation Improvements: Design, Construction, and
Testing, ASTM STP 1089, Melvin I. Esrig and Robert C.
Bachus, Eds., American Society for Testing and Materials,
Philadelphia, 1991.

ABSTRACT: This paper examines 6 stone column projects
constructed throughout the United States from 1985 to 1988
in which quick vertical load tests were performed to
predict performance of foundations and to compare this
prediction with the design prediction. The stone columns
were constructed through various types of soil including
sand, clay, silt, peat and construction debris.   The
configuration of the load tests varied from .76 m (30
inch) diameter to 1.8 m (6 foot) square steel plates with
loads from 89 kN to 916 kN (10 to 103 tons) on one to four
stone columns.

The proposed construction, design load, settlement
criteria and soil profile are reviewed for each project.
Special emphasis is placed on using the load test data to
predict the settlement of the planned foundations.   In
addition, the configuration and mechanics of the load test
set ups are presented.

KEY WORDS: load test, stone columns, vibro-replacement,
soil improvement, ground modification

The Stone Column Technique (Vibro-Replacement) is often used to
reduce settlement in soil profiles which cannot be densified in place
by vibration alone (Vibro-Compaction). The stone column reinforces
the cohesive soils by replacing a portion of the soil with a column of
higher modulus material. When evaluating the performance of Vibro-
Compaction, post treatment penetration testing and conventional
settlement calculations are performed to predict the performance
of the foundations. However, on Stone Column projects in cohesive

Mr. Hussin is the Senior Project Engineer and Mr. Baez is a
Project Engineer for GKN Hayward Baker Inc., 6850 Benjamin Road,
Tampa, Florida 33634

soils, post treatment penetration tests typically show little or no improvement in the native soil between the stone columns, resulting in the need for alternate methods of testing [1].

Load tests have been utilized to confirm predicted settlements of the foundation on the stone column improved sites. The tests are generally performed at the beginning of construction to confirm the design assumptions. If the soils have a low permeability and are susceptible to significant secondary settlement (i. e., soft clay or organics), several weeks or months may be required for the loaded area to undergo a high percentage of the settlement, even with the stone columns acting as drains [1]. However, for sites with moderate permeability and little secondary compression a quick load test may be used to predict long-term settlement of the proposed foundation.

The six quick load tests presented in this paper fall into three categories: (I) loading just the area of the stone column, (II) loading a stone column and the soil area tributary to the column, and (III) full scale footing load test. For each project, the load test data and theoretical settlement predictions are used to predict the settlement of the proposed foundations. The case histories are discussed below by category.

CASE HISTORY AND LOAD TEST DETAILS

Category I: Quick load test of stone column area

Case 1: Sound Suppressor Building, Langley AFB, Virginia

The 1,116 m^2 (12,000 ft^2) Sound Suppressor Building is a structure designed to house fighter aircraft during engine testing and dampen external noise. The structure is designed to bear on a grade beam reinforced mat with a design bearing pressure of 96 kN/m^2 (1 tsf).

The generalized soil profile is presented in Figure 1. The very loose silty fine sands present from a depth of 1.2 m (4 feet) to 4.3 m (14 feet) required improvement to permit construction of the facility an a shallow foundation with a maximum design settlement of 25 mm (1 inch).

A vibro-compaction program was specified calling for the soils between the probes to be densified to a depth of 4.9 m (16 feet) and a minimum relative density of 65%. The improvement program was performed using the wet method (since area for waste water settlement ponds was available) with an 80 horsepower vibrator penetrating on a 1.8 m (6 foot) triangular grid. A stone backfill was used to construct .76 m (30 inch) diameter stone columns. The stone was a 6 to 25 mm (.25 to 1 inch) diameter crushed granite. Post treatment Standard Penetration Test (SPT) and Cone Penetrometer Test (CPT) results revealed that minimal densification between the columns was achieved. Laboratory tests of the soils revealed that the fines (15 to 20 percent by weight) in the soils had a significant cohesion, which reduced the effectiveness of the vibro process.

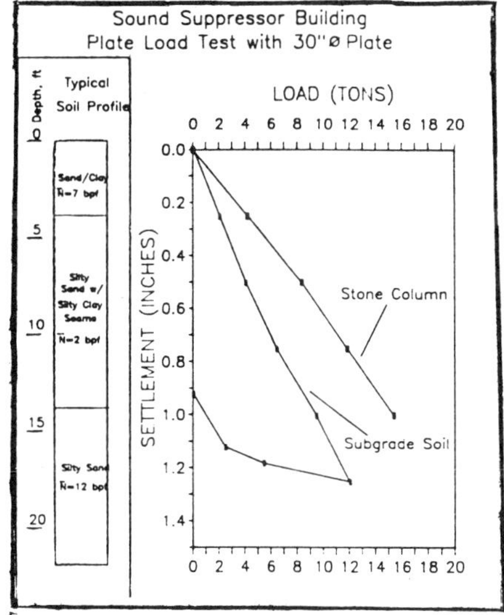

FIG. 1--Case 1, Soil profile and
load test results

FIG. 2--Case 1, Photograph of
load test results

Since the main criteria was that the structure's maximum total settlement was 25 mm (one inch), load tests were performed to predict the settlement. The load test results are shown in Figure 1. Three concentric 25 mm (1 inch) thick steel plates 0.3, 0.5, and 0.76 m (12, 21, and 30 inches) in diameter were placed on random stone columns and on the native soil at the mid point of the stone column grid pattern (similar set up as ASTM D 1194, Standard Test Method for Bearing Capacity of Soil for Standard Load on Spread Footings). The load was applied using a Cone Penetrometer Test (CPT) rig (Figure 2). The test rig's hydraulic system which usually pushes the CPT probe was used to apply the load. A rounded attachment was fixed to the end of a CPT extension rod and a socket was placed on the center of the plates. The settlement was measured using three dial gauges placed at third points around the perimeter of the plates. The gauges were attached to a 4.6 m (15 foot) reference beam which was fastened at its ends to stakes driven several feet into the ground until firm.

The load test on the stone column as applied in 20 percent increments to 130 kN (14.4 tons), the maximum load attainable with the CPT rig. Each increment was held for 15 minutes and the maximum load was held for 80 minutes. A plot of the settlement versus square root of time indicates that the primary compression was essentially complete at this time (Figure 3). Although the stone column is not a cohesive soil, the compression of the stone columns partially results from lateral bulging of the column which requires consolidation of the confining soil. Therefore, the authors felt that using the t_{90} method was reasonable. The load was then removed in the following steps: 60, 20, 10 and 0% of the maximum load. Rebound readings were taken 5 minutes after unloading.

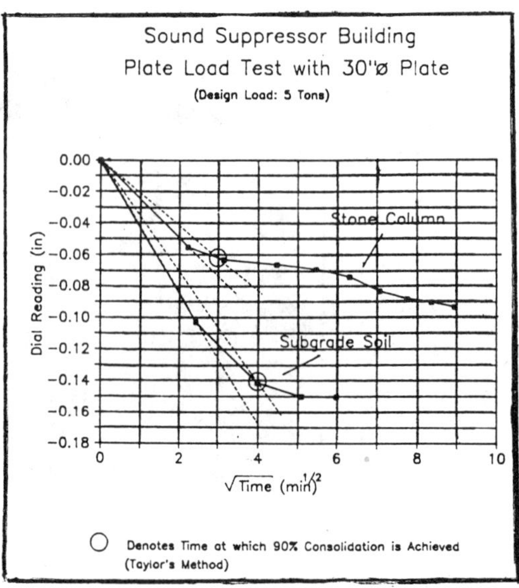

FIG. 3--Case 1, settlement vs. time for maximum load.
(1 foot = .30 m, 1 ton = 9 kN, 1 inch = 25 mm)

The soil load test located at the centroid of 3 columns was applied in the following percentages of the 103 kN (11.6 ton) total load: 12.5, 18.75, 25, 37.5, 50, 75 and 100 percent. Each load was held until the t_{90} settlement had occurred. The first three loads were held for about 15 minutes and the remaining loads for 30 to 40 minutes. The load was then removed to the following percentages of the maximum load 50, 25 and 0%.

Case 2: Capri Village Condominiums, Treasure Island, Florida:

Capri Village is a 3 story condominium and townhouse community, containing approximately 300 dwelling units. The buildings are wood frame structures designed on strip footings with a wall load of 30 kN/m (1 tlf).

Site investigations revealed a generalized subsurface profile (Figure 4) of three feet of surficial loose fine sand fill overlying a sandy peat/peaty fine sand zone varying in thickness from 0.6 to 1.2 m (2 to 4 feet). Beneath this, medium dense clean fine sands extended to depths in excess of 6.1 m (20 feet) below grade. The groundwater table was within 0.6 m (2 feet) of the surface. Foundation alternatives included piling, excavation and replacement or improvement of the sandy peat/peaty sand zone where necessary to allow shallow foundation support. Stone column installation was selected beneath all continuous strip footings to compact and strengthen the sandy soils and reinforce and replace the organic zones.

Stone columns were installed with an 80 horsepower vibrator using the wet method (since granular surface soils allowed run off water to permeate quickly) on 1.8 m (6 foot) centers to an average depth of 3.7 m (12 feet). The stone was a 6 to 25 mm (.25 to 1 inch) diameter crushed limestone. The column diameter was about .76 m (30 inches) in the upper and lower sands. However, additional time and effort was spent on the peaty layers to expand the columns. Post-construction CPT tests verified the continuity and density of the stone columns as well as the improvement of the granular soils to an average tip resistance value of about 10,000 kN/m^2 (100 tsf). However, to further evaluate the stone column load-carrying characteristics, individual stone columns were subjected to a .76 m (30 inch) diameter plate load test. The tests were applied using a CPT testing rig in a similar manner as in the Sound Suppressor Building (Case 1). Figure 4 shows a typical plot of the plate load test results.

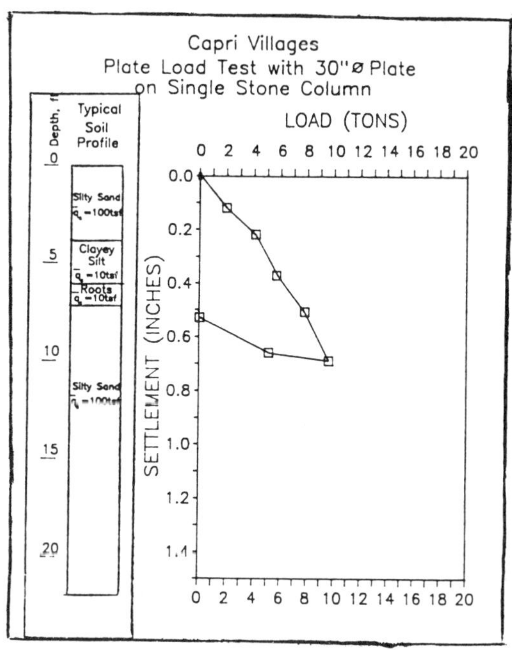

FIG. 4--Case 2, soil profile and load test results
(1 foot = .30 m, 1 ton = 9kN, 1 inch = 25 mm)

<u>**Category II:**</u>   <u>**Quick load test of stone column and tributary soil area**</u>

Case 3: Wheaton Plaza, Wheaton, Maryland:

Expansion to accommodate additional retail outlets at Wheaton Plaza Shopping Mall, in Wheaton, MD, had encroached on existing parking areas. To offset this loss of parking capacity, construction of a 3-level parking garage was planned at the ground level. The design bearing pressure was 295 kN/m^2 (3 tsf), and the maximum allowable settlement was specified as 25 mm (1 inch) under the design load.

The generalized subsurface profile (Figure 5) consisted of 6.1 m (20 feet) of silty, fine to medium micaceous sand fill, placed at the time of original mall construction, overlying decomposed mica schist. The fill density was erratic as a result of its uncontrolled placement. The water table was located at a depth of 11.6 m (38 feet) below grade. Improvement of the fill was required to reduce differential and total settlements.

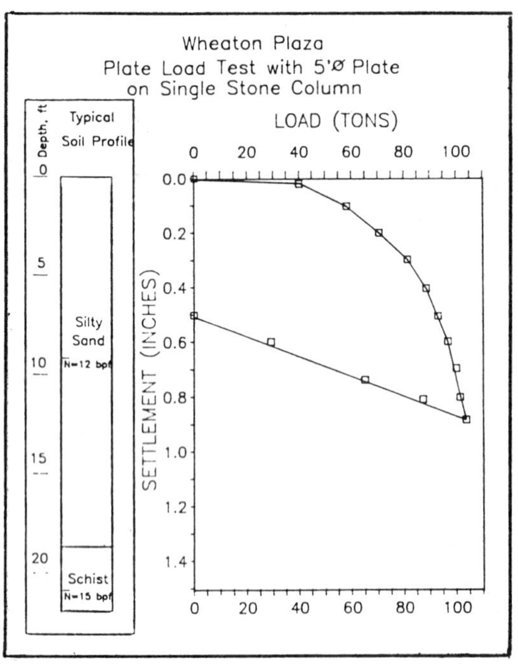

FIG. 5--Case 3, Soil profile and load test results
(1 foot = .30 m, 1 ton = 9kN, 1 inch = 22 mm)

The improvement program consisted of using the dry bottom-feed method (due to limited site area for handling waste water) with a 165 horsepower vibrator to construct .76 m (30 inch) diameter stone columns. The stone was a 25 to 38 mm (1 to 1.5 inch) diameter crushed granite. The number of stone columns per building column location varied from 1 to 7, based on footing size of .9 to 3 m (3 to 10 foot) square. The depth of the stone column was extended down to twice the largest footing dimension.

Prior to production, a full scale load test was conducted to confirm design predictions. Sixteen stone columns were installed at 4 building column locations. From these 16 columns, one was randomly selected and tested to 200% of the maximum design load. The results of the load test are presented in Figure 5.

The load test was applied to a 1.5 m (5 foot) diameter plate centered on a .76 m (30 inch) diameter stone column. The load was applied using a hydraulic jack with a calibrated pressure gauge. The reaction load was supplied by a dead load supported on a reaction frame. The plate deflection was measured by three dial gauges located on the perimeter of the plate and attached to a steel reference beam.

The loading was applied in general accordance with the procedure described in ASTM D 1143, Standard Method of Testing Piles Under Static Axial Compressive Load, Quick Load Test Method [2]. Load increments of 90 kN (10 tons) were placed to twice the design load for a maximum load of 930 kN (103 tons). Each increment of load was maintained for 1 1/2 minutes and the maximum load was maintained for 5 minutes. The load was then removed at one time.

Category III: Quick load test of a full scale footing on stone column reinforced soil.

Case 4: Kroger Shopping Center, Kokomo, Indiana:

The 8,147 m^2 (87,600 ft ^{2}) grocery store and adjacent strip shopping center is a single-story structure with design maximum column loads of 115 kN to 750 kN (13 to 85 tons) and a 20 kN/m^2 (0.2 tsf) floor load. The structure was designed to bear on shallow foundations designed for an allowable bearing pressure of 145 kN/m^2 (1.5 tsf) and a maximum total settlement of 25 mm (1 inch).

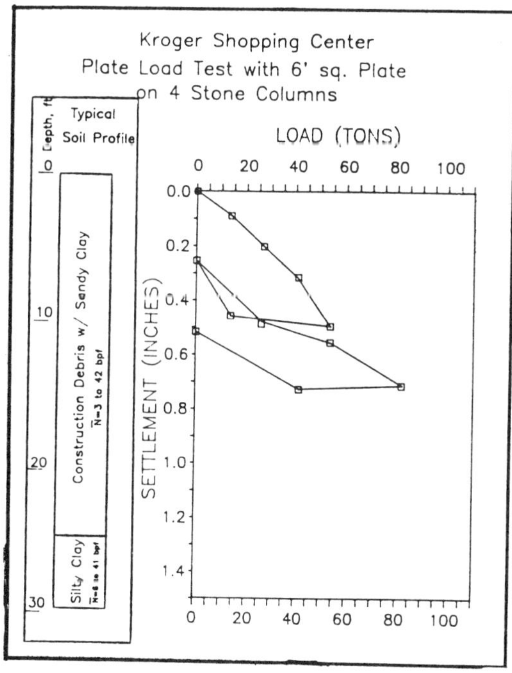

FIG. 6--Case 4, Soil profile and load test results
(1 foot = .30 m, 1 ton = 9 kN, 1 inch = 25 mm)

The site was found to be underlain by 1.5 to 9.1 m (5 to 30 feet) of sandy, silty, clayey fill with extensive construction debris. The fill was underlain by relatively dense fine sand and stiff clays (Figure 6). The fill placement was uncontrolled and standard penetration testing indicated erratic resistance values (N = 3 to 42). Shallow foundations on improved fill was the most economical alternative.

A stone column program was performed using a 165 horsepower vibrator and the dry bottom-feed method (due to limited site area to handle waste water) to fully penetrate the fill. One to four columns were constructed beneath the .9 to 1.8 m (3 to 6 foot) footing locations and on 2.4 m (8 foot) centers beneath load bearing walls. Stone columns were also installed on a 3 m (10 foot) square grid beneath the floor slab area. Average stone column diameters varied from .9 to 1.1 m (36 to 42 inches) in diameter. The stone was a 6 to 19 mm (.25 to .75 inch) diameter subangular gravel.

The load test was applied to a 1.8 m (6 foot) square plate centered on 4 stone columns spaced in a square pattern of 1.8 m (6 foot) on center. The load test results are presented in Figure 6. The load was applied with a calibrated hydraulic jack. The reaction load was supplied by a reaction beam attached to four rock anchors. The rock anchors were installed within 0.6 m (2 feet) of the plate. The anchors were about 15.2 m (50 feet) in length with only the bottom 6.1 m (20 feet) grouted and upper 9.1 m (30 feet) left as an open hole to avoid influencing the load test. The plate settlement was measured by 3 dial gauges attached to a steel reference beam (Figure 7).

FIG 7.--Case 4, Photograph of load test
(1 foot = .30 m, 1 ton = 9 kN, 1 inch = 25 mm)

The design load of 480 kN (54 tons) was applied in 25 percent increments, holding each for 15 minutes and reading the settlement every 5 minutes. The design load was held for 2 hours, reading the settlement every 10 minutes. The load was then removed in decrements of 25 percent of design load holding each for 15 minutes and reading settlement every 5 minutes. The zero load was held for 30 minutes. The load was then reapplied in increments of 25 percent of the design load to 150 percent of the design load. Each increment was held for 15 minutes reading settlement every 5 minutes. The maximum load of 720 kN (81 tons) was held for 2 hours, reading the settlement every 10 minutes. The load was then removed using the same procedure as the first unloading.

Case 5:   **Processing Building, Salem Nuclear Plant, New Jersey**

The processing Building is a single story 3,720 m^2 (40,000 ft^2) structure. The foundation design consists of spread footings with a design bearing pressure of 96 kN/m^2 (1 tsf).

The generalized subsurface profile consists of 10.7 to 12.2 m (35 to 40 feet) of silty fine sand and sandy silt hydraulic fill (see Figure 8). Due to the loose and soft nature of the fill, either a deep foundation or improvement of the fill was required.

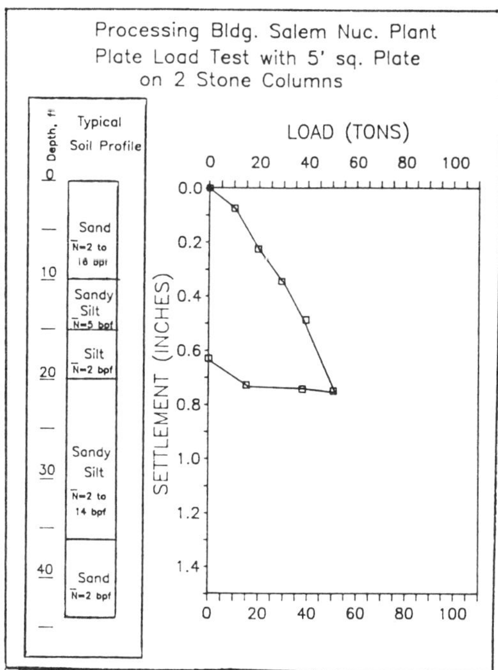

FIG. 8--Case 5, Soil profile and load test results
(1 foot = .30 m, 1 ton = 9 kN, 1 inch = 25 mm)

Soil improvement using stone columns was selected. A 165 horsepower vibrator was used to construct stone columns using the dry bottom-feed method (due to limited site area for handling waste water). The 1.1 m (42 inch) diameter columns were installed to an average depth of 5.5 m (18 feet) on 1.7 m (5.5 foot) spacing beneath wall and column footings and on a 2.6 m (8.5 foot) grid beneath the slab area. The stone was a 6 to 19 mm (.25 to .75 inch) diameter crushed limestone.

The load test was applied to a 1.5 m (5 foot) square plate with its diagonal centered over 2 stone columns. The load test results are shown in Figure 7. The load was applied with a 900 kN (100 ton) calibrated hydraulic jack with a load cell backup. The reaction load was supplied by a reaction frame supporting a dead load. The plate settlement was measured using three dial gauges attached to a steel referenced beam.

Twice the 220 kN (25 ton) design load was applied in 44 kN (5 ton) increments, holding each load for 15 minutes. The maximum load of 440 kN (50 tons) was held for 12 hours. The load was then removed in three decrements holding each for 15 minutes.

Case 6: Prison Facility, Marlin, Texas:

The new prison facility consists of 18 structures ranging from 56 m^2 (600 ft^2) entrance building to several 1,116 m^2 (12,000 ft^2) dormitory buildings. The structures were designed for shallow spread column and wall footings with 96 kN/m^2 (1 tsf) bearing pressure and a maximum total settlement of 25 mm (one inch).

The generalized subsurface profile consists of about 4.3 m (14 feet) of loose sandy silt and clayey silt, underlain by firm silty fine sand (see Figure 9). Both deep foundations and shallow foundations an stone columns were considered due to the compressibility of the loose silt.

Stone columns were selected and 165 horsepower vibrators were used to construct the 1.1 m (42 inch) diameter 4.9 m (16 foot) long stone columns using the wet method. Site area was available for settlement ponds for waste water. The stone was a 25 to 50 mm (1 to 2 inch) diameter crushed river rock. One to four stone columns were constructed below .9 to 1.8 m (3 to 6 foot) square column footings and a single row on 1.8 m (6 foot) center was constructed beneath load bearing walls.

The load tests were applied to a 1.8 m (6 foot) square plate centered on 4 stone columns placed in a square pattern of 1.5 m (5 foot) on center. The load test results are presented in Figure 9. The load was applied with a calibrated hydraulic jack. The reaction load was provided by a reaction frame supporting a dead load. The settlement of the plate was measured by four dial gauges attached to a steel reference beam.

Twice the design load was applied in increments of 25 percent of the design load. Each increment was maintained for 30 minutes and the maximum load was maintained for 1 hour. The load was then removed in 4 equal decrements with a 15 minute hold time for each.

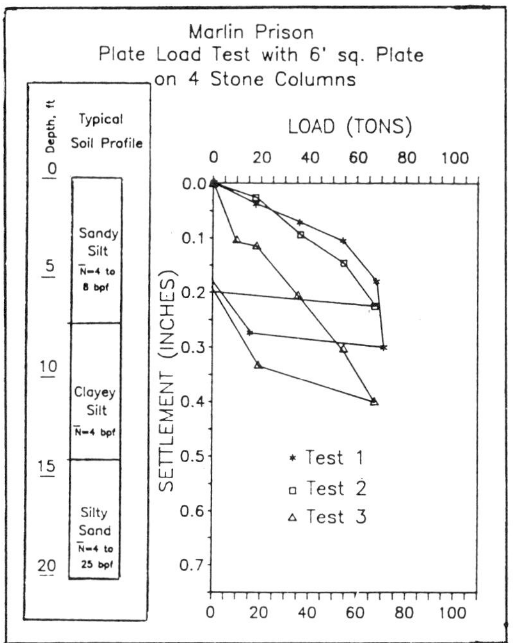

FIG. 9--Case 6, Soil profile and load test result
(1 foot = .30 m, 1 ton = 9 kN, 1 inch = 25 mm)

SETTLEMENT EVALUATIONS FROM QUICK LOAD TEST DATA

The quick load tests for the case histories presented were performed to confirm the predicted settlement of the planned foundations. In this section of the paper, methods for using this data to develop settlement estimates are presented for each category of load test.

I.   Quick load test on stone column area

In the first two case histories in which the stone column area alone was loaded, the test data was used to calculate an elastic modulus for the stone column (modulus equals stress divided by strain). This was appropriate since the compressible soil adjacent to the columns extended from the elevation of the plate load test to a depth of 2.1 to 3 m (7 to 10 feet), a relatively short distance. Because of this and the large difference in the modulus of the stone column and adjacent soil, the authors felt that a reasonable estimate of the stone column modulus could be determined by assuming the surface settlement of the column is principally due to the compression of this "unsupported" length of the stone column. The soils at the Sound Suppressor Building were cohesive silty sands in which the majority of the settlement should have occurred during the quick load test.

As for the stone columns at the Capri Village site, they were constructed so as to enlarge the column diameter in the organic layer. This increased diameter resulted in a high percentage replacement of the organics and could be considered as replacing the organic confining material around the .76 m (30 inch) diameter column with stone. This result greatly reduces the effect of the organic material on the performance of the stone column. Based on this and considering that little organic soils remained in the treated area, the great majority of the settlement of the stone column should occur during the quick test. Although some additional settlement would occur if the test was extended, the test should give a reasonable estimate of the column modulus value.

In addition, a load test on the native cohesive silty sand was also performed at the Sound Suppressor Building site. As mentioned above, the majority of the settlement occurred during the load test. The soil's elastic modulus was calculated by assuming an average stress of half the test load acting over a depth of twice the plate diameter. Therefore, using an elastic analysis: (modulus = stress x depth stressed divided by settlement), the conservative stone column modulus of 37,880 kN/m^2 (393 tsf) and soil modulus of 5,190 kN/m^2 (54 tsf) are calculated for the Sound Suppressor site. The soil modulus corresponds well to that predicted based on an average CPT value of 2,400 kN/m^2 (25 tsf) and a 2.5 conversion factor of tip resistance to modulus [5]. Using these values, a weighted average modulus was calculated based on the percentage of stone and soil beneath the foundation. The weighted average modulus of 10,290 kN/m^2 (107 tsf) was determined and the linear elastic analysis predicted a settlement of the building foundations of 28 mm (1.1 inches).

An analysis based on the GKN Keller in house design manual, "Stone Columns I" [3] was performed using these modulus values. This method predicts improvement factors based on effective area ratio and utilized Dr. Priebe's improvement factors [4]. The settlement of the foundation based on the untreated soil parameters alone is divided by this improvement factor to predict the foundation settlement on the stone column treated soil. Based on a soil modulus of 5,190 kN/m^2 (54 tsf), the untreated settlement of 56 mm (2.2 inches) is calculated. Based on an area treatment of .76 m (30 inch) diameter columns on a 1.8 m (6 foot) triangular grid, the improvement factor is about 2. This results in a post-treatment foundation settlement of 28 mm (1.1 inches), the same as that predicted by the linear elastic analysis.

Using the same analysis, the tests performed at the Capri Village site estimated the stone column modulus to be a conservative 29,430 kN/m^2 (306 tsf). The load test was not performed on the soil but the average CPT tip resistance (qc) of 1,440 kN/m^2 (15 tsf) in the organic soils resulted in a predicted modulus of 3,650 kN/m^2 (38 tsf) based on the modulus being 2.5 qc.

The weighted average modulus value of 16,540 kN/m^2 (172 tsf) results in a predicted post treatment settlement of 7.6 mm (0.3 inches) beneath the .6 m (2 foot) wide wall footings. The design was based on the GKN Keller method which predicted an untreated settlement of 64 mm (2.5 inches) and an improvement factor of 4.75, resulting in a post treatment foundation settlement of 7.6 mm (3 inches), the same as predicted by the load test.

II.   Quick load tests on a stone column and tributary area.

The Wheaton Plaza load test measured a settlement of 2.5 mm (0.1 inches) at the design load of 525 kN (59 tons). Assuming the profile to a depth of 3 m (10 feet) is compressed by one half the surface stress, the linear elastic analysis can be used to back calculate average modulus value of 173,050 kN/m^2 (1,800 tsf). The method can then be used to predict a settlement beneath a 3.4 m (11 foot) square footing of 5.6 mm (.22 inches).

III. Quick load tests on a full scale footing on stone column reinforced soil.

The full scale load tests for the last three case histories should reasonably predict the settlement of the foundation they model, as long as the duration of test is such that the majority of the primary consolidation has occurred. The maximum load is generally 150 percent of the design load. Increasing this to 200 percent seems unnecessary and can be expensive because of the large area loaded. In addition, the soil must be of a nature that secondary consolidation is minimal (i. e., not soft clays nor organics). Secondary consolidation will be minimal for the subject three case histories.

To evaluate if the test load was maintained for a sufficient time, the settlement versus the square root of time is plotted and the t90 analysis performed (presented in ASTM D 2435, One-Dimensional Consolidation Properties of Soils 11.6) [2].

SUMMARY AND CONCLUSION

The Category I load test is the least expensive quick load test and is useful in predicting settlement of foundations. Since the small diameter plate affects only a limited depth of soil beneath it, the soil which mainly contributes to the settlement of the foundation must be uniform in type and consistency. It is most critical that a firm layer with an underlying soft layer does not exist or the settlements will be greatly underestimated. A load test at the Sound Suppressor Building was performed at the ground surface, underlain by 3 feet of dense soil. This test experienced settlement of 20 percent of the test performed at a depth of 3 feet, the foundation bearing level.

In addition, it is important that the test loads are maintained long enough to allow the majority of the settlement to occur. In this respect, the soils cannot be such that secondary consolidation is a factor (i. e., same soft clays and organic soils). These load tests, although useful, do not actually measure the settlement of the foundation and rely on theoretical calculations to predict settlement.

The Category II load test is generally more expensive than the Category I test because of the greater load required, but technically models more closely the loading in the soil and stone column than the Category I test. However, the problem of having a test area which is smaller that the foundation load still exists. The same caution exists for this test as for the Category I test in so far as depth and uniformity of compressive soils and maintaining the loads for sufficient time. If the stone column reinforced compressible layer is thick, but uniform in nature and consistency, a finite element analysis can be performed to back calculate modulus values based on the load test data. Then the analysis can be performed to reasonably predict the settlement of the foundations.

The Category III tests are the most expensive and the technically preferred test since they exactly model the actual foundation loading. The primary consideration is that the design load and maximum test load are maintained for a sufficient time to permit the majority of the primary consolidation to occur. If secondary consolidation is a concern, the load test should be prolonged. The alternative is to use laboratory consolidation tests, the results of the quick load test and a finite element analysis to predict the total settlement of the foundations.

APPENDIX-REFERENCES

[1]   Barksdale, R. D. and Bachus, R. C., "Design and Construction of Stone Columns", for Federal Highway Administration, August 1982

[2]   "Annual Book of Standards, Section 4, Construction", American Society for Testing and Materials, 1986

[3]   Dobson, T. and Chambosse, G. "Stone Columns I, Estimation of Bearing Capacity and Expected Settlement in Cohesive Soils", GKN Keller Inc., In-House Publication

[4]   Schmertmann, J. H., Hartman, J. P. and Brown, P. R., "Improved Strain Influence Factor Diagrams", Journal of Geotechnical Engineering, American Society of Civil Engineers, August 1978

[5]   Priebe, H. "Abschaetzung des Setzungsverhaltens eines durch Stopfverdichtung Verbesserten Baugrundes." Die Bautechnik 5/1976

Donald R. Snethen[1] and Michael H. Homan[2]

DYNAMIC COMPACTION/STONE COLUMNS - TEST SECTIONS FOR CONSTRUCTION CONTROL AND PERFORMANCE EVALUATION AT AN UNCONTROLLED LANDFILL SITE

REFERENCE: Snethen, D. R. and Homan, M. H., "Dynamic Compaction/Stone Columns - Test Sections for Construction Control and Performance Evaluation at an Uncontrolled Landfill Site," _Deep Foundation Improvements: Design, Construction, and Testing_, _ASTM STP 1089_, Melvin I. Esrig and Robert C. Bachus, Eds., American Society for Testing and Materials, Philadelphia, 1991.

ABSTRACT: Completion of a portion of the highway loop around Tulsa, Oklahoma, required that State Highway 11 (SH-11) cross an old strip mine/uncontrolled landfill area located in north central Tulsa. A grade separation was required where SH-11 crossed Yale Avenue, a major city street. The entire project involves the bridge structure, approach embankments with a maximum height of approximately 9 m and the roadway paving. The foundation conditions varied from remnants of intact shale existing between the strip mined areas to layers of disturbed shale covering layers of trash varying in thickness from approximately 1 m to nearly 6 m. At both ends of the project, ground water was present.

Dynamic compaction was selected over other options (i.e. grouting or elevated roadway) to prepare the foundation material to support the embankments. As part of the dynamic compaction quality assurance program, three instrumented field test sections were constructed to verify the adequacy of dynamic compaction. In the portion of the project where the thickness of the trash layers was the greatest, the potential for additional settlement following dynamic compaction seemed large, so it was decided to support that section of the embankment on stone columns constructed using the dynamic compaction process. An on-site evaluation of different impact sequences was conducted to determine the most efficient method for stone column construction. Approximately 95 stone columns were constructed, using the selected procedure, in the area with the thickest trash. Following completion of the embankment, several

---

[1]Professor, School of Civil Engineering, Oklahoma State University, Stillwater, Oklahoma.
[2]Project Engineer, Law Engineering, Tulsa, Oklahoma.

locations were instrumented (settlement gages and piezometers) to observe the long term settlement performance of the embankment.

Dynamic compaction of the materials resulted in average settlement over the approximately 9 hectare site of 66 cm. Reasonable strength improvements were measured at each of the test sites. Settlement records show a maximum settlement of approximately 30 cm since completion of the embankment in Spring 1986. The major portion of the settlement occurred rather quickly (i.e. in a few months) with subsequent settlement being minimal.

KEY WORDS: deep foundation stabilization, stone columns, dynamic compaction, field test sections

INTRODUCTION:

The Gilcrease Expressway in Tulsa, Oklahoma, is an extension of existing State Highway 11 which connects the Tulsa International Airport with U.S. 75 and forms a portion of the proposed loop around Tulsa. Near Yale Avenue, the roadway crosses an abandoned coal strip mine. Subsequent to strip mining, the area was used as an uncontrolled sanitary landfill. The mine spoil extends to depths of up to 14 m below the existing grade. The existing ground surface elevation varies from approximately 198 m mean sea level (MSL) on the east end to 193 m MSL on the west end with the highest point of 204 m MSL located near the midpoint of the project. The roadway is a standard four-lane divided highway with an interchange and bridge at Yale Avenue. The maximum height of embankment at the interchange is approximately 9 m.

Prior to preparation for dynamic compaction the terrain in the project area consisted of a series of ridges and valleys formed by the strip mining operation. Randomly deposited trash was found throughout the site, primarily in the valleys and along trails. Large vegetation, such as trees and brush,were located west of Yale around the old strip mines.

A geotechnical investigation characterizing the site was performed by the Oklahoma Department of Transportation (ODOT) Materials Division [1]. The investigation included 67 test borings, two bulldozer cuts along the embankment centerline into the spoil bank ridges, and numerous field Standard Penetration Tests, Cone Penetration Tests and laboratory tests. The subsurface exploration confirmed the existence and extent of the strip mined area including the general location and thickness of trash layers, extent of spoil backfill, location of unmined ridges, and intermittent presence of ground water. Test data from the borings revealed that the mine spoil was a low to moderate plasticity clay with natural moisture contents near the plastic limit.

The centerline profile was generalized into three major subsurface profiles for design considerations, Figure 1. The first generalized subsurface profile, representing the central portion of the project, included approximately 16 m of mine spoil resting on intact shale with minimal trash and no continuous groundwater table. The second profile, representing the east portion of the project, included approximately a meter of mine spoil overlying approximately 2 to 6 m of trash resting on more mine spoil. No consistent groundwater level was indicated in this area. The third profile, representing the west portion of the project, included approximately 2 m of mine spoil overlying 2 to 7 m of trash with a groundwater level varying from 2 to 8 m below the ground surface.

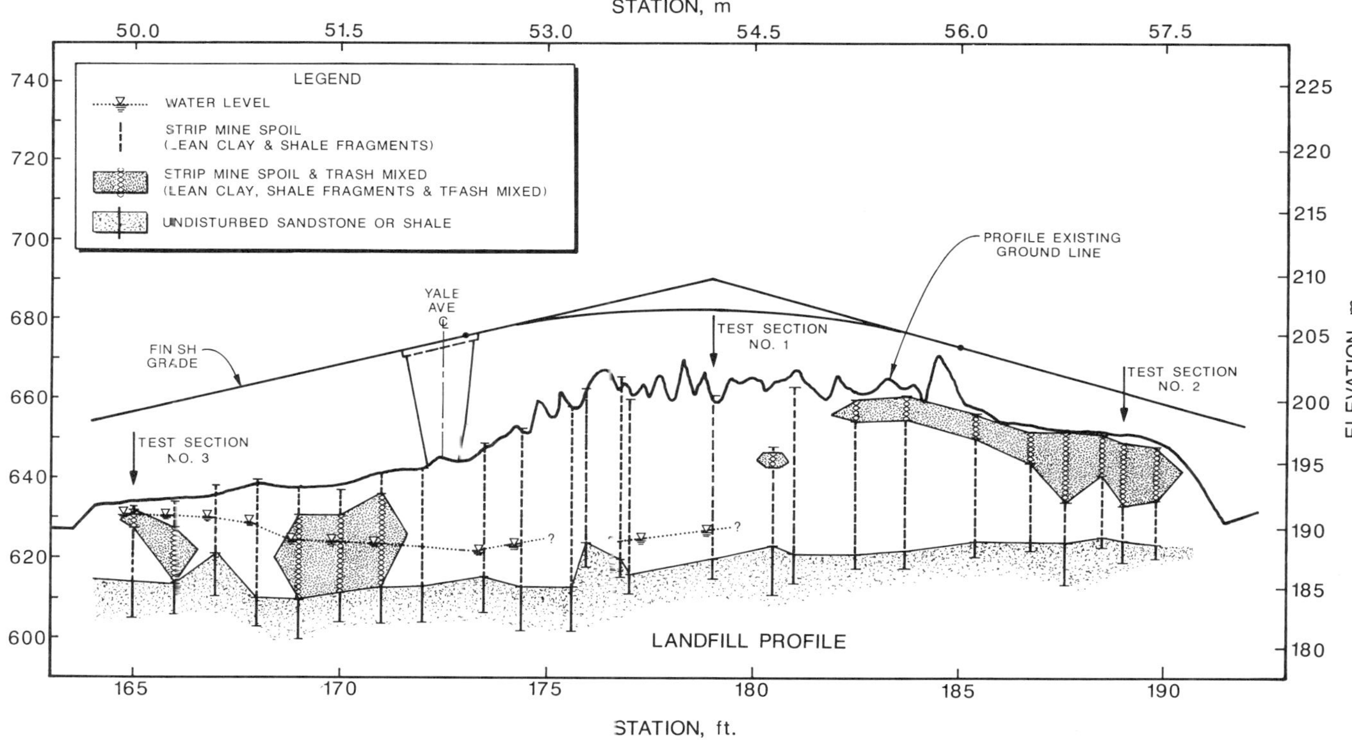

Figure 1a. Centerline Profile at Dynamic Compaction Project Site.

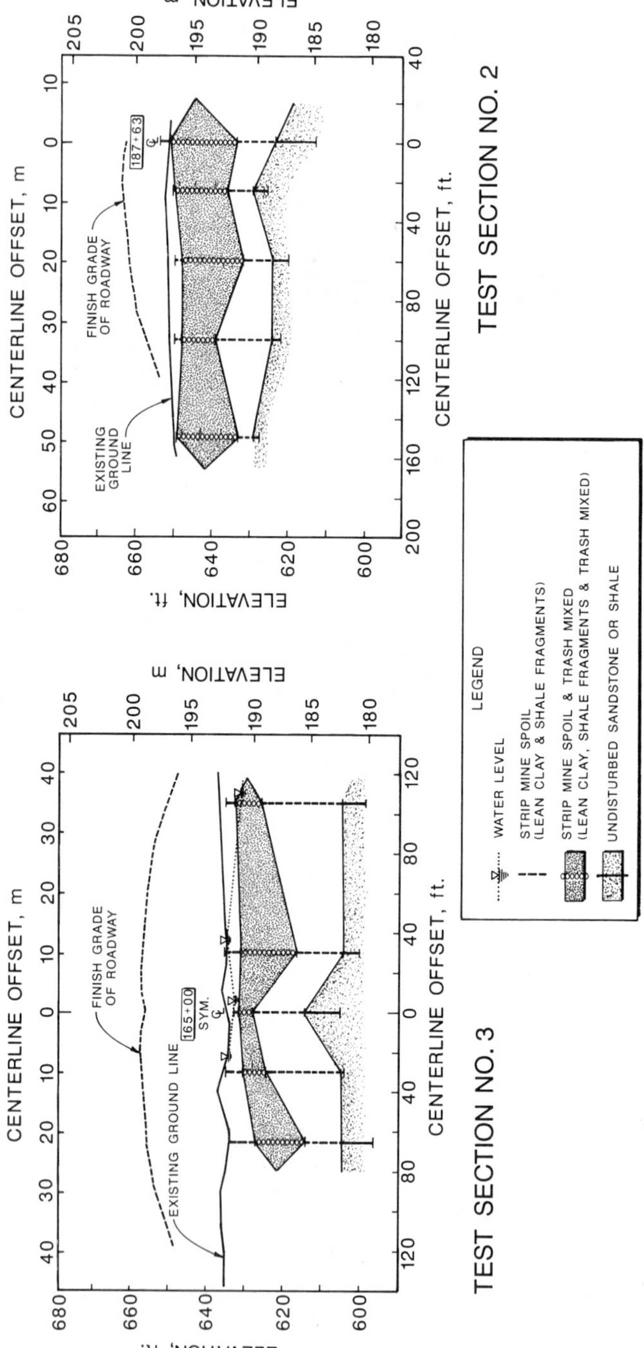

Figure 1b. Transverse Profiles at Stations 165+00 and 187+63.

Several options were considered to improve the foundation conditions to support the embankments and roadway. Specifically, the options included dynamic compaction, grouting, and constructing an elevated roadway founded on drilled shafts in the intact shale. Dynamic compaction was selected on the basis of construction feasibility and economy. Because of the variability and uncertainty associated with the subsurface conditions, ODOT required that three instrumented test sections be constructed to measure the effectiveness of dynamic compaction and to adjust the construction sequence as appropriate. The three test sections were located at the central, eastern and western portions of the project corresponding to the areas represented by the three idealized profiles previously described.

DYNAMIC COMPACTION - DESIGN/CONSTRUCTION CONSIDERATIONS

The design engineer, a consulting engineering firm from the Tulsa area, estimated that an average settlement of approximately 75 cm over the entire site would be required to achieve sufficient improvement of the foundation conditions to support the embankments and roadway. To achieve this average settlement, a total energy input of 312 m-tonne/m^2 was recommended with provisions for changes based on the results of the instrumented test sections.

The dynamic compaction construction sequence included the use of a 17.7-tonne round weight with a static contact pressure of 48 KPa and a drop height of 23 m. A square grid pattern with a spacing of 9 m was used for the specified 8 passes, with the grid pattern offset after each pass to achieve full coverage of the project site. The construction sequence required eight impacts per point for the first four passes (1-4) and six impacts per point for the last four passes (5-8). An ironing pass was required after the dynamic compaction to densify the near surface soils. This was conducted using a 2.1 m square weight weighing 7.0 tonnes and a drop height of 6 m and a 2/3 overlap of each impact point.

TEST SECTIONS - INSTRUMENTATION AND RESULTS

The three test sections were conducted using similar instrumentation and procedures [2, 3]; however, since the stone columns were used only In the portion of the project represented by Test Section No. 2 (i.e., Station 182 to 190) only the results from that test section are discussed.

Test Section No. 2 was located at Station 189+00. The soil profile at the test section consisted of approximately 1 m of mine spoil over nearly 6 m of trash resting on 1 m of mine spoil underlain by the intact shale. Prior to placing the instrumentation and doing the dynamic compaction, a 1 - m layer of crusher-run limestone (25 cm maximum size) was placed over the test area. Instrumentation at Test Section No. 2 consisted of two inclinometers and two electric piezometers. In addition, crater depths were monitored during compaction and surface elevation was monitored between passes (i.e., after leveling). Pre- and post-compaction continuous Standard Penetration Test borings were run. A plan view of Test Section No. 2 with the instrument locations and impact points is shown in Figure 2. At the time the piezometers were installed the ground water level was noted at a depth of 4.6 m. The test section was compacted using the same impact and pass sequence required for the dynamic compaction production work.

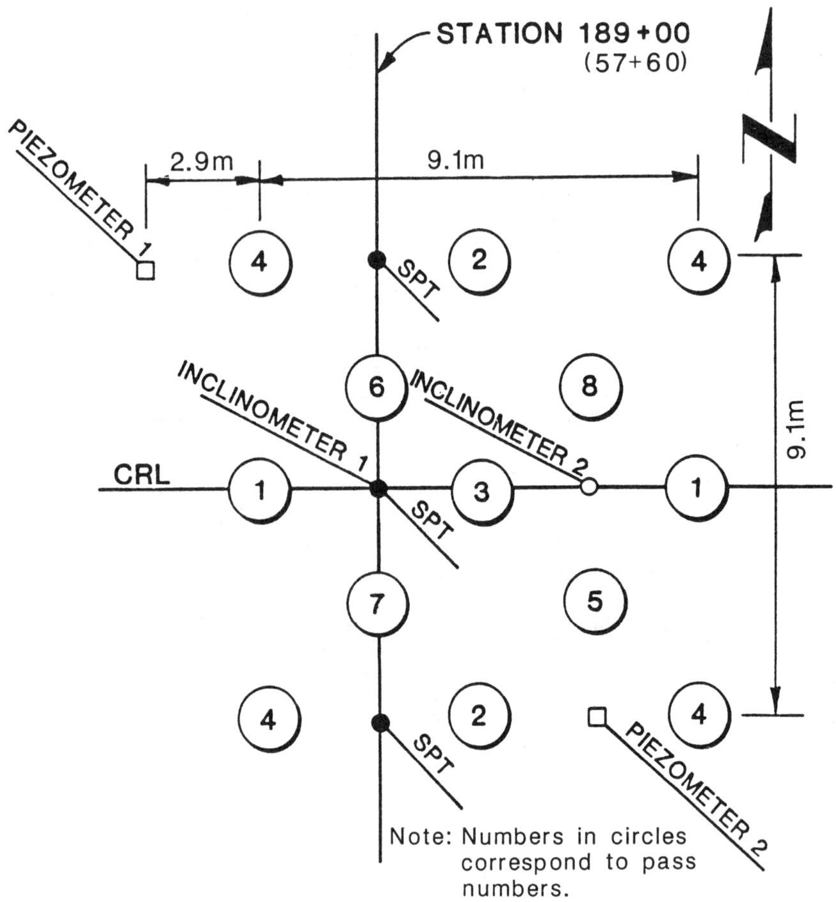

Figure 2. Instrumentation Layout and Impact Pattern, Test Section No. 2

One difference in construction procedure was the continuous monitoring of piezometers and crater depth after each impact and monitoring the inclinometers and surface elevation between passes. Only one of the 3 impact points in Test Section No. 2 received the full number of impacts. The other points received fewer impacts because of problems with the weight "sticking" in the crater. The crater depths were much greater in Test Section No. 2 because the wet condition of the trash increased the suction on the weight in the bottom of the crater making it more difficult to withdraw the weight. A summary of crater depths for all impact points at Test Section No. 2 is shown in Table 1.

TABLE 1 -- Summary of Crater Depth at Test Section No. 2

Pass No	Number of Impacts	Crater Depth (m)
1	8	3.8
1	6	4.0
2	6	4.1
2	6	4.1
3	7	3.5
4	3	2.6
4	6	3.6
4	6	3.4
4	6	3.7
5	4	3.0
6	3	2.4
7	4	2.4
8	4	2.9

Evaluation of the crater depth data from Test Section No. 2 indicated that the average crater depth was 3.4 m as compared to 2.3 m and 2.9 m for Test Sections 1 and 3, respectively. In addition, the incremental crater depth, a parameter typically used to control dynamic compaction, did not achieve the targeted minimum value of 0.15 m at any of the test section impact points.

The inclinometer data was not conclusive because of damage to the casings from the trash. The limited data indicated a "punching" shear failure under the impact of the weight. In other words, the trash was compressible enough to limit lateral movement around the impact point.

Piezometer data showed that excess pore water pressures during dynamic compaction would be minimal. Instantaneous peak values following impacting dissipated almost as quickly as they developed and the small residual values that followed the construction of the test section dissipated in approximately two weeks.

Typical Standard Penetration Test results before and after dynamic compaction are shown in Figure 3. The data shows that ground improvement at Test Section No. was not consistent.

The wet condition of the trash, large total crater depths and inconsistent strength improvement caused concern about the effectiveness of dynamic compaction to densify the trash layer. At this point the use of stone columns was considered beneath the central portion of the embankment between Stations 182+00 and 190+50. Stone columns seemed the most reasonable option because:
1.   They could be constructed using the equipment on-site without major interruption of the dynamic compaction [4].
2.   The inclinometer data indicated that there were minimal lateral movement around the craters (i.e. the stone columns could be "punched" to a firm layer below);

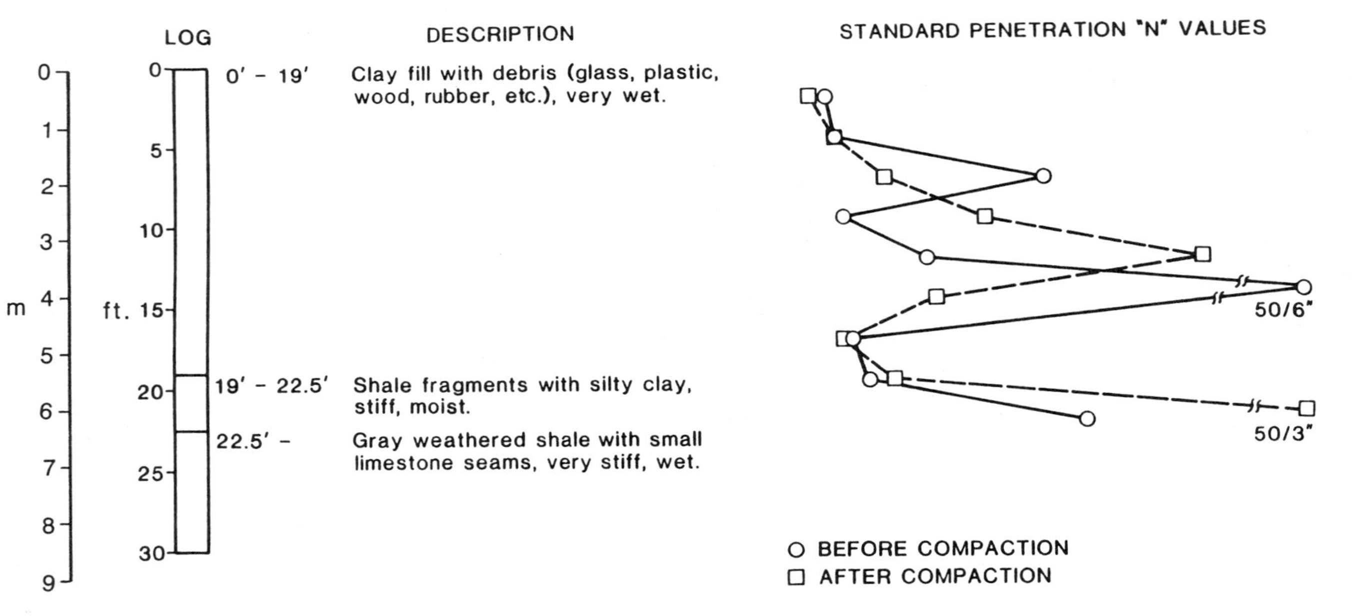

Figure 3. Standard Penetration Test Results, Along CRL, Test Section No. 2.

The decision to use stone columns was made without a firm construction sequence for the columns (i.e., impacts, limiting crater depths, backfilling). It was decided to construct three test stone columns using varying impact sequences to determine the most effective construction process. Using the total and incremental crater depth data from Test Section No. 2 as a guide, three test stone columns were constructed using the following sequences:

Test Column No. 1   --   three cycles of four impacts and one of six
                           impacts, backfilling after each cycle

Test Column No. 2   --   two cycles of six impacts, back-filling after
                           each cycle.

Test Column No. 3   --   three cycles of four impacts, backfilling after
                           each cycle.

During test column construction, crater depths were monitored after each impact. Typical total and incremental crater depths for Test Column 3 are plotted in Figure 4. Two general criteria were used to select the appropriate construction sequence; first, the individual cycles of impacts should result in a crater depth less than about 2.4 m and, second, the column should reach some level of "end" resistance at a depth of about 5 m. This second criteria was met by limiting the amount of incremental crater depth (i.e., change in elevation after each impact) to approximately 0.15 m.

Test Column No. 1 achieved both criteria but required an excessive number of impacts. Test Column No. 2 did not meet either criteria. Test Column No. 3 met both criteria and appeared to be the best balance between the number of impacts, number of cycles of backfill and construction of effective stone columns. A total of 95 stone columns approximately 2 m in diameter by 5 m long were constructed in four rows on 9 m spacing between stations 182+00 and 190+50 beneath the main lanes of the roadway. After construction of the stone columns, the dynamic compaction was completed on the remaining portions of the site.

EFFECTIVENESS OF STONE COLUMNS

Dynamic compaction (and stone column installation) was completed in the late summer 1985, and embankment construction was initiated in Fall 1985. The embankments were in place by late Spring 1986. During embankment construction the ODOT installed additional instrumentation to monitor embankment behavior. Specifically, two settlement plates were installed at Station 189+00 in the area of the stone column installation. A typical time - settlement record for one settlement plate and piezometer data from an adjacent piezometer is shown in Figure 5. The maximum settlement at Station 189+00 is just over 0.1 m for the two-year observation period. The closest location with similar subsurface profile (but less trash) and embankment height without stone columns and instrumented with settlement plates is near Station 165+00 (Figure 1b). The total settlement in that location was 0.15 m. It appears that both the dynamic compaction and stone columns were successful in reducing settlement.

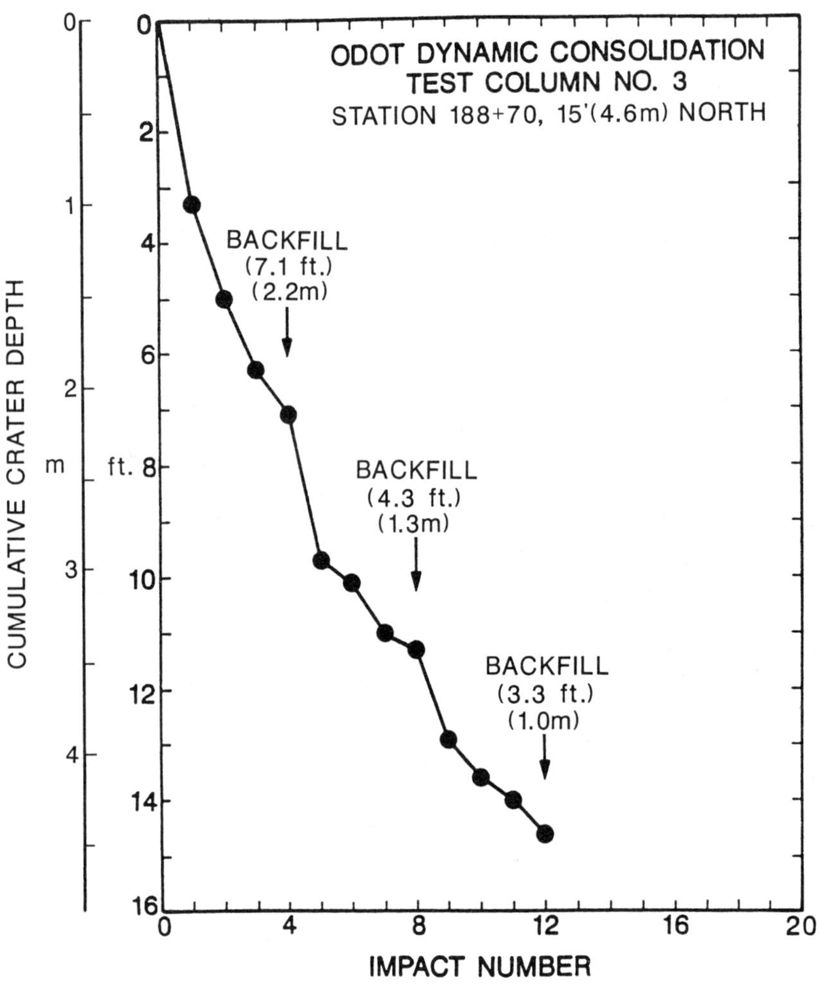

Figure 4a.   Cumulative Crater Depth Versus Impact, Test Column No. 3.

CONCLUSIONS:

Stone columns can be effectively constructed using dynamic compaction in areas such as old strip mines/landfills. The use of instrumented test section to control dynamic compaction is an excellent method to assess the need for additional foundation support and to provide basic information to assist in selecting the construction sequence.

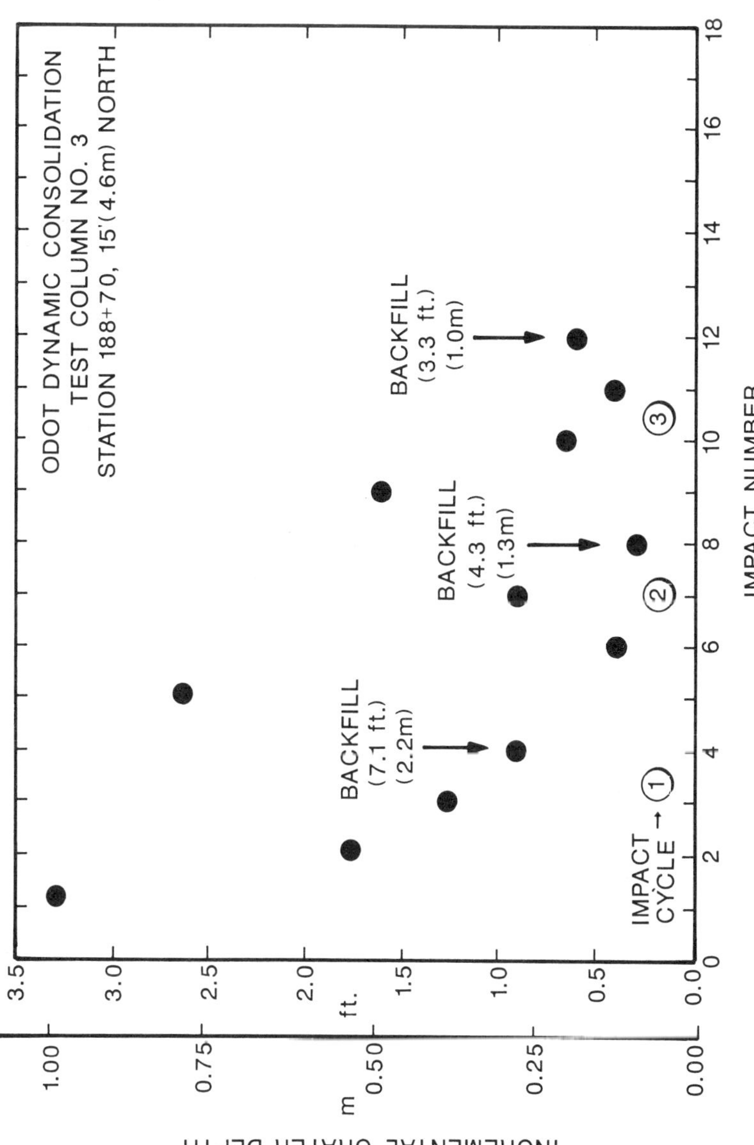

Figure 4b. Incremental Crater Depth Versus Impact, Test Column No. 3.

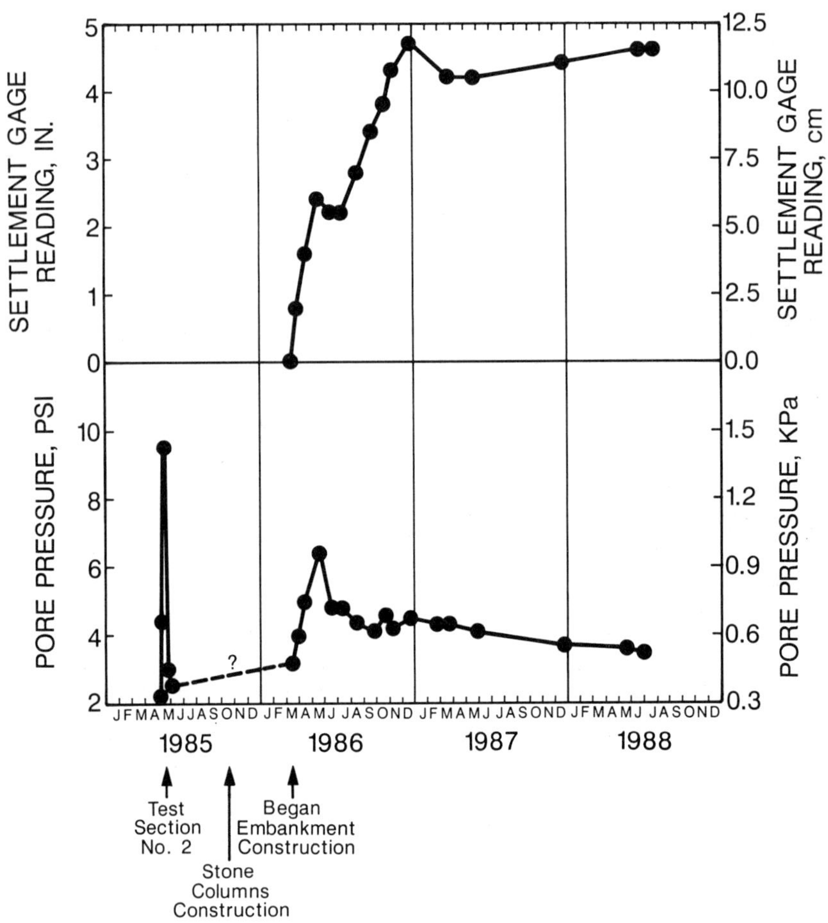

Figure 5. Settlement Gage and Piezometer Records, Test Section No. 2.

REFERENCES:

1   Plans and Specifications, Project No. F390(27), State Highway 11, Tulsa County, Oklahoma Department of Transportation, Oklahoma City, Oklahoma.

2   Snethen, D.R., and Homan, M.H., "Evaluation of Dynamic Compaction in Tulsa, Oklahoma", report to Oklahoma Department of Transportation, prepared by School of Civil Engineering, Oklahoma State University, Stillwater, Oklahoma, January, 1986.

3 Homan, M.H., "Monitoring and Evaluation of Dynamic Compaction for the Gilcrease Expressway, Tulsa, Oklahoma," M.S. Thesis submitted to Graduate College, Oklahoma State University, Stillwater, Oklahoma, July, 1985.

4 Barksdale, R.D., and Bachus, R.C. "Design and Construction of Stone Columns, Volume I," FHWA Report FHWA/RD-83-026, Federal Highway Administration, Washington, D.C., December, 1983.

Ken S. Watts and J. Andrew Charles

THE USE, TESTING AND PERFORMANCE OF VIBRATED STONE COLUMNS IN THE
UNITED KINGDOM

REFERENCE: Watts, K. S. and Charles, J. A., "The Use, Testing,
and Performance of Vibrated Stone Columns in the United Kingdom,"
Deep Foundation Improvements: Design, Construction, and Testing,
ASTM STP 1089, Melvin I. Esrig and Robert C. Bachus, Eds.,
American Society for Testing and Materials, Philadelphia, 1991.

ABSTRACT: Vibrated stone column ground improvement techniques
are being used extensively for low rise housing and light
industrial units in the United Kingdom. Much of the work is on
filled sites. There is a need for simple and inexpensive forms
of testing. The advantages and limitations of different types
of in situ tests are considered. A brief account is given of
field studies to evaluate the effectiveness of vibrated stone
columns at two sites with different ground conditions.

KEYWORDS: vibroflotation, stone columns, ground improvement,
site investigation, field test, load test, settlement

INTRODUCTION

In the United Kingdom (UK) scarcity of good building land and the
desirability of re-using land in inner cities has increasingly led to
construction on ground that would previously have been regarded as
unsuitable or at best marginal for the use of shallow spread
foundations. Ground treatment techniques can often be used to reduce
differential foundation settlements to acceptable values. The
techniques most commonly adopted are the various deep vibratory
processes collectively described as 'vibro' in which ground treatment
is effected by a powerful torpedo-shaped vibrating poker. Vibro
includes the processes of vibrocompaction/vibroflotation and vibrated
stone columns.

These ground treatment techniques were introduced intc the UK from
Germany about 30 years ago and initially many applications were
associated with civil engineering works. However, for the last 20 years

Mr. Watts is a Senior Scientific Officer and Dr. Charles a Senior
Principal Scientific Officer at the Building Research Establishment,
Garston, Watford WD2 7JR, United Kingdom.

vibro has been used extensively for low-rise buildings, particularly
housing and light industrial units and this is now the main use for
these methods in the UK. Projects range from a few treatment points
beneath strip footings for a single detached dwelling to the treatment
of large areas with a uniform pattern of treatment points. Well over
half the work is in fills with some work in soft clays. The vibrated
stone column technique has been used on many urban redevelopment sites.
It has been estimated that the annual value of vibro contracts in the
UK is around £12 million [1].

A survey carried out by the Building Research Establishment
(BRE) on the use of vibro in the UK showed that there has been little
systematic monitoring of the performance of structures built on ground
treated by vibro, despite its widespread use [1]. BRE is currently
carrying out a programme of field studies to monitor vibro at a number
of sites. Both natural soil and fill sites are being investigated and
the work includes the use of load tests on treated and untreated ground
to compare with observations of settlement of the structures built on
the treated ground. The effectiveness of stone columns in soft clay
soils has previously been investigated by BRE in laboratory tests [2].

SOIL TESTING

Where heterogeneous fills are to be treated, small scale laboratory
testing of the fill may be of limited value and in situ testing methods
may be more appropriate. In situ tests may be carried out as part of
the site investigation to determine the soil profile, to characterise
soil properties and to estimate bearing capacity or settlement of the
untreated ground. In situ tests carried out both before and after
treatment may give an indication of the effectiveness of ground
treatment in improving soil properties. Load tests on treated ground
can give a direct estimate of the performance of the structure which
subsequently will be built on the ground.

Penetration Testing

A wide range of in situ testing techniques are now available eg
standard penetration test (SPT), cone penetration test (CPT), dynamic
probing (DP), flat dilatometer (DMT), Menard pressuremeter (MPM).
International reference test procedures are available for SPT, CPT and
DP [3]. Not all these methods are suitable for all soil types and none
may be suitable in some heterogeneous fills.

CPT, DP and DMT have been used by BRE to provide a comparison of
treated and untreated ground. These techniques have been used in both
cohesive and granular soils, however the comparison is largely
qualitative. After vibrated stone columns have been installed, the
ground is non-uniform with significant differences in properties
between stone column and surrounding soil. Also the properties of the
soil will vary with the distance from the stone column.

Penetration tests have been used to investigate the change in soil
properties with distance from a stone column. In a natural sandy soil

treated by the wet vibroreplacement technique, BRE carried out CPT and DMT testing before treatment and after treatment at different distances from a stone column. Values of cone end bearing resistance, q_c and dilatometer parameters p_0 and p_1 were found to have increased by up to 100 % within 0.5 m of the column following treatment. There was little improvement at distances greater than 2 m from the column.

Geophysical Testing

Geophysical measurements can be used to characterise the soil conditions prior to ground treatment and to estimate the deformation properties of the soil. The measurement of any overall improvement in soil properties due to the treatment is difficult because the treated area is affected by the introduction of the dense stone forming the columns. BRE has used surface shear wave velocity measurements to estimate settlement and has compared these predictions with load tests on treated and untreated ground. The accuracy of these estimates appears to be limited to immediate and relatively short term settlement.

Load testing

Two important factors are the size of the loaded area and the duration of the test. The test can only be used to directly predict settlement of a building on vibrated stone columns if a representative area is loaded ie an area including at least two columns and the soil between them. Also with most fills it is long term settlement rather than immediate settlement or bearing capacity that is of concern. This means that a quick test where the load is applied by a hydraulic jack using the weight of a vehicle or construction plant as reaction may not be very helpful. A test using kentledge and lasting several weeks is needed. However as vibro is used on many very small developments, full scale load tests involving the use of steel or concrete kentledge may be prohibitively expensive. A simple, inexpensive test which will indicate the likely long term performance of a structure is required. It should be noted that load tests only predict settlement due to applied loads. The structure built on treated ground could be damaged by settlement due to other causes eg. biodegradation of organic matter, collapse compression on inundation etc. A load test may give little indication of the extent to which treatment has removed susceptibility to these other hazards.

Plate Loading Test

The plate loading test is carried out as a routine control procedure. A 600mm diameter plate is placed on top of a column and the load deformation behaviour is determined during a quick loading and unloading cycle. The load is provided by a hydraulic jack using the weight of a vehicle or crane as reaction. The test may give some indication of workmanship and uniformity, but cannot be used for design or to predict long term movements of structures which stress a large number of columns and the ground between them.

Area or Zone Test

To predict movements of structures built on treated ground it is necessary to load a representative area that includes a number of columns and the ground between them in the same way that the structure will apply load to the treated ground. It is necessary to maintain the load for a reasonable period of time to obtain an indication of the rate of settlement in the long term subsequent to the immediate response to the application of the load. Such tests are called area or zone tests and usually use kentledge to apply the load. Typically a concrete slab is cast over a number of columns and loaded to 1.5 times or more the working load. The test can be adapted to suit the specific treatment and foundation design and can simulate higher foundation loads. The major disadvantage is the cost which may well be prohibitively high on many small sites.

Portable Footing Test

A test has been developed using a portable footing attached to a heavy road vehicle. This enables two or three columns and the ground between them to be tested rapidly for immediate response. Special vehicles are being developed to apply much larger loads in this manner. The cost will be related to the length of time the test load is in place.

Skip Test

A simpler and cheaper form of area test appropriate for typical housing loads on shallow fill has been developed [4]. A small area is loaded by a rubbish container or 'skip' filled with sand. Lightweight structures with strip footings stress the ground significantly typically only to depths of 1.5 m to 2.5 m. Consequently, it is relatively simple to test load the fill to reproduce the actual stress level and distribution with depth. A model concrete footing can be cast over two or more columns or alternatively a pre-cast or steel model footing can be used. Larger stresses can be applied by placing a second skip on top of the first, but are limited typically to those induced by two-storey housing.

Measurements of settlement can be made by precise levelling. The test should be maintained as long as significant settlements are being recorded, probably a minimum of one month. Settlements can be plotted against the logarithm of time and extrapolated to estimate the settlement during the life of the structure.

FIELD STUDIES

The effectiveness of ground improvement using vibro techniques has to be judged by the long-term performance of the structures built on the treated ground. Although there appear to be few reported instances of unsatisfactory performance, there are very few documented case histories where settlements have been measured. The effectiveness of

vibro in reducing or eliminating total and differential settlement of a
structure can only be assessed by detailed monitoring of the structure
during and subsequent to completion. The behaviour of untreated ground
under loading conditions similar to those imposed by the structure
gives an insight into the effectiveness of the treatment. BRE is
currently carrying out a programme of field studies of the use of
vibrated stone columns at a number of sites in the UK, two of which are
summarised below.

Sand with Peat Layer

Vibro has been used extensively for low rise public sector housing
in Manchester in the natural sandy deposits found over large areas to
the south and west of the city. BRE was given the opportunity to study
the use of vibro at a housing development and to monitor the settlement
of a number of semidetached and terraced house blocks built on the
treated ground [5].

The development consists of fifteen semidetached and terraced two
storey house blocks. Alluvial deposits of sand and sand and gravel
found over the entire site were underlain by firm clay at about 4 m to
5 m below ground level. A peat layer up to 0.35 m thick was located at
depths varying between 0.45 m and 1.35 m over much of the site. BRE
levelling stations [6] were installed in six of the blocks to monitor
the long term settlement of the houses built on treated ground.

Ground treatment was carried out using the wet vibroreplacement
technique. Treatment consisted of single rows of stone columns along
the line of all load bearing walls and a small number under the
ground floor slabs. The full depth of alluvial sand was treated to
depths varying between 2.8 m and 4.2 m, the average column length and
diameter for the whole site being 3.4 m and 0.6 m respectively.
Figure 1 shows a typical layout for the stone columns and footings.

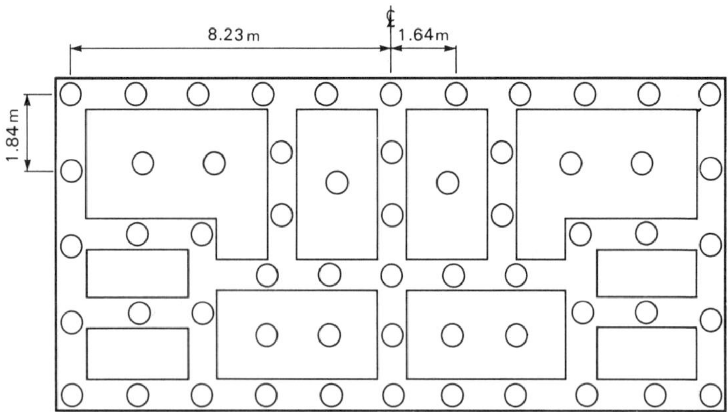

Fig. 1 Typical layout of stone columns and footings
(columns not to scale)

An in situ load test was performed on untreated ground at a
location where soil conditions were similar to the main site, including
a shallow layer of peat 0.2 m thick and 1.3 m below existing ground
level. A pre-cast concrete pad, 0.75 m wide x 2.25 m long x 0.2 m deep,
modelling the house foundations, was bedded 0.9 m above the peat layer.
An empty steel skip was placed on the pad and filled with sand which
applied 50 kN/m^2 to the loading pad. This was a typical foundation
pressure for load-bearing walls in the house blocks. The load test was
in place for four months during which time a total settlement of 4.3 mm
was measured, of which 1.5 mm was identified by a magnet settlement
gauge as compression within the 0.2 m thick peat layer. Figure 2 shows
the sand-filled skip during the load test.

Fig. 2   In situ load test using a sand-filled skip
on precast footing

CPTs were carried out next to the load test and in untreated ground
on the housing site. They indicated that the sand immediately below the
load test was significantly denser than at a similar depth on the
developed site. Calculations [7],[8] using the CPT data from next to
the load test predicted 3.5 mm settlement over a four month period. The
analysis was not designed to estimate the settlement of the peat. It
would seem, therefore, that the CPT was able to provide a method of
estimating the settlement of structures on this site. Similar
calculations using data relevant to the developed site were used to
estimate the settlement of houses had no treatment been carried out.

Over the first two year period of monitoring, the average
settlement of all the monitored house blocks was between 2.7 mm and 4.3
mm. In general, those house blocks founded above the peat settled the
most. Figure 3 shows the average settlement of three of the blocks
monitored. Block 1 was founded where the peat layer is 0.35 m thick,
block 7 where it is 0.1 m thick and block 12 where no peat was found

during the site investigation. Estimates of settlement of houses on untreated ground based on CPT data were much larger. Taking into account the compression of the relevant thickness of peat beneath each individual house block, estimates of up to 11 mm were computed for a similar time period. It was concluded, therefore, that settlements were less than half of what they would have been without ground treatment.

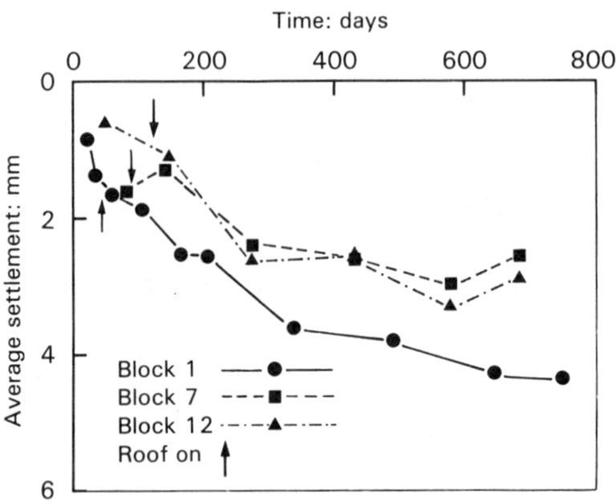

Fig. 3   Average settlement of house blocks

Clay Fill

A former gravel pit at Abingdon, near Oxford was backfilled with miscellaneous clay fill some 20 to 30 years ago. For a number of years the southern half of the site was used as a storage compound for heavy civil engineering plant, while the northern part had been grassed over and had remained undisturbed since backfilling. Two trial pits in the southern part of the site showed tarmac, hardcore fill and lean mix concrete in the top 1.0 m, underlain by 0.6 m of firm clay fill. Beneath this were 2.2 m of soft clay fill and 0.5 m of sandy gravel. The bottom of the original excavation was at 4.3 m and was a firm natural clay deposit. In the northern area, two trial pits revealed a firm clay fill up to 1.6 m thick overlying 2.0 m of soft clay and silt fill. This fill contained organic material.

A two storey steel framed structure with concrete blockwork infill panels has been built on the site. All structural columns are supported on reinforced concrete foundation pads. The pads are linked by reinforced concrete ground beams to support the blockwork infill. The building has one wing on the southern half of the site and the other on the northern half. The two wings are joined across their western ends by a structure of similar construction.

Prior to construction the fill was treated using the dry vibrodisplacement technique. Stone columns were constructed through the full depth of the fill. One, two, three or four columns, depending on structural loading, were placed at the location of each foundation pad with single rows of columns under the ground beams. A few treatment

points were positioned under the ground floor slabs. The foundation
plan layout and location of treatment points is shown in figure 4.

Levelling stations were installed on the upper surface of seven of
the main foundation pads on the north wing and two on the south wing. A
deep levelling datum was installed outside the fill area. Monitoring of
settlement of the foundations began before the steel frame was erected.
Levelling was transferred to BRE levelling stations in the adjacent
blockwork panels when the foundations were covered.

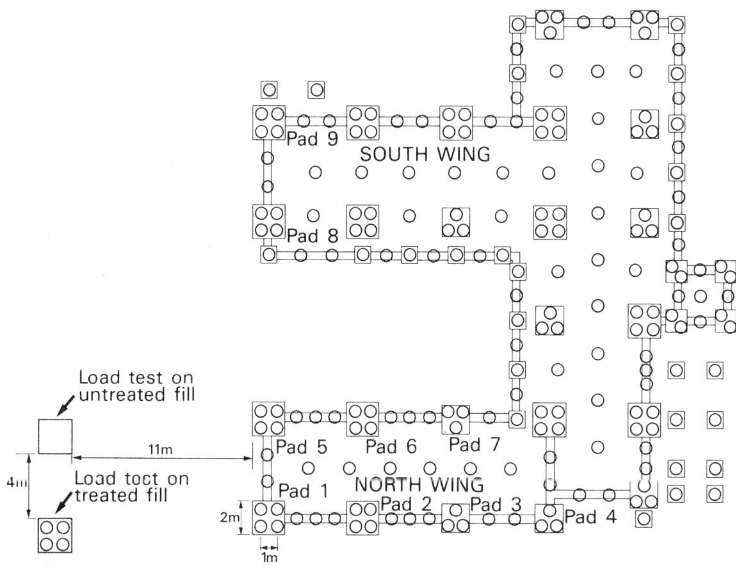

Fig. 4 Foundation plan and location of treatment points

Two full scale load tests were carried out on fill close to the
north wing (figure 4). An additional group of four stone columns were
constructed during ground treatment and a 2.0 m square concrete pad was
cast over them to model a foundation pad. An identical pad was cast on
adjacent untreated ground. Each pad was loaded with two skips, one on
top of the other, and their settlement was measured over a six month
period. The combined weight of the two skips on each pad was
approximately 21 tonnes, thus an additional 50 kN/m^2 was applied to the
treated and untreated fill. This represented 60% of the estimated
structural loading imposed on foundation pads 1, 2, 5 and 6 of the
actual structure. Figure 5 shows the treated ground and the sand filled
skips in place. Sampling was carried out next to each load test and the
properties of the stiff and softer clay fill samples were measured in
the laboratory. The settlement of each skip test had no ground
treatment been carried out was estimated from laboratory data. Values
of m_v, the coefficient of volume compressibility, were calculated from
oedometer tests carried out on 75 mm diameter specimens obtained from
selected depths within the fill. An elastic distribution of vertical

stress due to the additional load was assumed and estimates of the settlement of each pad were computed.

Fig. 5a Treated ground below load test

Fig. 5b Sand filled skips in place

The settlement of the building has been measured over a fourteen month period since the foundations were constructed. During that time

the monitored part of the south wing has settled an average of 7 mm.
The seven points monitored on the north wing over the same period have
settled between 12 mm and 21 mm with an average settlement of 15 mm.
The average settlements of each wing of the building is plotted from
the beginning of construction in figure 6.

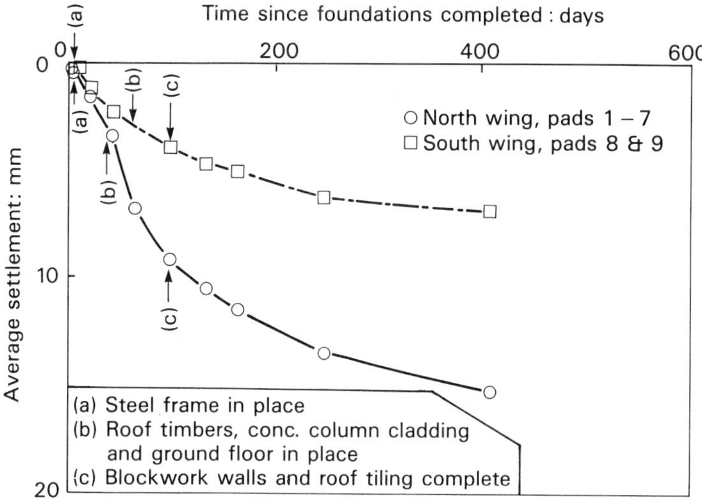

Fig. 6   Average settlement of foundation pads

The load test founded on untreated ground settled 13 mm whilst the
pad on the four stone columns settled 18 mm. Predictions from
laboratory tests for untreated ground adjacent to each test gave
12 mm and 19 mm respectively. This highlights the variability of the
fill and suggests that the columns constructed under one of the pads
had little effect in reducing its total settlement.

The south wing of the building, located on fill previously used for
heavy storage, has settled only half as much as the north wing. There
has also been a significant amount of differential settlement taking
place on the north wing. The load tests were designed to model
foundation pads 1, 2, 5 and 6 under the north wing of the building
which were loaded to 85 kN/m^2 by the structure. These four pads have
all settled by 13 mm whilst the load test on untreated ground is likely
to have settled 22 mm had the full 85 kN/m^2 been applied during the
test.

It would seem that vibro has not eliminated significant
differential settlements in the variable clay fill. However, the total
movements are relatively small and a major proportion of the settlement
was built out during construction of the flexible steel frame building.
There is no evidence to date of any distress to the completed
structure.

CONCLUSIONS

1. Vibro is being used extensively in the UK for housing and light industrial developments usually on filled ground.
2. Determination of soil properties required for design calculations is difficult in heterogeneous fills.
3. In situ penetration tests and geophysical tests may be of use in the characterisation of soil properties in some soil conditions.
4. Typical projects are small and simple, inexpensive methods are required for carrying out load tests on the treated ground.
5. Area or zone load tests are an appropriate although sometimes prohibitively expensive way of predicting the long term performance of structures built on ground treated by vibro. The skip test can be an effective and economic alternative in some circumstances.
6. Field studies indicate that although vibro may reduce settlement in some situations, but does not eliminate it.

ACKNOWLEDGEMENTS

The work described in this paper forms part of the research programme of the Building Research Establishment, United Kingdom and is published by permission of the Director. The main client for the work is the Construction Industry Directorate of the Department of the Environment. Permission to carry out the study at Manchester was granted by the City Architect, Manchester City Council. The study at Abingdon was carried out with the permission and co-operation of ARC Conbloc. The authors are grateful to Bauer Foundations for their assistance in carrying out the field work. Mr.A.P.Butcher carried out and interpreted the cone penetration tests at the Manchester site. Dr.C.P.Abbiss carried out the geophysical testing and interpretation. Mr.I.R.Holton, Mr.D.Burford and Mr.K.McElmeel assisted with the field work.

REFERENCES

[1] St John,H.D., Hunt,R.J. and Charles,J.A., "The use of 'vibro' ground improvement techniques in the United Kingdom", BRE Information Paper, 5/89.
[2] Charles,J.A. and Watts,K.S., "Compressibility of soft clay reinforced with granular columns", Proceedings of 8th European Conference on Soil Mechanics and Foundation Engineering, Helsinki 1983, 1, pp 347-352.
[3] "Report of the ISSMFE Technical Committee on Penetration Testing of Soils - TC 16 with Reference Test Procedures CPT - SPT - DP - WST", International Society for Soil Mechanics and Foundation Engineering, 1989, Swedish Geotechnical Institute.
[4] Charles,J.A. and Driscoll,R.M.C., "A simple in situ loading test for shallow fill", Ground Engineering, 14, No.1, 1981, pp 31, 32, 34, 36.

[5] Watts,K.S., Charles,J.A. and Butcher,A.P., "Ground improvement for low rise housing using vibro at a site in Manchester", Municipal Engineer, 6, June 1989, pp 145-157.
[6] Cheney,J.E.,"Techniques and Equipment using the Surveyor's Level for Accurate Building Movement" in Field Instrumentation in Geotechnical Engineering, Butterworth, London, 1973, pp 85-99.
[7] Schmertmann,J.H.,"Static cone to compute static settlement over sand", ASCE Journal of Soil Mechanics and Foundations Division, Vol. 96, SM3, 1970, pp 1011-1043.
[8] Schmertmann,J.H., Hartman,J.P. and Brown,P.R., "Improved strain influence factor diagrams, ASCE Journal of Geotechnical Engineering Division, Vol. 104, GT8, 1978, pp 1131-1135.

Ryouichi Babasaki, Kichio Suzuki, Satoshi Saitoh, Yoshio Suzuki, and Kazuya Tokitoh

CONSTRUCTION AND TESTING OF DEEP FOUNDATION IMPROVEMENT USING THE DEEP CEMENT MIXING METHOD

REFERENCE: Babasaki, R., Suzuki, K., Saitoh, S., Suzuki, Y., and Tokitoh, K., "Construction and Testing of Deep Foundation Improvement Using the Deep Cement Mixing Method," _Deep Foundation Improvements: Design, Construction, and Testing_, ASTM STP 1089, Melvin I. Esrig and Robert C. Bachus, Eds., American Society for Testing and Materials, Philadelphia, 1991.

ABSTRACT: The construction of an office building (hereafter referred to as "Building N") was planned in Kagoshima City in western Japan. The building site consisted of loose, alluvial sand called Shirasu, which is subject to liquefaction during earthquakes. It was decided that a newly developed soil improvement technique would be adopted as the foundation construction method. The technique employs the Deep Cement Mixing Method to construct a grid of hardened soil. This grid controls the generation of excess pore water pressure by reducing the occurrence of shear deformation in the native soil within the grid, thereby preventing liquefaction during earthquakes.

This paper discusses the design and construction of the improved soil foundation for Building N.

Keywords: liquefaction, design, improved soil foundation, numerical analysis, centrifuge, laboratory mixing test, trial construction

BUILDING N AND ITS FOUNDATION SOIL

Fig. 1 shows the plan and elevation views of Building N and its improved soil foundation. The building is a three story, reinforced concrete building. It has a continuous footing foundation, supported by soil that has been improved in a grid configuration.

The boring log for the foundation soil is given in Fig. 2. The soil is composed of saturated alluvial Shirasu with an SPT-blows number of less than 10 down to approximately GL-20.0m, the depth to which boring was conducted. Soil composition to a depth of GL-5.0m is a sandy layer mixed with pumice and gravel, below which is a layer of fine sand to a depth of GL-20.0m. Using a simplified method of liquefaction analysis (as given in the Recommendation for Design

R. Babasaki, K. Suzuki, and S. Saitoh are chief research engineers; K. Tokitoh is a research engineer; and Y. Suzuki is a research manager at TAKENAKA TECHNICAL RESEARCH LABORATORY, 5-14, 2-chome, Minamisuna, Kotoh-Ku, Tokyo, Japan.

of Building Foundations of the Architectural Institute of Japan), it was determined that liquefaction was possible in one section down to a depth of GL-12.5m. For this reason, soil improvement in a grid configuration was undertaken below the continuous footing foundation of the building to a depth of GL-13.5m.

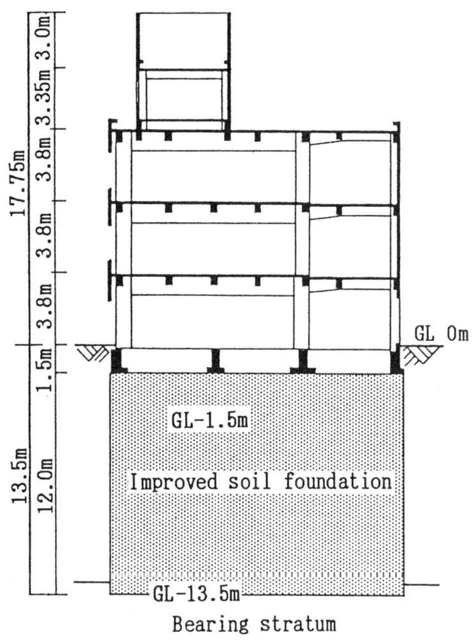

(a) Elevation of improved soil foundation and structure

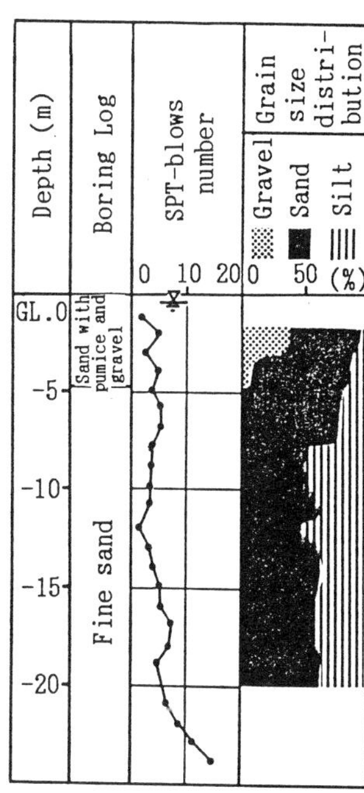

Fig. 2 -- Soil profile

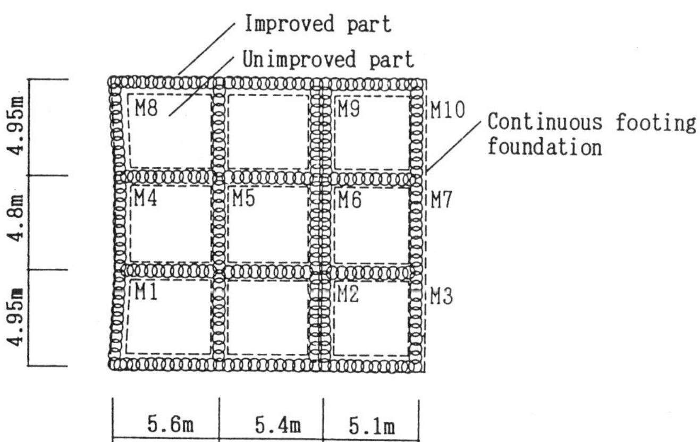

(b) Plan of improved soil foundation

Fig. 1 -- Plan and elevation views of Building N

METHOD OF DESIGN FOR IMPROVED SOIL FOUNDATION

Fig. 3 shows the design sequence for an improved soil foundation. A brief summary of each of the items is given below.

Task 1: A liquefaction assessment method is used to determine if any soil layers are prone to liquefaction.

Task 2: The allowable compressive stress of the improved soil is established using the following factors: the designed standard compressive strength of the improved soil; the abatement coefficients of the improved soil; and the safety factor.

Task 3: The length, width, and depth of the improved soil foundation, the space between grid walls, and the thickness of the grid walls are determined. The depth of the improved soil foundation is set at one meter below the depth of the soil layer that has been found to be prone to liquefaction. The space between grid walls is set according to two conditions: 1) the space necessary to prevent liquefaction, as determined by laboratory model shaking tests; and 2) the space necessary to ensure a uniform distribution of the building's vertical load throughout the area of the improved soil foundation, as measured at the foundation's bottom.

Task 4: The allowable bearing capacity of the bearing stratum is calculated as a spread foundation, factoring in the length, width, and depth of the improved soil foundation and the soil constants of the native ground.

Task 5: The vertical bearing capacity is determined on the bottom surfaces of both the continuous footing foundation and the improved soil foundation. The compressive stress of the improved part is assessed at the bottom surfaces of the continuous footing foundation and the improved soil foundation.

Task 6: The stability of the building and the improved soil foundation during earthquakes is assessed by simulating the application of horizontal external force (inertia force). Vertical bearing capacity is determined in the same way as described in Task 5 above. Compressive stress and shearing stress of the improved part are assessed at the bottoms of the continuous footing foundation and the soil improvement foundation. Sliding of the building and the improved soil foundation is also assessed at the bottoms of the continuous footing foundation and the improved soil foundation.

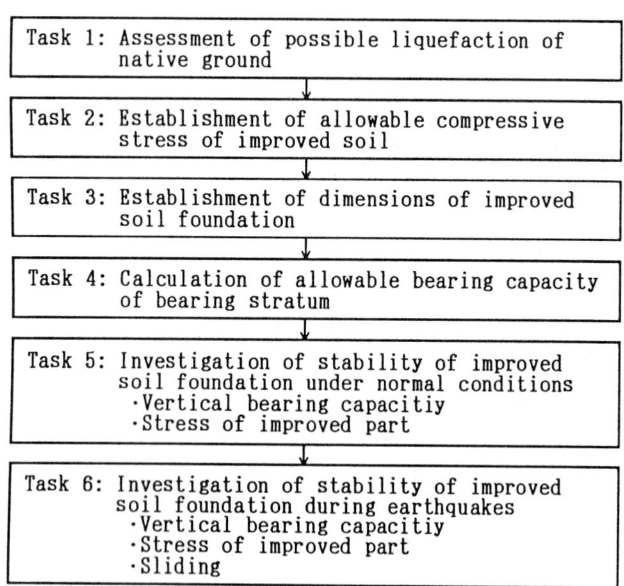

Fig. 3 -- Design sequence for ground improvement

VERIFICATION OF DESIGN METHOD FOR IMPROVED SOIL FOUNDATION

Numerical Analysis

To assess the resistance of the grid-configured improved soil foundation to liquefaction, seismic response analysis was conducted using a 3-dimensional linear FEM model. As shown in Fig. 4, analysis was conducted on a 1/4 section of a full FEM model. Analysis was carried out for unimproved ground and for an improved soil foundation. The ground was designated as the SOLID element, and the pillars and beams of the building were designated as the BEAM element. The building's mass was added as a lumped mass at the joints of the pillars and beams. Initial stress analysis and dead load analysis (including the weight of the building) were conducted first, followed by dynamic analysis. Pre-analysis of the building only was conducted and the analytical constants for the pillars and beams were established. The input earthquake motion was based on seismic waves measured on the grounds of the Takenaka Technical Research Laboratory (Tokyo) at a depth of GL-60.0m during an offshore earthquake that took place near Tokyo in December 1987 (Fig. 5). The maximum acceleration and duration of the input earthquake motion were set so that maximum acceleration on the ground surface of the unimproved ground reached 200 cm/s^2.

Fig. 6 shows the liquefaction resistance coefficient (FL value) in the depth direction for the elements listed in Fig. 4. In contrast to the FL value of less than 1.0 obtained in the loose, sandy unimproved ground to a depth of GL-12.5m, the FL value for the improved soil foundation was approximately 1.5, demonstrating the effectiveness of the grid-configured improved soil foundation in preventing liquefaction.

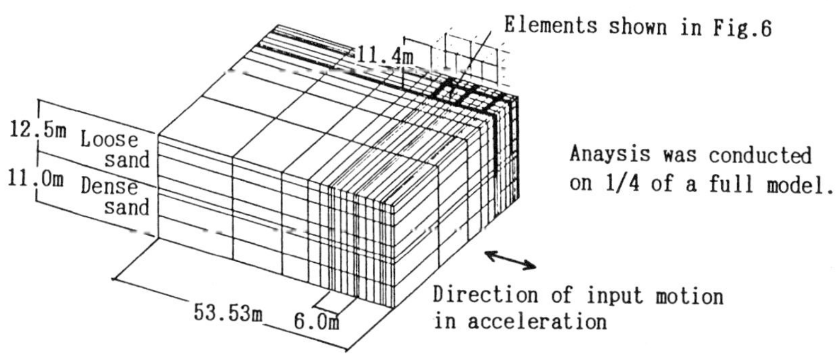

Fig. 4 -- FEM model

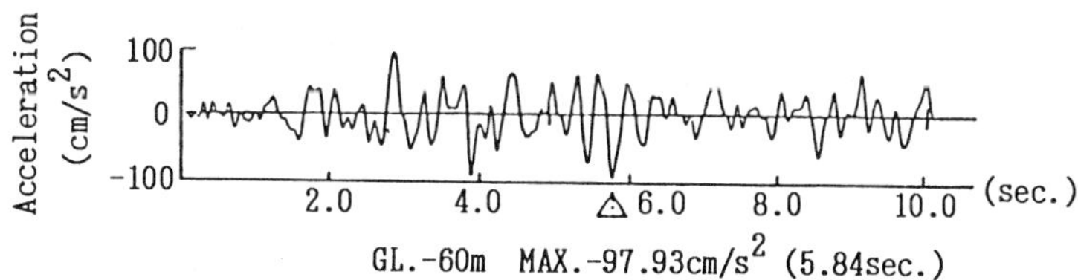

Fig. 5 -- Input motion in acceleration

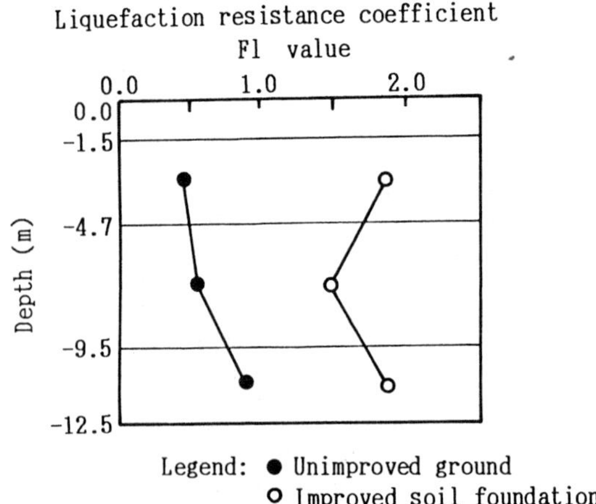

Legend: ● Unimproved ground
         O Improved soil foundation

Fig. 6 -- Relationship between depth and
liquefaction resistance coefficient

Centrifuge Model Tests

Loading test on the improved soil foundation: A centrifuge model test was conducted using facilities at Chuo University to verify the bearing capacity of the improved soil foundation for Building N.

Fig. 7 shows the model, which was constructed on a scale of 1/80 within a container measuring 70cm in length, 54cm in width, and 50cm in height. It was prepared within a block of Shirasu that was sampled intact from the building site at a depth of GL-5.0m to GL-5.5m. Cement, bentonite, water, and Shirasu were mixed together to make a cement slurry that was used to construct the improved part of the improved soil foundation. The same slurry was also used to make specimens for unconfined compression tests. After confirming that the specimen had achieved the designed standard compressive strength of 2 MPa, the loading test was conducted in a centrifugal field of 80g (g is gravity).

Fig. 8 shows the results of the loading test. A load of up to approximately 6 times the vertical design load (13280 kN) of the actual improved soil foundation of Building N was applied. Based on the results shown in Fig. 8, it was confirmed that no yield load was evident within the area where load was applied, and that the improved soil foundation had enough bearing capacity to support the vertical load as originally designed. In addition, inspection of the model of the improved part after the loading test was completed revealed no damage.

To verify the safety of the design method, the stability of the actual improved soil foundation for Building N was checked by measuring settlement during the building's construction. Fig. 9 shows the amounts of settlements of the measurement locations shown in Fig. 1 that occurred from the time the continuous footing foundation was completed to the time the building was completed. Actual settlement during this period ranged between 3 and 7mm, which is less than half the 15mm settlement predicted through calculations using the coefficient of subgrade reaction, given by the load settlement curve shown in Fig. 8.

Model shaking experiments: Model shaking experiments were conducted to evaluate the effectiveness of the grid-configured improved soil foundation in controlling liquefaction. Two models were constructed, one improved and one unimproved, at 1/100 the size of the actual foundation within containers measuring 40cm in length, 18cm in width, and 27cm in height, as shown in Fig. 10. Pore water pressure was measured in both models. To obtain a comparable time scale between the model and the actual site in terms of the pore fluid's seepage characteristics, a glycerin solution with a viscosity 100 times that of water was used in the model. The experiment was conducted in a centrifugal field of 100g.

Fig. 11 shows the input motion in acceleration. Fig. 12 shows the relationship between the maximum excess pore water pressure ratio (Δ u/σ v'), and the ratio L/H, where L is the space between grid walls and H the thickness of the layer prone to liquefaction. In contrast to the maximum excess pore water pressure ratio of 1.0 obtained in the center of the unimproved ground, the ratio for the improved soil foundation was approximately 0.5, demonstrating that the grid-configured improved soil foundation method was effective in controlling liquefaction.

LABORATORY MIXING TESTS

Laboratory mixing tests were conducted to determine what kinds of cement must be added in what amounts to achieve the designed standard compressive strength. Two soil samples were obtained separately for the tests: the first from a sandy soil layer containing pumice and gravel at a depth of less than GL-5.0m; the second from a deeper layer of fine sand. Specimens measuring 35mm in diameter and 80mm in height were made by mixing

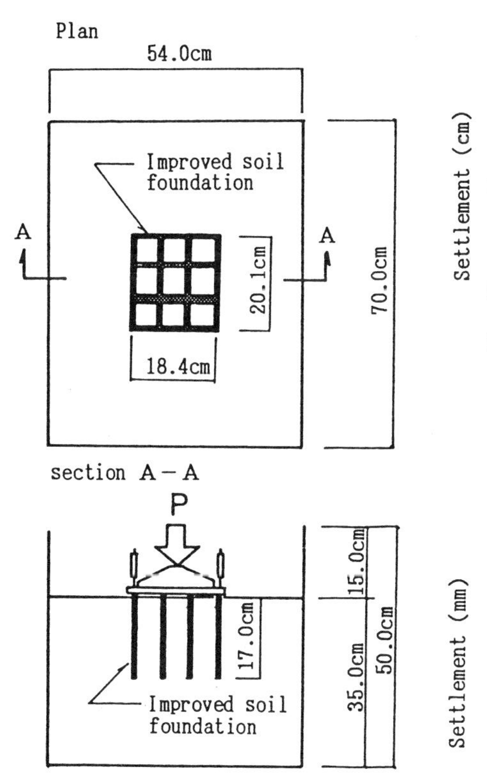

Fig. 7 — Improved ground model of Building N

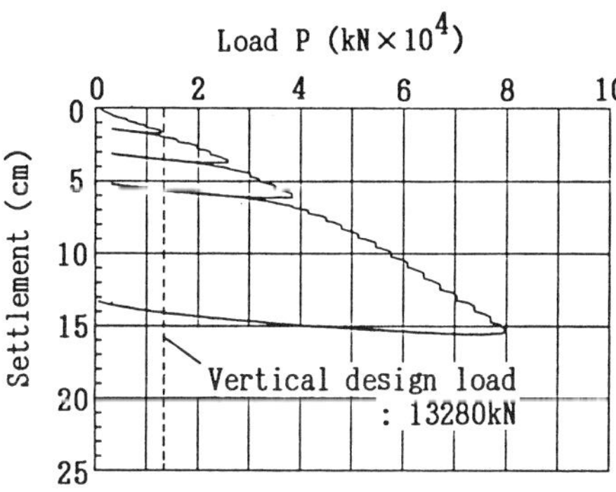

Fig. 8 — Relationship between load and settlement

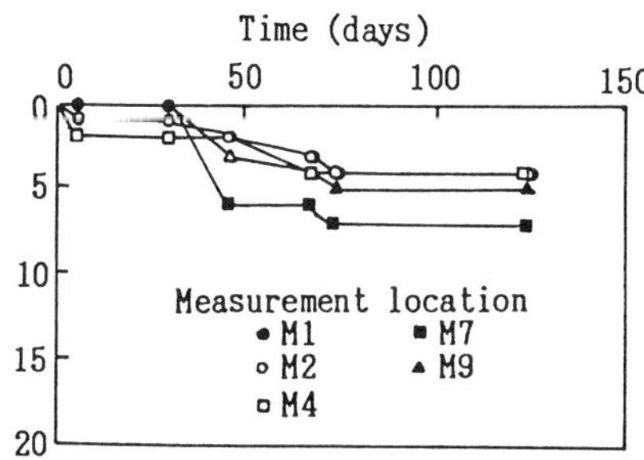

Fig. 9 — Relationship between time and settlement of Building N

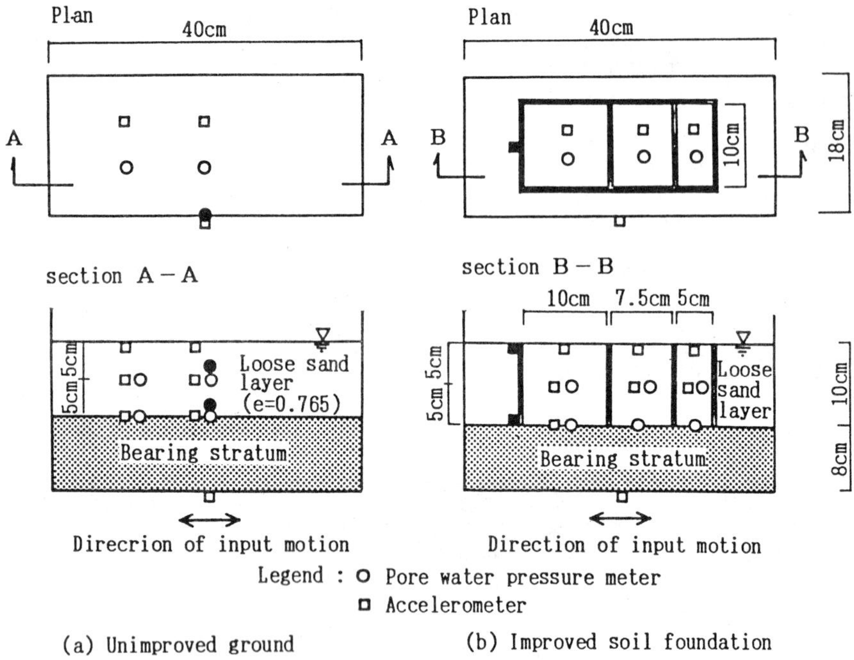

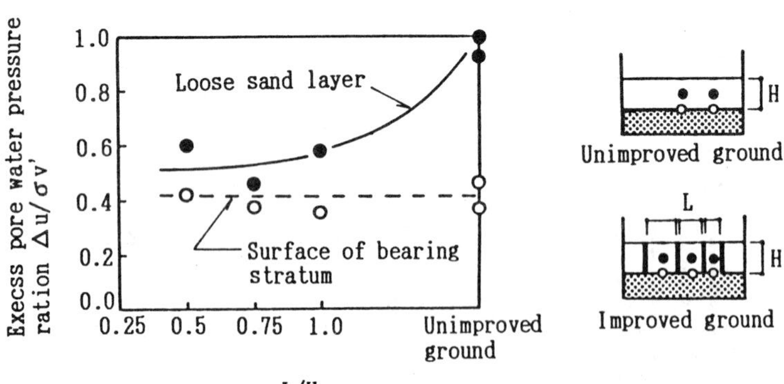

Fig. 10 -- Model ground

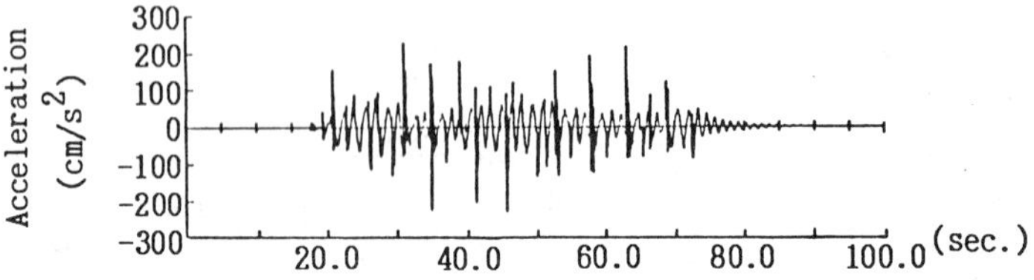

(a) Unimproved ground   (b) Improved soil foundation

Fig. 11 -- Input motion in acceleration

Fig. 12 -- Relationship between maximum excess pore water
pressure ratio $\Delta u / \sigma v'$ and L/H

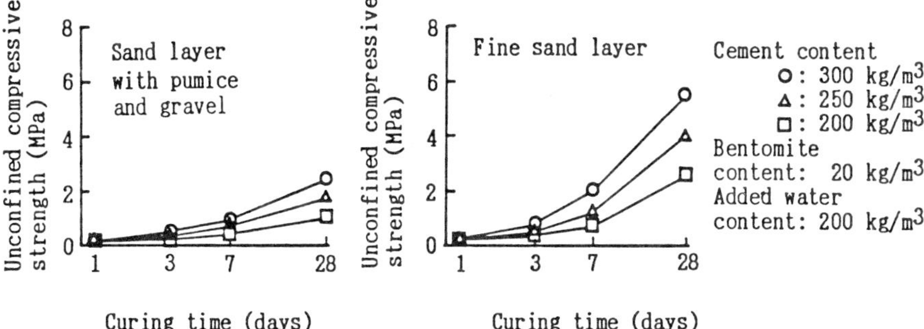

Fig. 13 -- Relationship between curing time and unconfined compressive strength of improved soil made with blast furnace cement

cement (blast furnace cement and ordinary Portland cement), bentonite, and water to the Shirasu. Unconfined compression tests were conducted after the specimens had been cured for a specified period of time. The main conclusions derived from these tests were as follows:

1) Blast furnace cement is more effective than ordinary Portland cement for use in improved soil foundations.
2) A layer of fine sand at a depth of more than GL-5.0m is more effective than a layer of sandy soil at a depth of less than GL-5.0m that contains pumice and gravel for use in improved soil foundations.
3) To achieve the designed standard compressive strength of a specimen with blast furnace cement that has been cured for 28 days, 200kg of cement must be added to every cubic meter of fine sand, and 300kg must be added to every cubic meter of sandy soil that contains pumice and gravel.
4) The unconfined compressive strength for specimens that had been cured for 28 days was 4 times that of specimens that had been cured for 3 days shown in Fig. 13.

TRIAL CONSTRUCTION

Trial construction of an improved soil foundation was undertaken on the actual site to determine the mixing specifications for the cement, bentonite, and water, and to determine the working specifications for the mixing machine. The same machine that would be used in actual construction was used for the test: a three-axis soil-cement mixing machine (Photo 1). Fig. 14 shows the arrangement of the improved units made in the trial construction. Each unit was constructed according to the construction and mixing specifications shown in Tables 1 and 2. During trial construction, control measurement was applied to the depth of the mixing machine, its penetration and withdrawal speeds, and the quantity of the cement slurry discharged.

Fig. 15 shows the relationship between the depth and the unconfined compressive strength of improved soil sampled from the improved part three days after trial construction was completed. The unconfined compressive strength at a depth of GL-5.0m or less was less than that found at a depth of more than GL-5.0m, indicating a tendency similar to that found in the laboratory mixing tests. As the figure shows, the amount of cement needed to achieve the designed standard unconfined compressive strength of 2 MPa after curing for 28 days was different for the two layers above and below a depth of GL-5.0m. Since the laboratory tests showed that the unconfined compressive strength for samples that had been cured for 28 days was 4 times that of samples that had been cured for 3 days, it was concluded that 300kg of cement were required per cubic meter for the upper layer, and 200kg were required per cubic meter for the lower layer to achieve the designed standard compressive strength.

Construction of the actual improved soil foundation was carried out on the basis of the results of the trial construction. The mixing and construction specifications for the actual foundation are given in Table 3.

CONCLUSION

A newly developed soil improvement technique for preventing liquefaction during earthquakes was used to construct the foundation of Building N. This required a new design method, which was verified through numerical analysis and centrifuge model tests. In addition, laboratory mixing tests and trial construction for the actual foundation were conducted to bring this project to successful completion.

Photo 1   Mixing machine

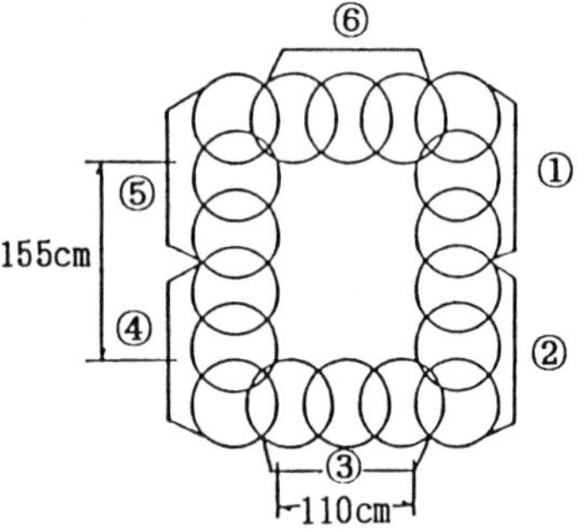

(a) Plan of improved units

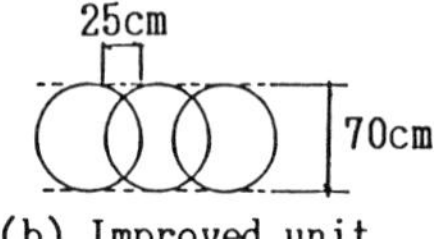

(b) Improved unit

Fig. 14 -- Arrangement of improved units

Table 1 -- Construction specifications

Items	Unit number	
	1,2,3,5,6	4
Depth of improvement (m)	GL-13.5	
Penetration speed (m/min.)	0.5	
Withdrawal speed (m/min.)	1.0	1.5
Rotating speeds of blades (r.p.m.)	25	

Table 2 -- Mixing specifications

Unit number	Cement content (kg/m^3)	Bentonite content (kg/m^3)	Added water content (kg/m^3)
1	202	10	202
2,4,5,6	253	10	202
3	296	10	197

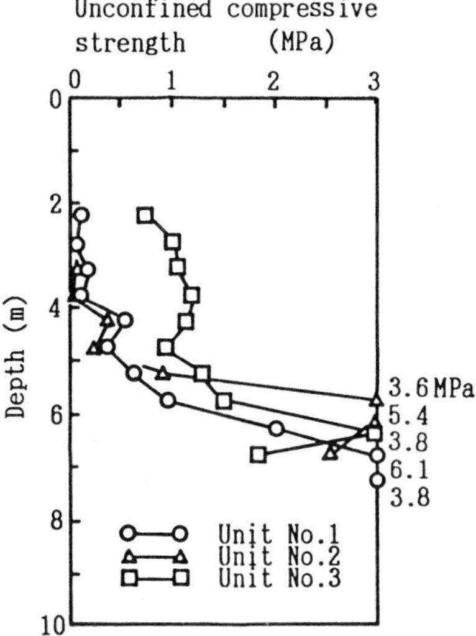

Fig. 15 -- Relationship between depth
and unconfined compressive strength of
improved soil at curing time of 3 days

Table 3 -- Mixing and construction specifications for actual foundation

Layer	Mixing specifications			Construction specifications				
	Cement content (kg/m^3)	Bentonite content (kg/m^3)	Added water content (kg/m^3)	Method of slurry injection	Depth of improvement (m)	Penetration speed (m/min.)	Withdrawal speed (m/min.)	Rotating speeds of blades (r.p.m.)
Sand layer with pumice and gravel	300	10	200	Injection when penetrating	GL-13.5	0.5	0.5	25
Fine sand layer	200					0.5	1.0	

ACKNOWLEDGMENT

The authors would like to thank Professor Fujii of Chuo University for his help in conducting the centrifuge model tests.

Michael J. Byle, Paul M. Blakita, and Ernest Winter*

SEISMIC TESTING METHODS FOR EVALUATION OF DEEP FOUNDATION
IMPROVEMENT BY COMPACTION GROUTING

REFERENCE: Byle, M. J., Blakita, P. M., and Winter, E., "Seismic Testing Methods for Evaluation of Deep Foundation Improvement by Compaction Grouting," <u>Deep Foundation Improvements: Design, Construction, and Testing</u>, <u>ASTM STP 1089</u>, Melvin I. Esrig and Robert C. Bachus, Eds., American Society for Testing and Materials, Philadelphia, 1991.

ABSTRACT: Improvement of subsurface soils by the injection of grout inclusions, termed compaction grouting, has been used since the mid 1950's. For improvement verification, past practice has generally been to test the compaction grouted area with SPT or CPT methods. However, since these procedures test only a single point: too little testing may not accurately represent the non-homogeneous nature of the improved soils while testing in sufficient numbers for accurate statistical evaluation may be uneconomic. An alternative to these methods is direct transmission seismic testing. This technique is particularly well suited to evaluate inclusion-improved soils since the seismic velocity is measured throughout the zone of improvement. To date, little information has been published on this method. The purpose of this paper, therefore, is to demonstrate the use of seismic test procedures, to evaluate data accumulated from two test sites and to present recommended guidelines for the use of these methods.

KEYWORDS: compaction grouting, crosshole seismic, downhole seismic, uphole seismic, Poisson's ratio, relative density

Ground modification methods have become widely accepted in recent years. Compaction grouting has been used as a ground modification technique since the early 1950's. This and similar techniques are based on the densification of the surrounding soils by displacement during the introduction of an inclusion in the soil mass. The inclusion in compaction grouting is accomplished by injection of a mortar-like grout under high pressure [1].

---

* Michael J. Byle, P.E. is an Associate with Schnabel Engineering Associates, 8460C Tyco Rd., Vienna, VA 22180; Paul M. Blakita, E.I.T., is a Project Engineer with GKN Hayward Baker Inc., 1875 Mayfield Rd., Odenton, MD 21113; Ernest Winter, P.E., is a Principal of Schnabel Engineering Associates, Suite 250, 10215 Fernwood Rd., Bethesda, MD 20817.

The introduction of an inclusion causes a non-uniform distribution of soil densification. The soils nearest the inclusion, or grout bulb, will undergo the greatest displacement and will achieve the highest density, while those at a distance will be less affected. Thus, although pressures and the volume of grout injected can be measured, the average amount of increased density or performance across the treatment area cannot be directly interpolated from these measures.

Verification that the anticipated improvement has been accomplished is a common problem with this type of ground improvement technique [2]. To date, engineers have of necessity relied on localized testing from which deductions may be made as to overall improvement. However, unlike traditional techniques, seismic methods provide data from which average improvement values may be calculated with relative accuracy and within the economic constraints of most projects.

Performance Testing

Performance testing procedures such as plate load tests may not easily be undertaken on a large enough scale to evaluate improvement in the entire improved zone. Common load testing practice using a 0.3 m to 0.6 m (1 to 2 ft) diameter plate generally tests only the upper 1.2 m to 1.8 m (4 to 6 ft) of the soil. The ultimate performance test, construction of a completed structure with settlement monitoring, is often the only performance test available within the economy of most projects.

SPT, CPT and DMT

Other common testing practices include the drilling of sample borings using the Standard Penetration Test (SPT), undisturbed tube sampling with laboratory density testing, Cone Penetration tests (CPT) or Flat Dilatometer Test (DMT). These do give a measure of soil improvement, but only at the single location where they are used. These tests, unless done in very large quantity sufficient for statistical analysis, may give a very misleading estimate of actual soil improvement. Both CPT and DMT procedures are limited by the inability of this type of test equipment to penetrate the improved soils [2].

Seismic Methods

Seismic methods are based on the rate of transmission of seismic waves through the soil. The velocity of seismic waves through soils is a function of basic soil properties such as modulus, density and Poisson's Ratio [3], [4]. These methods have traditionally been used to evaluate the stratigraphy of soils and rock and have only recently been used in the evaluation of ground modification [2], [5], [6]. Seismic methods have an advantage in that the seismic wave speed measured between any two points will be some average of the wave speeds for the materials between those two points. This holds promise for evaluating the average improvement of a soil mass rather than simply spot values as with many other test methods.

Other Methods

Other methods of evaluating ground modification that may hold promise include nuclear density logging of boreholes and ground penetrating radar. These methods are relatively new and have as yet not come into common usage.

SEISMIC METHODS FOR SOIL EVALUATION

Seismic methods are based on the characteristically constant velocity of seismic waves travelling through a uniform medium. There are two principal types of seismic waves: dilatational (compression or p-) waves and distortional (shear or s-) waves. The p-wave is a wave of energy which acts to alternately compress or dilate the soil. The s-wave is a wave of energy which acts to distort the soil in shear.

The rate of travel of these waves is dependent on fundamental properties of the soil: the shear modulus, G; the mass density of the soil, ρ; and Poisson's ratio, ν [3]. The compression wave velocity is given by the following relationship:

$$V_c = \sqrt{(\lambda + 2G)/\rho} \tag{1}$$

where λ = Lame's Constant, and the shear wave velocity as:

$$V_s = \sqrt{(G/\rho)} . \tag{2}$$

Since Lame's constant, λ , is related to the shear modulus by:

$$\lambda = \frac{2\nu G}{1 - 2\nu} \tag{3}$$

Poisson's ratio can be determined by the ratio between V_c and V_s from the following relation:

$$\frac{V_c^2}{V_s^2} = \frac{2 - 2\nu}{1 - 2\nu} \tag{4}$$

Empirical equations for angular grained materials have been developed by Richart et al. [3] to relate shear wave velocity and shear modulus to void ratio and overburden as follows:

$$V_s = 159 - (53.5)e \quad \bar{\sigma}_o^{0.25} \tag{5}$$

and

$$G = \frac{1230\ (2.97 - e)^2}{1 + e} \qquad (6)$$

where: e = void ratio and $\bar{\sigma}_o$ = overburden pressure. By assuming an initial unit weight, the void ratio can be determined from eq. 5 for a measured shear wave velocity. The effect of any small error in the initial unit weight assumed on the calculated void ratio from equation (5) is minor since the fourth root of the surcharge is used in this equation. Dry unit weight γ_d is calculated from the void ratio as:

$$\gamma_d = \frac{\gamma_w\ G}{1 + e} \qquad (7)$$

where G_s = specific gravity (2.8 for micaceous sands); γ_w = unit weight of water 9.8 kN/m^3 (62.4 pcf) [4].

DATA ACQUISITION AND COMPACTION GROUTING PROCEDURES

<u>Test Sites</u>

The test sites for this study were very similar townhouse buildings that had been built over 3.0 m to 8.5 m (10 to 28 ft) of uncompacted sandy silt fill at locations roughly 4.8 km (3 miles) apart. Displacement of the homeowners during repairs was undesirable. Compaction grouting was a means of improving the soils insitu without underpinning and could be done while the buildings remained occupied. The fill material was of local origin, primarily residual soil taken from nearby cut areas. Standard Penetration Test blow counts for the fill were typically around N = 2. The underlying natural deposits consisted of residual silty sands and decomposed rock. The parent rock is a schistose gneiss of the Wissahickon Formation. This formation typically weathers into a micaceous sandy silt. Typical soil properties are given in the table below:

Table 1
Average Fill Properties

	Unit Weight (kN/m^3)	Blow Count (top 8 m) Before Grouting (N min)	(N avg)	USCS Class	Blow Count After Grouting (N min)	(N avg)
Site 1	14.8	1 blow for 18" (.45 m)	6	SM	6	10
Site 2	15.6	2	4	SM	-	-

*(1 kn/m^3 = 6.37 psf)

Site 1: The building at site 1 was a four level townhouse building with a brick veneer finish. The lower level was below grade at the front and daylighted to the rear at grade. The building had been in place approximately nine months, during which time cracking and settlement of the front stoops and walls had occurred.

The test area was located at one end of the building in a relatively flat area. The existing loose fill was at its deepest [approx 8.4 m (23 ft)] in this area and the location was relatively free of obstructions to the grouting and to the seismic testing. Figures 1 and 2 show the plan view and section of site 1, respectively.

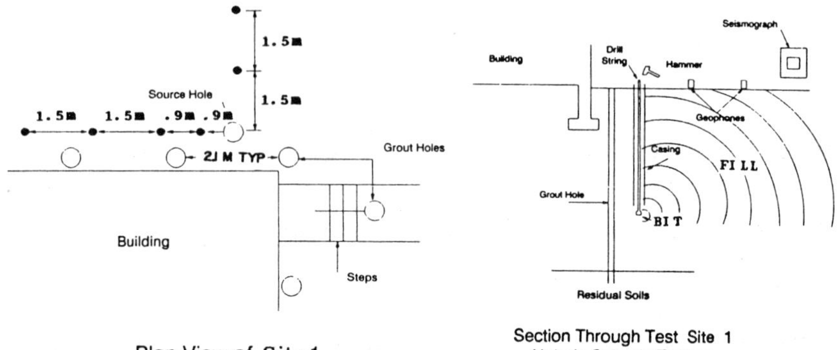

Plan View of Site 1

Section Through Test Site 1
Uphole Seismic Testing

Site 2: This site was similar to site 1 in all respects except that the building was a three level townhouse building without brick finishes. The seismic test section was selected at the rear of one townhouse unit where soil borings indicated the deepest fill area. Figures 3 and 4 show the plan view and section for site 2, respectively.

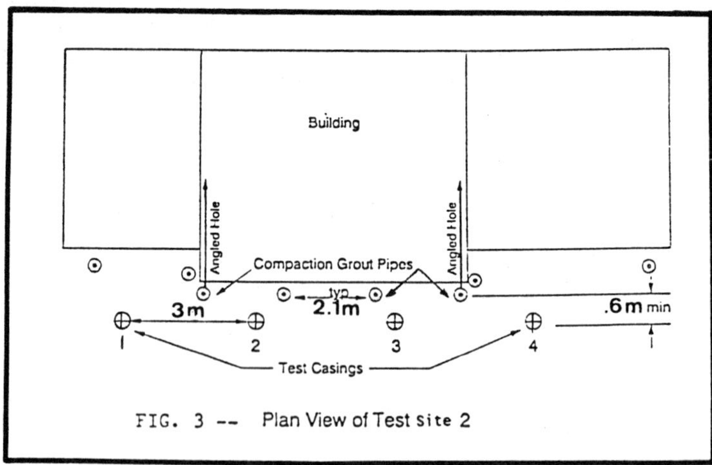

FIG. 3 -- Plan View of Test Site 2

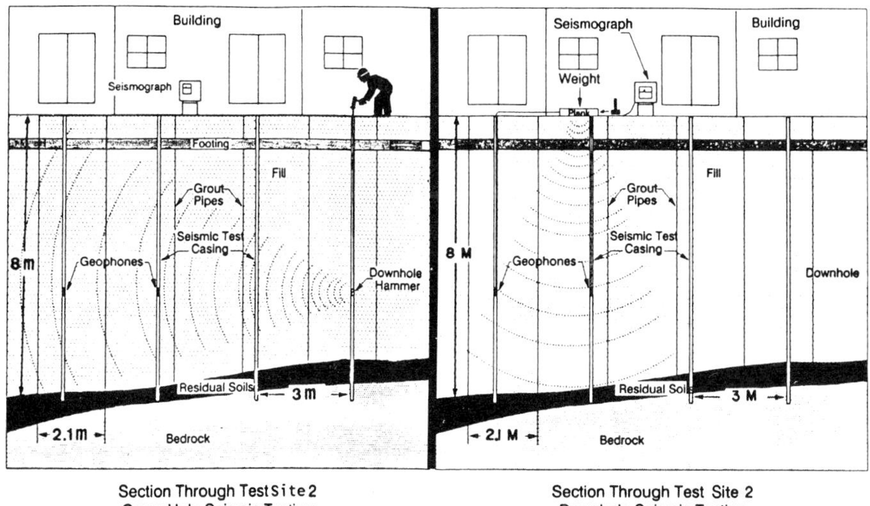

Section Through Test Site 2   Section Through Test Site 2
Cross Hole Seismic Testing   Downhole Seismic Testing

Figure 4

Testing Program

<u>Site 1:</u> The testing program at site 1 consisted of before and after testing of the SPT resistance and the seismic wave velocities. The SPT values were measured in a single boring at approximately the same location before and after the grouting. The test location was selected between two proposed grout injection points to limit the influence of direct contact with grout.

The seismic testing by the uphole method was conducted concurrent with the SPT test. A schematic representation of the test location and the arrangement of test equipment is shown in Figure 2. The initial test hole was drilled with the same pneumatic track-drill as was used to drill the grout holes. The signal source was the center bit extended below the bottom of a 75 mm (3-inch) I.D. steel casing. The top of the drill rod was struck with a 5 kg (12 lb) sledge hammer with a piezo-electric trigger switch attached. The arrival of the seismic waves at the ground surface was monitored with geophones at 5 locations parallel and perpendicular to the building and line of grouting. The geophones were monitored with a 12-channel seismograph with triggering by the hammer switch. The test was performed at 1.5 m (5 ft) intervals from the ground surface to a depth of 7.6 m (25 ft). The signals were recorded on paper tape for later analysis in the office.

After grouting was completed, the test hole was redrilled with a soil drilling rig at a slight offset to the original test location. The standard split spoon driven through hollow stem augers was used as a source of seismic waves. The equipment arrangement was as described above.

<u>Site 2:</u> At site 2, crosshole and downhole seismic methods were used. The test arrangement used four predrilled holes with light duty PVC casing grouted in place with a slurry of weak cement bentonite grout. The holes were drilled with the pneumatic track drill to a depth of 7.6 m (25 ft) at a spacing of 3 m (10 ft) between holes.

The crosshole test utilized all four of the holes. A signal source was placed in Hole 1, with geophones placed in the remaining holes to monitor s-wave and p-wave arrivals. The signal source was a mechanically wedged shear block hammer which was activated by striking the top of a connecting rod at the ground surface. The hammer was designed to seat in the PVC casing at any depth. The geophones were triaxial geophones with a packer attached, which could be inflated to achieve tight contact with the casing. Monitoring was again provided with a conventional 12-channel seismograph. The cross-hole test was performed at 1.5 m (five ft) intervals from the ground surface to a depth of 7.6 m (25 ft) before and after grouting.

The downhole test was performed initially in Hole 3 and after grouting in both Holes 2 and 3, since the grouting process had pinched off Hole 3 below 3.5 m (11.5 ft). The signal source for the down-hole arrangement used a heavy plank, approximately 1.2 m (4 ft) long, placed on the ground surface and held in place by a deadweight (vehicle wheel). The plank was struck alternately on either end to generate polarized shear waves. The shear wave reversal by this procedure allows easier identification of the shear wave arrival. Vertical blows were also used to generate strong
p-waves to better delineate the p-wave arrival. A triaxial geophone with packer was set at various depths within the hole being tested to measure arrivals at 1.5 m (5 ft) intervals to 7.6 m (25 ft) below the ground surface.

Grouting Program

Because of similarities in soil conditions and structure design, the compaction grouting procedures used at each site were nearly identical. The same drilling procedures, grout mix design and ready mix supplier were used. Hole spacings and treatment depth criteria were also similar between sites.

Design Consideration: Because of the presence of some finer grained soils, (silty fine sands, clay lenses, etc.) and the lack of confinement due to the light structural loads and shallow footings, grout hole spacings were limited to no more than 2.4 m (8 ft) in areas of known soft soils. In most cases, grout pipes were installed at 1.8 m to 2.1 m (6 to 7 ft) spacings. At all locations, the grout pipes were installed through the fill soils to bedrock or 0.6 m (2 ft) into residual soil.

Grout volumes were estimated to be 8 - 10% of the treated soil volume. The soil volume was based on a compactive zone of influence being a column of soil slightly larger in diameter than the pipe to pipe spacing. This provided a compacted soil column of 1.8 m to 2.4 m (6 to 8 ft) in diameter with target grout injection volumes of 0.28 to 0.47 cu.m (3 to 5 cu.ft/ft) of hole. The average volume injected was 0.28 cu.m (3 cu.ft per ft) of hole.

Grout Mix Design: Due to site constraints, on-site batching of the grout was not practical and therefore ready mix was used. Use of ready mix offers unique challenges in high pressure pumping because of a lack of natural fines in the fine aggregate common to ready mix plants. These fines provide water retention and lubricity thereby increasing the pumpability of the mix. To compensate for the low percentage of natural fines, the grout mixture used a slag/cement blend at higher than normal quantities. Each 0.77 cu.m (1 cu.yd) of grout consisted of 1,136 kg (2,500 lbs) of concrete sand, 364 kg (800 lbs) of slag cement and approximately 150 l (40 gals.) of water to produce a 2.5 to 5 cm (1 to 2 in) slump material.

<u>Casing Installation:</u> The 75 mm (3 in) I.D. flush joint casing was installed using an air track drill by simultaneously driving and drilling using a retractable bit inside and just ahead of the casing tip. This method overcuts the hole just enough to drive the casing without damage but still maintains a tight casing to soil seal.

Most holes were installed at a 10 to 15 degree angle under the building so that a densified column was produced under the structure without drilling through the footing. Between units at site 2, a steeply angled (35 to 40 degrees below horizontal) hole was drilled to provide densification below the load bearing party walls without having to drill from inside the basements.

<u>Compaction Grouting:</u> Compaction grouting was performed with a diesel hydraulic piston grout pump conveying the transit mix grout through 75 mm (3 in) high pressure grout hoses. The pump was capable of grouting pressures of 42 kg/sq.cm (600 psi) with a variable output from 0.4 l/min (0.2 cu.ft per minute) up to 30 l/min (15 cu.ft per minute).

As pumping proceeded, grouting pressures, injected volumes, and structural movement were continuously monitored and recorded. Grout injections continued at a given 0.9 m (3 ft) stage until a refusal condition was reached. Refusal was considered to be one of the following conditions:

a. 0.65 cu.m/m (7 cu.ft/ft) of grout was injected at a pressure of 7 kg/sq.cm (100 psi) or higher
b. Grout flow ceased at an injection pressure of 42 kg/sq.cm (600 psi)
c. Ground or structural movement was detected 3 mm (1/8 in)
d. A maximum of 0.93 cu.m/m (10 cu.ft/ft) was injected

Upon reaching a refusal condition, the casing was extracted 0.9 m (3 ft) and the process repeated. This grouting/withdrawal sequence was carried out from a depth of 0.6 m (2 ft) into residual soil to just below the existing footing.

Ground heave and structural movement monitoring was performed with a laser level and audible targets along with an optical level sighting on targets affixed to the structure. The optical level was used to verify target elevations and to sight locations which were in the "shadow" of the laser. Tolerance was set at 3 mm (1/8 in).

Quality Assurance tests included slump tests on each ready mix load delivered and casting of 50 mm (2 in) cubes. Grout log sheets were maintained for each hole and detailed pressures, injected volumes and reasons for refusal at each stage. Grout take summaries for sites 1 and 2 are presented in figures 3 and 4.

Test Results

The results of the seismic testing show a general improvement of the soil in the areas grouted. The uphole seismic method conducted in the open drill hole below the bottom of a drill string gave the least useful results while the crosshole testing gave the most meaningful and repeatable results. The downhole testing proved easiest to perform with meaningful and consistent results.

Uphole Method: The uphole method was used only at site 1 and had difficulties peculiar to the methods and equipment used. The "noise" generated by the signal rod either rattling around inside the casing or binding in the casing caused stray pulses from various points above the bit to arrive at the geophone before the desired pulse. Interpretation of the wave traces for the uphole was complicated by the presence of these early arrivals. The noise was so extreme as to make most records unusable.

Crosshole Method: The crosshole method was done at site 2 only, by the procedures described previously. A major factor affecting its use at site 2 was the presence of adjacent mechanical equipment. The testing program was adjusted to avoid times when this mechanical equipment was active. The wave records were generally clear and p-wave arrival was easily obtained. Shear wave arrival was slightly more difficult to evaluate and had to be inferred from wave shape since the down-hole hammer, and consequently signal polarity, was not reversible. The p-wave and s-wave arrivals were measured from the signal hole at three listening holes.

The arrivals at some elevations in the second and third listening holes showed evidence of refraction by the underlying bedrock and through the adjacent building foundations. Because of the upward sloping rock surface, the p-wave arrived at the third and fourth listening holes before the first at depths below 6.1 m (20 ft). This demonstrated the importance of using closely spaced holes when refracting media are nearby.

A second factor affecting the use of the cross-hole procedure is that of possible damage to the test holes. Holes numbered 3 and 4 were both damaged by the grouting process which pinched them off below 3.7 m and 6.1 m (12 and 20 ft), respectively. This did not severely limit the testing at this site, but for critical applications, provision should be made for re-installing the plastic casings, or initially installing them midway between grout points rather than adjacent to them.

Plots of shear wave arrival, Poisson's ratio and unit weight calculated from the data before and after grouting are presented in Figures 6, 7 and 8, respectfully. Figure 6 shows a slight increase in the shear wave velocity on the order of 15% between depths of 2.9 m to 6.6 m (8 to 18 ft). Figure 7 shows no significant change in Poisson's ratio with grouting as calculated from equation 4.

Figure 8 presents the unit weight calculated from equations 5 and 7.

These were calculated using an assumed overburden based on 1,602 kg/cu.m (100 pcf) unit weight and a specific gravity of solids of 2.8. These results indicate a 15 to 25% increase in calculated dry unit weight of the grouted soil between depths of 2.9 m to 6.6 m (8 to 18 ft). These values cannot be taken as actual density of the soil because they include the effect of grout inclusions. However, this does indicate a significant improvement in the area grouted. This corresponds well to the areas of maximum grout take. Grouting profiles for this area are given by Figure 5.

Downhole Method: This method was also used at site 2. The method was affected by mechanical equipment and pipe damage similar to the cross-hole method, but was much easier to interpret. By reversing the polarity of the shear wave, the two wave traces could be overlaid, with the point of divergence

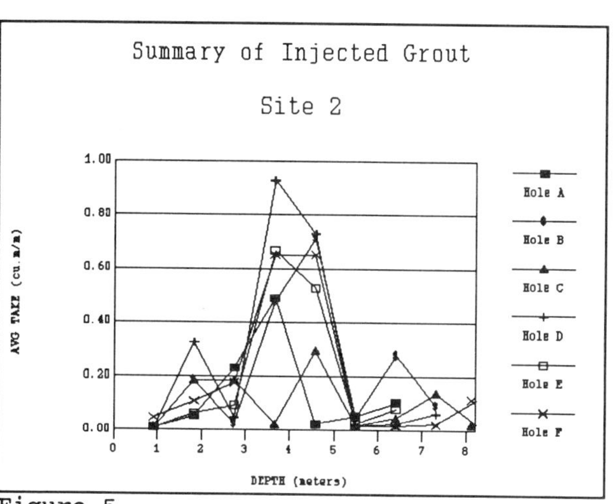

Figure 5

denoting the point of shear wave arrival [8]. This method was not at all affected by refraction from the underlying rock, and by selecting the signal point carefully, the influence of the building foundations could be greatly limited if not eliminated. Some interference from waves traveling down the casing was experienced but was small and required additional interpretation.

The shear wave velocities by this method are averaged from the surface to the depth of the geophone. Average velocity is plotted versus geophone depth in Figure 9. This figure shows a significant increase in shear wave velocity of 30 to 90 percent with the most substantial increases in the zone from depths of 2.0 m to 6.9 m (9 to 19 ft). The very high shear wave velocity of 241 m/sec. (790 ft/sec) at 2.9 m (8 ft) depth the immediate area of the pipe.

Calculated Poisson's ratio is presented in Figure 10. This figure shows that Poisson's ratio has remained relatively constant at about 0.4, slightly higher than indicated by the crosshole. The calculated average densities are (Figure 11) shown to increase from 35 to 40 percent after grouting. Except for the unusually high calculated density of 20.4 kN/m^3 (130 pcf) at the 2.9 m (8 ft) depth, the results generally indicate a reasonable 35% increase in dry density in the zone between 3.0 m and 6.9 m (9 to 19 ft) deep.

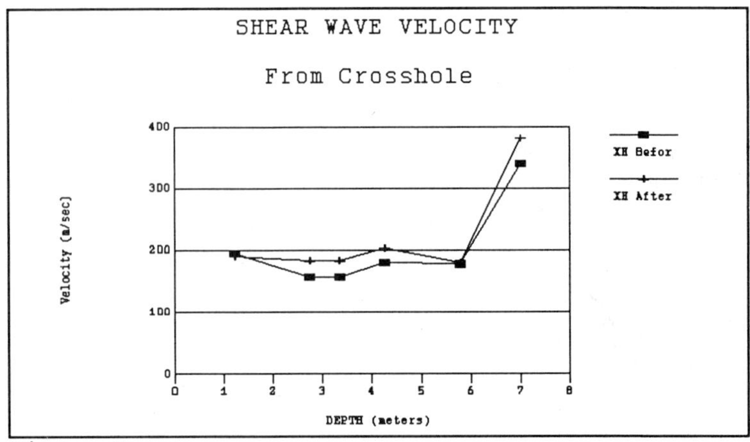

Figure 6

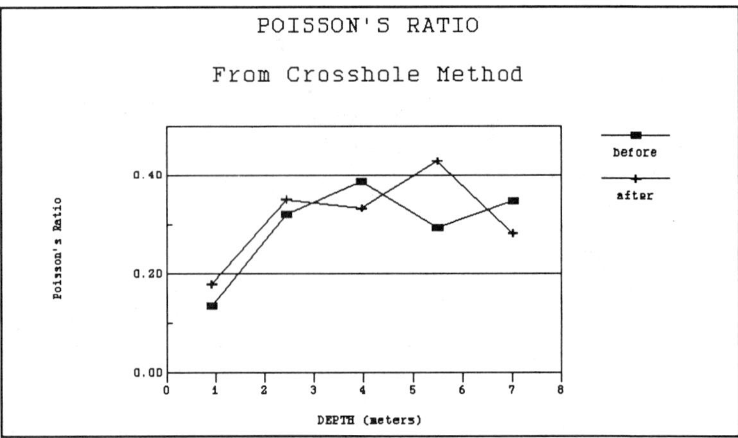

Figure 7

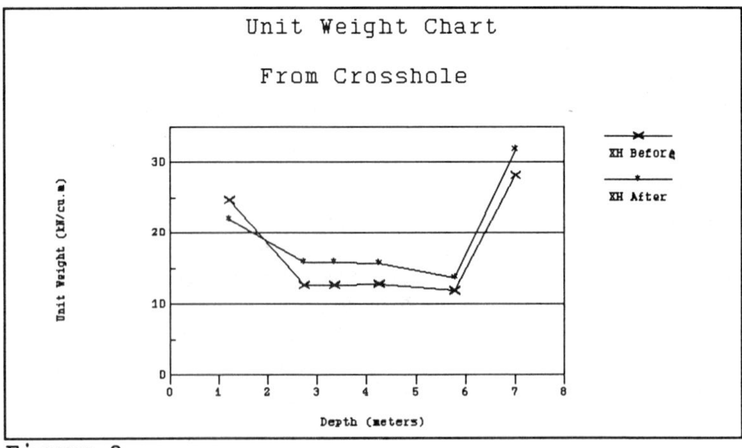

Figure 8

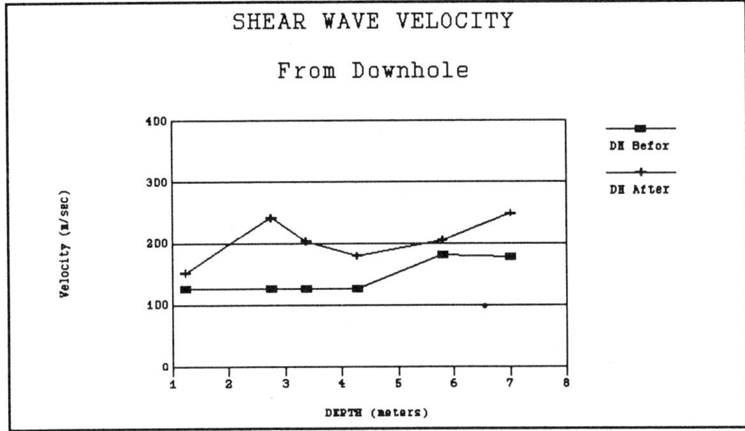

Figure 9

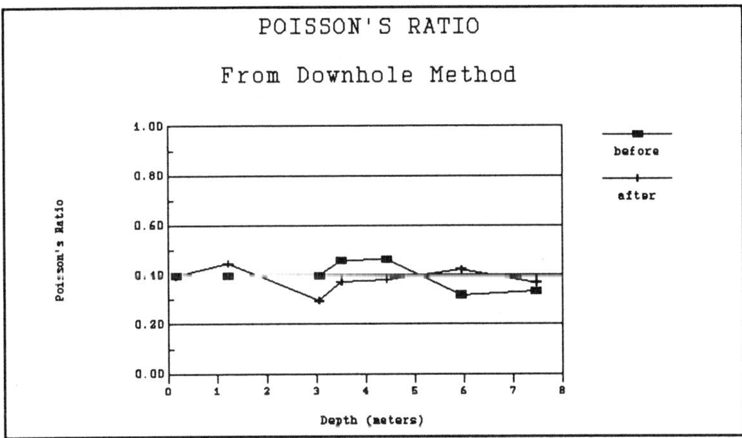

Figure 10

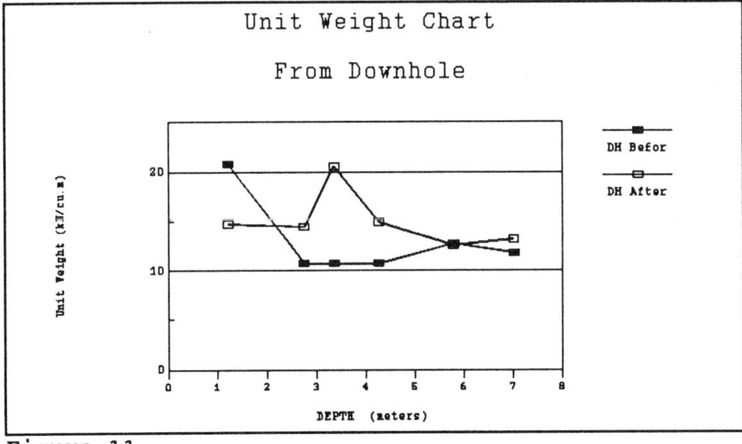

Figure 11

Discussion and Conclusions

The results of the testing presented herein show the crosshole and downhole seismic methods to be reproducible and reliable methods for measuring the improvement of soils by introducing inclusions of grout. The method should be suitable for use in other ground modification procedures where a change in the mechanical properties of the soil is induced by the inclusion. The method has the benefit of evaluating average properties over larger areas than most single point tests. The equipment and expertise for these seismic methods is readily available and relatively inexpensive.

The methods do have limitations and each should be practiced only by those experienced in the interpretation of seismic data and knowledgeable in the factors influencing the test procedure. A significant limitation is that casing sufficient to withstand all of the forces induced by ground modification will likely interfere with, or short circuit the transmission of seismic waves, particularly for the downhole method. Damage to the test holes is more likely with some ground modification procedures than others and test holes should be sited accordingly. Provisions should be made to accommodate the likelihood that some test holes will be damaged and will require replacement.

The downhole method is preferred where no surface obstructions are present within the area of the test. Where slabs or other structures are present, such as when testing is to be done inside or beneath a building, the crosshole procedure should be considered. The test holes should be spaced closely for the crosshole test, with spacing approximately equivalent to the inclusion spacing. Additionally, test hole spacing should be less than twice the distance from the hole to foundations or other solid objects, 0.9 m to 3 m (3 to 10 ft), to reduce the influence of refractions.

The uphole method is not considered an appropriate method for the evaluation of ground improvement. The uphole method has only very marginal economies over the downhole procedure and provides much poorer data.

Seismic methods provide a good qualitative measure of average improvement. Though density values can be obtained by these methods, the calculated densities will be colored by the influence of inclusions. Test locations and methods can be selected to minimize these effects or to examine significantly large areas to obtain an averaging effect. However, in this case, calculated void ratios and densities should be considered as indicators only and not taken as true values. Where inclusions comprise a relatively small percentage of the soil volume and have dynamic properties similar to those of the soil, calculated densities should be fairly accurate but should be correlated with some physical tests for verification.

Seismic methods hold much promise and, with further research, may be shown to be very effective in applications for quantitative evaluation for ground improvement. Additional research is needed to evaluate the density correlations for different materials and to better define the influence of small inclusions on the calculated values.

REFERENCES

[1] Brown, D. R. and Warner, James, Compaction Grouting ASCE JSMFD Vol. 99, No SM8, pp 589-601.
[2] Welsh, Joseph P., In Situ Testing for Ground Modification Techniques, Proc. In Situ '86, ASCE GT Div., Blacksburg, VA, 1986, pp 322-335.
[3] Richart, F. E., Jr., Hall, J. R., Jr., and Woods, R.D., Vibrations of Soil and Foundations, Prentice Hall, NJ, 1970.
[4] Stokoe, K. H. and Woods, R.D., In Situ Shear Wave Velocity by Crosshole Method, ASCE JSMFD, SM5, May 1972, pp 443-459.
[5] Woods, R. D. and Partos, A., Control of Soil Improvement by Crosshole Testing, Proceedings of 10th Intl. Conference on Soil Mechanics and Foundation Engineering, Rotterdam, 1981, pp 793-796.
[6] Partos, A., Woods, R.D., and Welsh, J. P., Soil Modification for Relocation of Die Forging Operations, Proc. Grouting in Geotechnical Engineering, ASCE, New Orleans, LA, February 1982, pp 738-758.
[7] Bowles, J.E., "Soil Volume and Density Relationships" in Foundation Analysis and Design, 3rd Ed., McGraw-Hill Book Company, NY, 1982, pp 13-16.
[8] Baily, A. D. and Alstine, C. L., Downhole Shear Wave and Soils Study, Nimbus Instruments Publication, Sacramento, CA., 1984.

Michel P. Gambin

LATERAL STATIC DENSIFICATION AT MONACO
DESIGN, CONSTRUCTION AND TESTING

REFERENCE: Gambin, M. P., "Lateral Static Densification at Monaco Design, Construction, and Testing," _Deep Foundation Improvements: Design, Construction, and Testing_, ASTM STP 1089, Melvin I. Esrig and Robert C. Bachus, Eds., American Society for Testing and Materials, Philadelphia, 1991.

ABSTRACT: The so-called "Compaction Grouting" technique is now being used in France under various names. Due to a large research programme launched a few years ago, Soletanche was in a position to develop a more scientific approach to this process when it is used as a soil improvement method.

This paper describes the various rules used for design with their background and gives an example of a lateral static densification job.

The positive role of microshearing in sandy soil and that of solid mortar columns in clayey soils is described.

KEYWORDS: soil improvement, compaction grouting, densification, reinforcement, cavity expansion, potential of liquefaction, settlement.

Lateral Static Densification is the combination of 2 types of soil improvement techniques : sand densification and soil reinforcement. It is designed to give an overall increase in soil parameters such as shear strength and E-modulus, in sandy, silty or even clayey soils, saturated or not saturated, at depth.

The background of conventional techniques of deep soil improvement such as vibrocompaction, Ménard dynamic consolidation (1), controlled explosions, coupled with our knowledge regarding stone columns, stone piers (2), etc ... helped Soletanche to convert the

M.P. Gambin is head of the soil improvement department at Soletanche Entreprise, P.O. Box 511, 92005 Nanterre, France.

American compaction grouting   technique (3) into a scientifically controlled procedure renamed Lateral Static Densification.

We called it Static Densification, because stresses are slowly applied, as opposed to vibratory or dynamic stresses used in other methods, and Lateral Static Densification, because soil deformations which induce densification are mostly created in the horizontal plane.

The mechanism of lateral static densification can be theoretically analyzed on the basis of the latest developments regarding the expansion of cylindrical cavities in soils and also the reinforcement of soils by inclusions.

A test section carried out in the Principality of Monaco (French Riviera) and various jobs since performed either in Monaco or elsewhere made it possible to prove the validity of our theoretical approach.

I. THE TECHNIQUE

Preloading and/or dynamic processes (such as vibrocompaction or dynamic compaction) are not the most appropriate methods to improve a soft or loose formation overlaid by a much stronger thick layer (see Fig. 1).

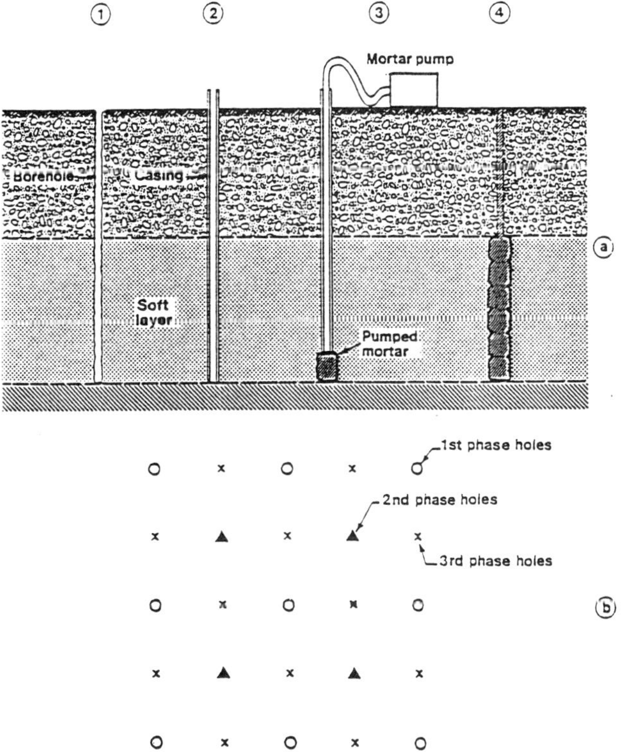

FIG. 1 - Lateral Static Densification Principle
    a) Soil cross section and sequences
        nota: Stages 1 and 2 are usually combined
    b) Plan view

Indeed, when preloading is carried out, its effectiveness is not as great as expected, due to a broader stress distribution on top of the soft deposit. The same phenomenon would occur under a dynamic compaction impact, the stiffer layer then acting as a screen. If vibrocompaction is considered a large amount of energy is wasted vibrating large diameter pipes through the upper layer.

Why not, then, drill to the full depth, with full face bits and 100 mm dia casing, leave the casing and connect it to a specific mortar pump with a 1-10 m3 per hour pumping capacity at a 0.5-8 MPa pressure ? The casing is then pulled up 0.5-1.5 m at a time and a mortar, very viscous and with a high friction angle is pumped into the cavity. Pumping rate is such that the mortar will not segregate and create hydraulic fracturing in the surrounding soil. Mortar stays in the shape of bulky cylinders on top of each other.

The treatment procedure generally involves several passes of drill holes so as to correctly improve the whole volume of soil.

Up to this point of the description, nothing really differs from the compaction grouting. What is new is as follows:

a) the final soil parameter values are derived from the specifications given by the Consulting Engineer for the proposed structure such as footing bearing capacity and maximum allowable differential settlement,

b) the effect of the cylindrical mortar inclusions is involved as well as the soil densification between the cylinders,

c) the volumes of mortar to be pumped are calculated before treatment from the original soil parameters and the final values as derived above,

d) the mortar column diameter is assessed, based on the original soil parameters and pumping capacity,

e) a pattern of drill holes is set out, based on the requirements of the previous points.

II. THE THEORETICAL APPROACH

The theoretical approach can be described in the same sequence as above.

2.1. Final bearing capacity and settlements.

According to the sand fraction included in the soil layer to be treated, the inclusion will or will not play a major role. We shall analyze the 2 extreme cases: clean sand (or sand and gravel) and clayey soil. For most of the treatment cases an intermediate situation is faced where the specialist contractor has to use his own judgment.

2.1.1. In the event of a clean sand layer the placement of the mortar columns will create an overall densification which can be easily visualized.

Let us look at the soil behaviour around the first column of mortar. The analysis of a clean loose sand response is given step by step in fig. 2 as the mortar column grows in diameter. Axes are in Almansi coordinates (4) for the first diagram

$$a = \frac{1}{2} \frac{\Delta V}{V}$$

for the radial deformation and

$$\mu(a) = 2 \frac{d\,(aV)}{dV}$$

for the volume change (which is negative in the case of densification).

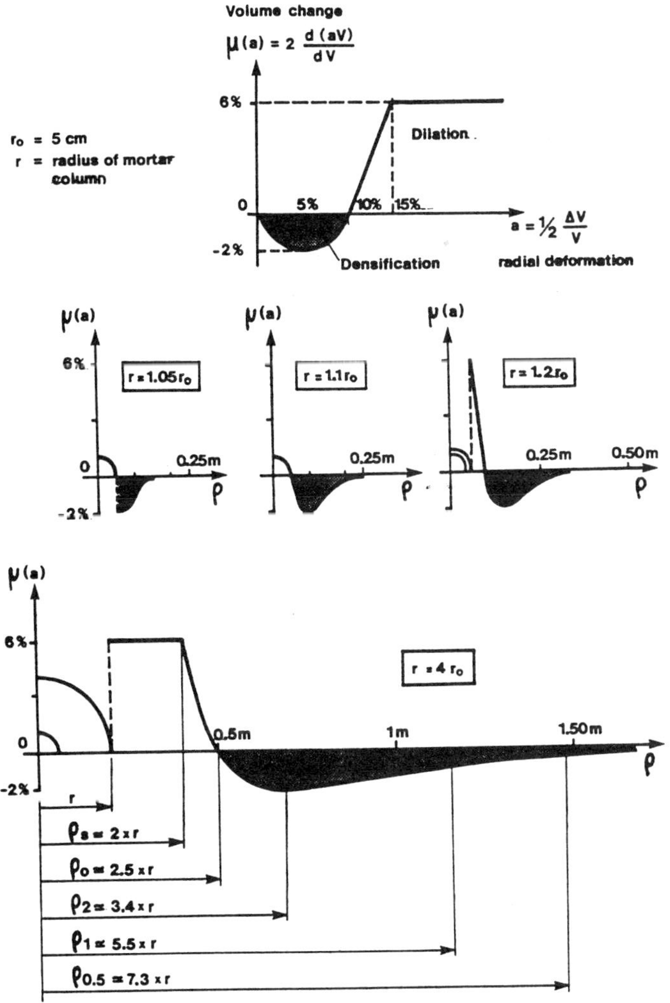

FIG. 2 – Densification and dilation around a growing mortar cylinder in soil (5)

On the second series of diagram one can see how as soon as the mortar column diameter reaches 1.2 times the drill hole diameter dilation appears on the wall of the soil in contact with the mortar. However this dilation in a medium which was first densified helps increase the outside densification. This densification extends to 7.3 times the radius of the mortar column when the latter is 4 times the borehole radius. The subscript of " ρ " on the last diagram gives the densification ratio in percent. The mean densification ratio in this specific example is 1.25 % within the cylinder 1.63 m in radius (including the ring volume now in dilation).

When a second or third column is placed, densification will increase according to a mechanism which is still under analysis but which can be visualised by superimposition of the densification curves.

Densification and later, dilation is obtained by microshear strains within the soil, induced by the expansion of the cylindrical cavity created by the pressurized mortar. The effect of direct compression only appears when mortar columns are closer and this is less effective. The predominant role of microshear strains during the process makes it very valuable to decrease the potential of liquefaction of soils in earthquake prone areas.

Now, what is the effect of this apparently small amount of densification on bearing capacity and differential settlements ? As everyone knows bearing capacity in sand is directly related to allowable settlements. In the line of previous works (6) we give in table 1 some results for a given sand. As one can see for a very slight variation of the void ratio or the specific gravity the allowable bearing capacity exhibits a very large increase when the target is to keep a differential settlement between adjacent footings limited to 20 mm: Terzaghi's rule.

TABLE 1

PARAMETER	Loose state	%	Medium dense state	%	Dense state
Relative density (Dr)	56		69		86
$e = e_0 - Dr\,(e_0 - e\ min)$	0.536		0.504		0.464
$n = {}^e/1 + e$	0.349		0.335		0.317
$\gamma = 2.7 - n \times 1.7$	2.107	1.1	2.130	1.4	2.161
$\gamma d = (1 - n) \times 2.7$	1.758	2.2	1.798	2.5	1.844
allowable bearing capacity (kPa)	200		300		500

It must be noted that up to this point we have neglected the effect of the inclusions in our calculation.

2.1.2. In clayey soils this inclusion effect will become predominant because the clayey soil densification is very slight.

Mortar colums will act as follows (see Fig. 3): a force Qt acts on their top, transmitted by the upper layer, a force Fn acts on the upper section of their shaft as a negative skin friction, a force Fp acts on the lower section of their shaft as a positive skin friction and a force Qp acts on their point. Each one of these forces can be estimated as a function of the displacement of the corresponding section of the inclusion with respect to the surrounding soil.

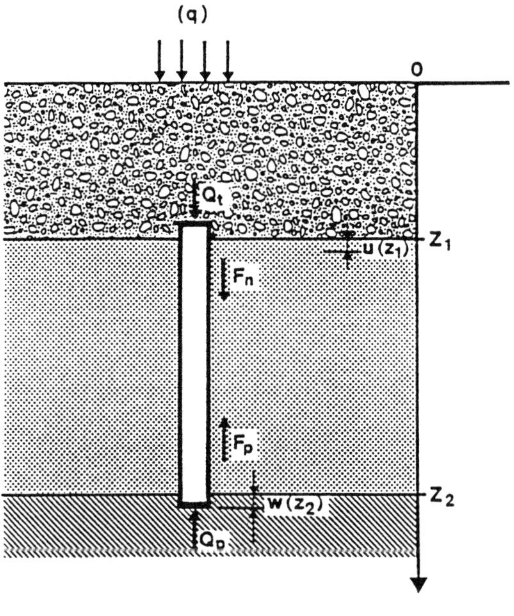

$$Q_t + F_n = F_p + Q_p$$

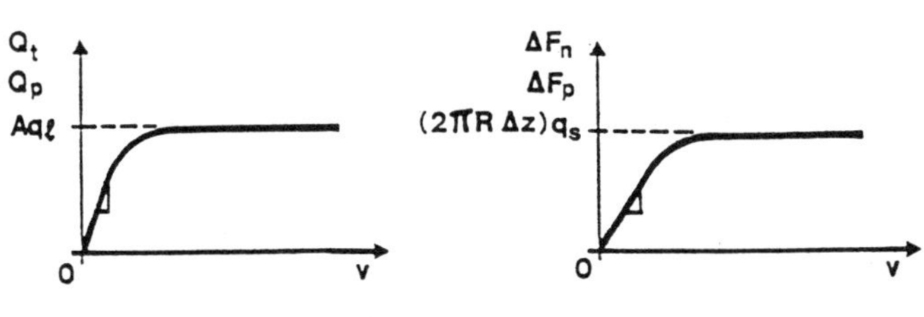

FIG. 3 – Soil-inclusion interactions

Our approach is mostly based on the Menard pressuremeter test (7–8) which is the most often used in-situ test in France (80 % of State owned structures are solely designed on pressuremeter tests results).

If $u(z)$ and $w(z)$ are respectively the settlement of the inclusion and that of the soil at depth (z), the relative displacement between the inclusion and the soil at the depth z is:

$$v(z) = w(z) - u(z)$$

and every unknown is a function of $v(z)$.

The relationships to be used involve:

. a limiting stress value q_1 under a pile point or above a pulled-up flat anchor and the positive or negative ultimate skin friction q_s along the pile shaft (7-9).

. idealized stress strain relationships either below a pile point, or above a flat anchor or along a pile shaft (10-11).

Then it is possible to determine the 4 functions:

$$Qt = f_1(v) \qquad \text{at } z = z_1$$

$$Qp = f_2(v) \qquad \text{at } z = z_2$$

$$\Delta Fn = f_3(v) . \Delta z$$

$$\Delta Fp = f_4(v) . \Delta z$$

which have similar forms (see Fig. 3).

A computer program which is an extension of a former program (12) to estimate the pile settlement on the basis of pressuremeter tests results makes it possible to determine Qt, Qp, Fn, Fp as well as $u(z_1)$ and $w(z_1)$ and to compare them with the $u_o(z_1)$ value without reinforcement.

An effectiveness ratio $(u_o - u)/u_o$ can be used to compare various inclusion spacings or diameters.

2.2 Assessment of mortar volumes

These volumes are calculated with a different approach according to the mechanism which is prevalent.

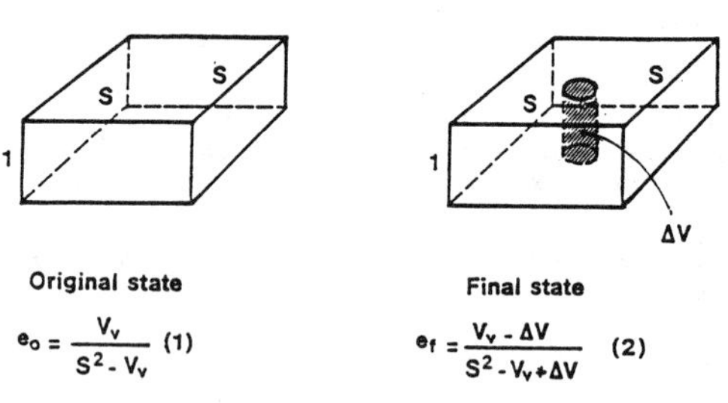

FIG. 4 - Calculation of mortar volume in loose-sand

2.2.1. In the event of a loose sand layer (see Fig. 4), assume the original void ratio is e_o and the final value is e_f, as obtained from above mentioned calculations. Volume of inclusion ΔV must be

$$\frac{\Delta V}{V} = \frac{e_o - e_f}{(1 + e_o)(1 + e_f)}$$

This formula would be valid if the final e_f value were to be constant throughout in the medium. However this is not the case due to the soil compressibility. Consequently, the designer will take a mean ΔV value given by

$$\frac{e_o - e_f}{(1 + e_o)(1 + e_f)} < \frac{\Delta V}{V} < \frac{e_o - e_f}{1 + e_o}$$

2.2.2. In the event of a clayey layer, we have shown in para 2.1.2. how a single computer program can yield every mortar column parameter.

2.3. Assessment of the mortar column diameter

According to the soil characteristics one can experience a pumping refusal. Consequently it is interesting to analyze the expansion of the mortar column in the soil.

First of all it must be kept in mind that the soil reaction around the mortar column cannot be greater than the limit pressure measured by the Menard pressuremeter. As head losses in the hose between the pump and the casing and in the casing itself are never more than 1 MPa for a 50 m length, the pressure mostly dissipates within the mortar column due to the high friction of the material (see Fig. 5).

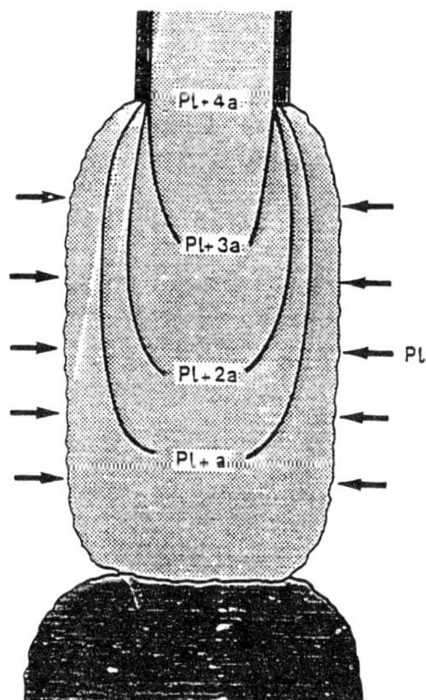

FIG. 5 - Cross section of a mortar cylinder during expansion
Contour lines are pressure equipotentials

Assume p the mortar pressure at the casing outlet and p_1 the soil limit pressure, V_o and V respectively the original volume of the cavity and the mortar column volume, if can be demonstrated that

$$\frac{p}{p_\ell} = \left(\frac{V}{V_o}\right)^n$$

with

$$\frac{\sin \phi_M}{1 + \sin \phi_M} < n < \frac{4}{3} \frac{\sin \phi_M}{1 + \sin \phi_M}$$

where ϕ_M is the friction angle of the mortar. For $\phi_M = 40°$, a usual value,

$$0.39 < n < 0.52$$

Consequently with a good approximation we can write

$$p \simeq \frac{d}{d_o} \times p_\ell$$

where d_o and d are respectively the borehole diameter and the final mortar column diameter.

If the holes were drilled in 100 mm diameter, the final mortar column can be only 600 mm if the pump cannot reach a pressure greater than 6 times the limit pressure of the soil.

2.4. Assessment of the drill hole lay-out

In the event of a clayey soil this lay-out is obtained by the computer programme used to design the reinforcing columns.

In the event of a loose sand layer we still have to set out the drill holes.

We know $= \Delta V/V$ from para 2.2.1

d/d_o from para 2.3

consequently we can easily find the spacing S between drill holes after the last phase, given by:

$$d \sqrt{\frac{\pi(1+e_o)}{e_o - e_f}} < S < d \sqrt{\frac{\pi(1+e_o)(1+e_f)}{e_o - e_f}}$$

to take into account the soil compressibility.

2.5 Assessment of number of phases

This can only be done through a test section as it is not yet easy to predict the excess pore water pressure building-up during construction. Pore water pressure transducers will be set before treatment and it will be easy to find out whether all the columns can be constructed in one phase or not.

III. THE FONTVIEILLE ZONE D JOB AT MONACO

This job was carried out in 1985 after a successful test section funded by Soletanche (13). This test confirmed Soletanche expectations and helped them to prepare the design for their first job in the vicinity.

3.1. Background of the project

The Fontvieille area in the Principality of Monaco was reclaimed in the early 70's to provide some more space for urban development between the Alps range and the sea. After reclamation the typical soil cross section (see Fig. 6) was as follows:

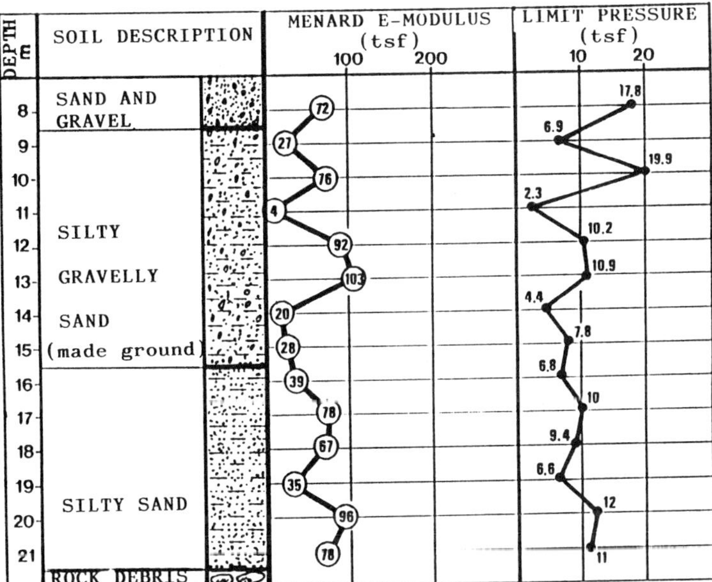

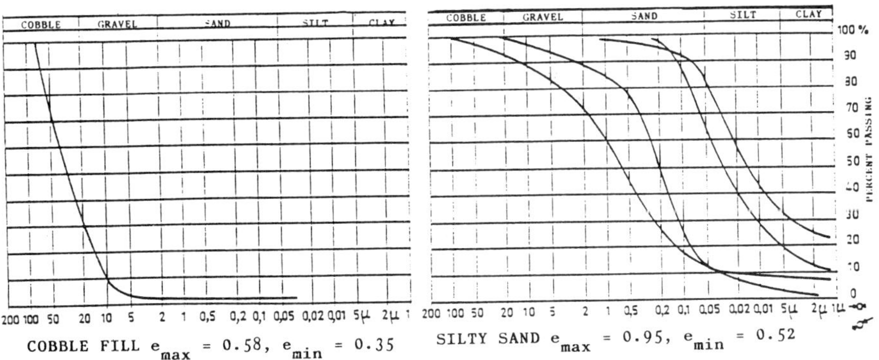

FIG. 6 - Typical geotechnical cross section at Fontvieille zone D (Ménard pressuremeter tests) and grain size curves (1 tsf = 100 kPa)

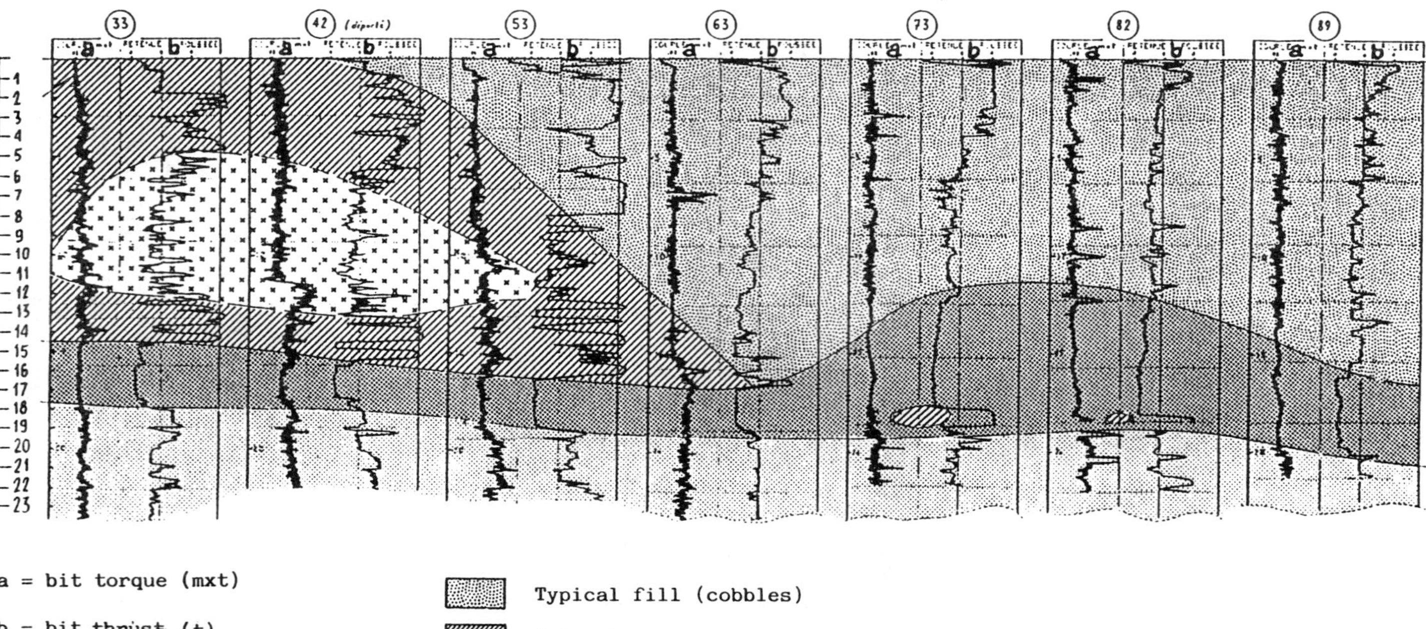

FIG. 7 – Typical longitudinal cross section derived from Enpasol Mark II combined parameters logs.

- sand dumped from barges, including gravel and cobbles and some
silty pockets, up to 15 m deep,

- silt and sand (original sea bed), 2-3 m thick,

- undersea talus (rock debris).

The French Riviera being an earthquake prone area, it was
necessary to check the relative density of these formations and to
densify them according to the various types of projects involved.

In the late 70's, after a rest period, the upper layer of the sand
and gravel fill was compacted by the Menard dynamic technique, except
close to the sea front where deep vibratory methods were used. To
compact the deeper layers, heavy preloading by embankments up to 16 m
high was carried out in open spaces.

In zone D, on a 8,900 m2 surface area surrounded by housing blocks
and other concrete structures, the principality planners were proposing
community premises. These involved a church, a post office, fire and
police stations, on top of a 2 storey parking garage. Loads would vary
between 30 and 90 KPa due to some extra fill for landscaping. Piling
was considered too expensive, especially since a former temporary dyke
made of rip-rap was crossing the site. However a raft could not be
anticipated without deep soil treatment: static settlements of 30 to 90
mm were expected and liquefaction potential was not negligible.

3.2 The soil treatment tender

The Public Work Agency of Monaco launched a tender for soil treatment to:

- decrease foundation beds compressibility,
- lower the non homogenity of their response to loads,
- reduce the potential of liquefaction to an acceptable value.

The contractor had to guarantee:

- under structural loads
  . a total settlement of less than 40 mm
  . differential settlements of less than 1/1000

- under live loads and landscaping earth loads
  * for non critical buildings, (mostly parking garages),
    . additional total or differential settlements of the same
      magnitude,
  * for critical structures (such as the church and the office
    buildings),
    . additional total settlement of less than 30 mm,
    . differential settlement of less than 0.7/1000.

- under a 0.2 g earthquake, the stability of each building.

Acceptance tests were set up as follows:

- regarding static stability:

Minimum Ménard E-modulus: 3 MPa
Ménard limit pressure pl-po greater than 0.6 MPa
Ménard E-modulus floating mean greater than 8 and 10 MPa

for non critical and critical structures respectively.

Cross hole VL and G floating mean greater than 2000-2100 m/sec and 1800-1900 MPa respectively:

Minimum VL and G : 1600 m/sec and 1000 MPa.

- regarding stability under seismic condition:

$$N > 21 \text{ at } 12 \text{ m depth}$$
$$N > 25 \text{ at } 20 \text{ m depth}$$

as long as $D_{50} > 0.3$ mm.

3.3 The proposed treatment

Due to their own experience gained during the above mentioned test section Soletanche submitted a bid on a design and built basis. The proposal was based on the following treatment by lateral static densification:

- primary and secondary columns in the non critical zones with a final spacing equal to 3.6 m in a square pattern and a mean cumulated $\Delta V/V$ equal to 3.2 %,

- primary, secondary and tertiary columns in the critical zones with a final spacing equal to 2.5 m in a square pattern and a mean cumulated $\Delta V/V = 4.8$ %.

Two tests sections were carried out in each type of zones to successfully confirm these expectations.

For the whole job, continuous monitoring involved:

. Drilling parameters recording with the Soletanche Enpasol Mark II (14-15) on 10 % of the holes (i.e. on a 10 x 10 m square pattern),
. Recording quantity of mortar placed and applied pressure for each hole pass,
. Checking for possible soil heave.

On the site 100 holes were monitored with the Enpasol Mark II which makes it possible to derive combined parameters from which geotechnical cross sections can easily be drawn. On fig. 7, the location of the temporary dyke previously mentioned is clearly visible.

During the test sections a correlation was found between the Ménard E-modulus between 4 and 12 MPa and a combined parameter β (relatively similar to Sommerton parameter) for cobble fill and silt-sand formations (see Fig. 8). As it was possible to relate the $\Delta V/V$ to the deficiency of the measured E-modulus values against either 8 or 10 MPa at a given depth it became apparent that $\Delta V/V$ was also a function of the deficit of β to E = 8 or 10 MPa.

Consequently the instruction sheets could be automatically derived for each Enpasol monitored hole. An interpolating process was used for the intermediate holes.

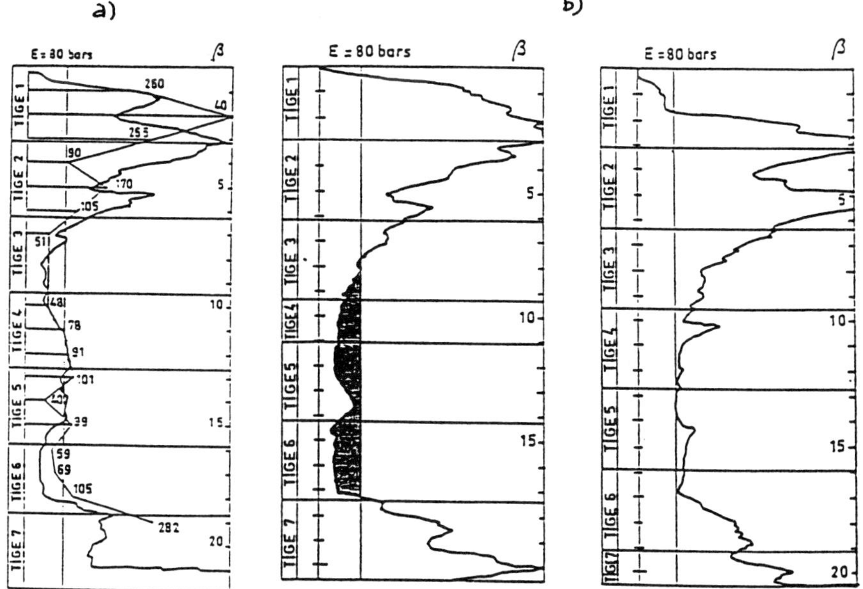

FIG. 8 - a) Linear correlation between Ménard E-modulus and the parameter (Enpasol Mark II)

b) The β parameter (Enpasol Mark II) before and after treatment. The hatched zone expresses the deficit of E-modulus from which $\Delta V/V$ can be estimated at each pass

3.4 The results

A total of 770 columns were placed, involving 3,000 m3 of mortar, leading to an average $\Delta V/V$ = 3.9 %

In the silt-sand natural formation $\Delta V/V$ was 3.5 % in non critical zones and 6.6 % in critical zones. If the volume of mortar is related to the surface area, it corresponds to a height of 260-480 mm in non critical zones and 480-530 mm in critical zones.

Acceptance tests included:

36 Enpasol monitored holes,
12 Ménard pressuremeter holes,
 8 SPT soundings
 7 seismic cross hole tests.

In fig. 9 a statistical analysis of the Ménard E-moduli before and after treatment is presented. As one can see:

- in the non critical zones the E-modulus median value exhibits a 100 % increase,

- in the critical zones, this increase is 128 %,
- after 2 passes there are no E-modulus less than 3 MPa,
- after 3 passes there are no E-modulus less than 4.5 MPa,
- a third pass does not increase the highest values of E, on the contrary dilation effects start to become apparent.

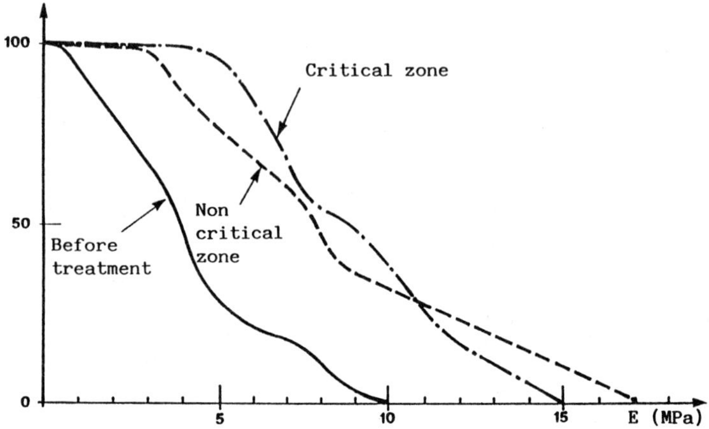

FIG. 9 - Statistical analysis of E-moduli before and after treatment

3.5 Expected settlements

From the Ménard E-moduli results it is possible to estimate the expected settlements.

Two series of calculations could be done, one only involving the densified materials, the other one taking the mortar column effect into account.

First, in table 2, we present the various mean values as measured for the geotechnical parameters.

TABLE 2 - Soil parameters

Depth	Soil	Before treatment Ménard E-modulus (MPa)	After treatment Ménard E-modulus (MPa)	Limit pressure after treatment (MPa)
0- 7 m	Dynamically compacted fill	–	–	2
7-15 m	Cobble fill	4	8 - 9	1.5
15-18 m	Silt and sand	5	8 -10	1.5

higher E-moduli refer to critial zones (3 passes)

In the non critical zone, mean mortar column diameters are respectively 0.59 m and 0.76 m in cobble fill and in silt and sand.

In the critical zones, mean mortar column diameters are respectively 0.50 m and 0.725 m.

At the time of this job approximate rules were taken regarding the mobilization of skin friction and soil reaction against column top and point. A more sophisticated approach is now being used, drawing on all available French experience (16).

From table 3 one can see that taking into account the mortar column effect means a decreases of the settlement by 50 % or more. Columns support 40-45 % of the tributary load.

TABLE 3 - Anticipated settlements

Zones	Loads (kPa)	Settlement without treatment (mm)	Settlement without column effect (mm)	Settlement with column effect (mm)
non critical	22	21	11	5.6
	50	48	26	13
critical	87	84	39	16

α factor (according to Ménard) was taken as 1/3 in cobbles and 1/2 in silt and sand

Soil reaction at the top and at the point steadily increases from 9 and 5 % respectively for the lower loading to 21 and 12 % for the middle loading to 28 and 13 % for the highest loading.

A typical split of the various forces acting on one column is represented of fig. 10.

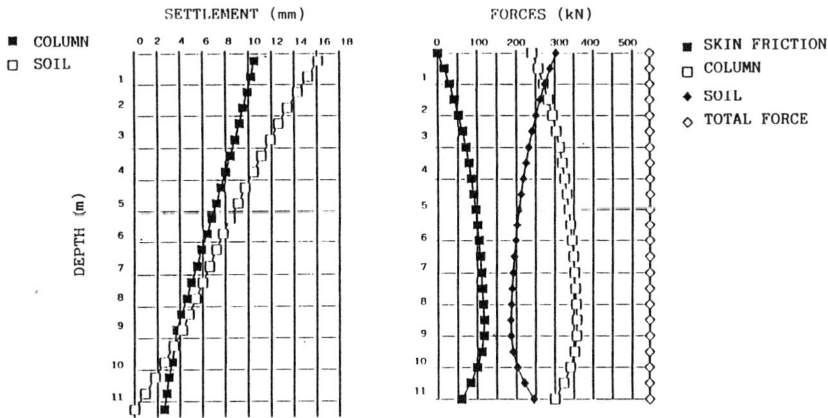

FIG. 10 - Split of settlement and the various forces, between the column and the soil (5)

The building was completed in mid 1987 and at that time observed settlements were less than 10 mm, which is not surprising since live loads were not fully active.

3.6 Conclusion for this job

It finally appears that although the Client's consulting engineer found Soletanche too audacious by placing such small quantities of mortar, we could have placed even less and still have complied with the specifications.

IV. GENERAL CONCLUSION

Application of the most recent theories on cavity expansion in soils and on reinforcement effects of vertical mortar columns has helped Soletanche to set up cost effective design rules for compaction grouting and to make it a more refined technique for soil improvement under the name of lateral static densification.

The rules involve:

- the analysis of soil strength and deformation parameter as a function of void ratio,
- the response law of soils exhibiting a negative dilatancy around a cylindrical cavity in expansion,
- a correlation between the mortar cylinder size and the ratio of the pressure developed by the pump to the limit pressure measured in the surrounding soil by the pressuremeter,
- the mechanics of reinforcement by mortar columns.

The role played by microshear strains in the densification process makes this technique very useful in decreasing the potential of liquefaction of soils in earthquake prone areas.

The Monaco Fontvieille zone D job is a example among many others. This reclaimed land located in a seismic zone was treated by this technique. The bid was on a design and built basis. The job progressed according to the design criteria, pumping instruction sheets being computerized on the basis of a combined drilling parameter deficit. No acceptance test failed at the time of handing over the site, although mortar inclusion did not exceed 3.8 % in the non sensitive zones (mostly parking garage) and 4.5 % in the sensitive zones (office buildings and church). Leveling survey during construction and after commissioning showed that settlements were well within acceptable limits.

These conclusions lead Soletanche to further use and expand this technique worldwide.

REFERENCES

(1) Gambin M.," Ménard Dynamic Consolidation", _Sols Soils_ No 29, 1979, pp 27-43 (Also : ASCE Ground Reinforcement Seminar, Washington D.C., Jan).
(2) Gambin M., "Puits ballastés à la Seyne sur Mer", _Proceedings International Symposium on Soil Reinforcement_, ENPC, Paris, Oct 1984, pp 139-144 (English version available from the Author).

(3) Warner J., "Compaction grouting, the first thirty years" <u>Grouting in geotechnical engineering</u>, ASCE, New Orleans, Feb., 1982, pp 694-707.

(4) Baguelin F., Jezequel F., Shields D., <u>The Pressuremeter and Foundation Engineering</u>, Trans Tech Publication, Switzerland, 1978.

(5) Bouchelaghem A., PhD Thesis, Ecole Nationale Supérieure des Mines, Paris (in preparation).

(6) Thorburn S., "Building structures supported by stabilized ground", <u>Ground Treatment by Deep Compaction</u>, ICE, London, 1976, pp 83-94.

(7) Menard L., "Calcul de la force portante des fondations sur la base des résultats des essais pressiométriques", <u>Sols Soils</u> No 5, June 1963, pp 9-32.

(8) Menard L., "The interpretation of pressuremeter test results", <u>Sols Soils</u> No 26, 1975, pp 7-43.

(9) Bustamante M., Gianeselli L., "Prévision de la capacité des pieux isolés sous charge verticale". <u>Bull. Liaison Labo P. et Ch.</u>, n° 113, May- June 1981, pp 83-108.

(10) Gambin M., "Calcul du tassement d'une fondation profonde en fonction des résultats pressiométriques", <u>Sols Soils</u> No 7, Dec. 1983, pp 11-31.

(11) Franck R., Zhao S.R., "Estimation par les paramètres pressiométriques de l'enfoncement sous charge axiale de pieux forés dans les sols fins", <u>Bull. Liaison Labo P. et Ch.</u>, n° 119, May-June 1982, pp 17-24.

(12) Marchal J., "Calcul du tassement des pieux à partir des méthodes pressiométriques", <u>Bull. Liaison Labo des P. et Ch.</u>, n° 52, May 1971, pp 22-25.

(13) Gambin M., "Le compactage statique horizontal", <u>Comptes rendus Colloque Franco-Soviétique sur l'amélioration des sols</u>, Moscou, Oct 1985 (available as Rapport des Labos LPC, No GT 20 published 1987), pp 7-14.

(14) Pfister P., "Recording drilling parameters in ground engineering", <u>Ground Engineering</u>, London, April 1985.

(15) Pfister P., Hamelin J.P., "Computer aided soil investigations with drilling parameters", <u>Proceedings Fifth International Conference on Numerical Methods in Geomechanics</u>, Nagoya, Japan, April 1985, pp 1715-1720.

(16) Bustamante M., Franck R., Gianeselli L., "Prevision de la courbe de chargement des fondations profondes isolées", <u>Proceedings 12th International Conference on Soil Mechanics and Foundation Engineering</u>, Rio de Janeiro, August 89, vol 2, pp 1125-1126.

Note : Sols Soils papers in French have an expanded English summary

George A. Munfakh

DEEP CHEMICAL INJECTION FOR PROTECTION OF AN OLD TUNNEL

REFERENCE: Munfakh, G. A., "Deep Chemical Injection for Protection of an Old Tunnel," _Deep Foundation Improvements: Design, Construction, and Testing_, _ASTM STP 1089_, Melvin I. Esrig and Robert C. Bachus, Eds., American Society for Testing and Materials, Philadelphia, 1991.

ABSTRACT: The foundation of a 19th Century brick-lined railroad tunnel was reinforced by a deep chemical stabilization scheme which allowed successful construction of two new tunnels about 7 ft. (2 m) below its invert. After evaluation of several protection schemes which included ground freezing, reinforcement with micro piles and conventional structural underpinning, chemical grouting of the foundation soils was selected based on technical and economical merits. Laboratory and field grouting tests were performed in conjunction with the grouting design to select the appropriate type of grout and determine the grouting specifications. The successful application of the chemical grouting scheme allowed construction of the new tunnels with a minimum impact on the existing tunnel.

KEYWORDS: deep chemical injection, soft-ground tunneling, soil coherence, ground settlement, brick-lined tunnel, laboratory and field grouting tests, grouting specifications, grouting operation.

INTRODUCTION

Chemical stabilization is a well accepted method of ground treatment, particularly for construction of highways over unsuitable soils. Stabilizing a soil at the surface is usually easy to accomplish and relatively

Dr. Munfakh is vice president and head of the Geotechnical Department, Parsons Brinckerhoff Quade & Douglas, Inc., One Penn Plaza, New York, NY 10119.

inexpensive. When at depth, however, the task becomes more difficult to implement and to monitor.

Chemicals are applied at depth, by injection or by deep mixing methods, to bind the soil particles together, generally resulting in reduction in the soil's plasticity and increase in its strength and coherence (or cohesion). The most common methods of deep chemical stabilization are penetration grouting (by cement or chemicals) and lime columns.

This paper presents a case history where deep chemical injection was used effectively and economically to protect a 19th Century structure. It describes the design and construction aspects of this ground treatment method, and discusses the laboratory and field testing performed to evaluate its effectiveness.

PROJECT DESCRIPTION

The Lexington Market Section of the Baltimore Metro consists of 1700-ft-long (515 m) twin, single-track subway tunnels driven in soil using the compressed-air shield method. The spacing between the two tunnels is 33 ft (10m) center to center. The 18-ft (5.5 m) - diameter tunnels were driven directly under a number of existing structures including a 90-year-old brick-lined Baltimore & Ohio (B&O) Railroad tunnel (Fig. 1). A 7-ft (2.1 m) clearance existed between the crowns of the new tunnels and the invert of the old tunnel.

The B&O Railroad tunnel is a horse-shoe shaped, brick lined tunnel constructed in the 19th Century. To maintain its serviceability and minimize the impact of construction of the new tunnels beneath it on its structural integrity, a number of protection schemes were considered including chemical stabilization.

SUBSURFACE CONDITIONS

Fig. 2 illustrates the subsurface conditions at the intersection of the new and old tunnels. Generally, granular soils of the Cretaceous age extend from the ground surface to depths of 53 to 58 ft (16 to 17.5 m),about 10 to 15 ft (3 to 5 m) below the invert of the B&O Railroad tunnel. These deposits are underlain by residual cohesive material of decomposed rock. The new tunnels were driven with their inverts in the cohesive residual material, but with some 1 to 6 ft (0.3 to 1.8 m) of Cretaceous soil exposed at the crown. The granular soils at the site were generally dense with SPT values generally between 35 and 50. The cohesive soils were stiff to hard. Grain size analyses of three Cretaceous soil samples collected from the layer sandwiched between the new and old tunnels were almost identical showing

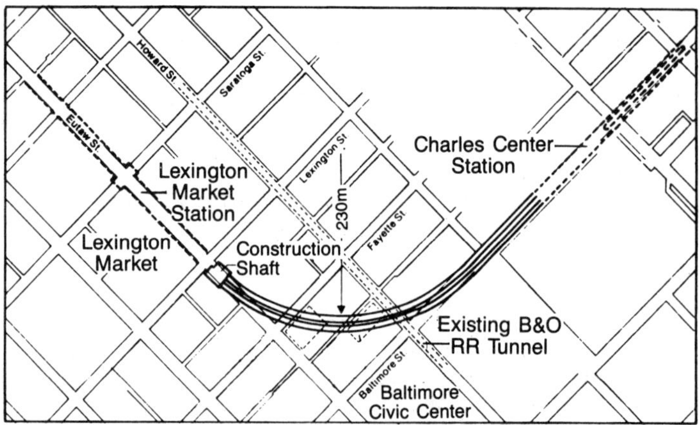

Fig. 1.   Project location plan

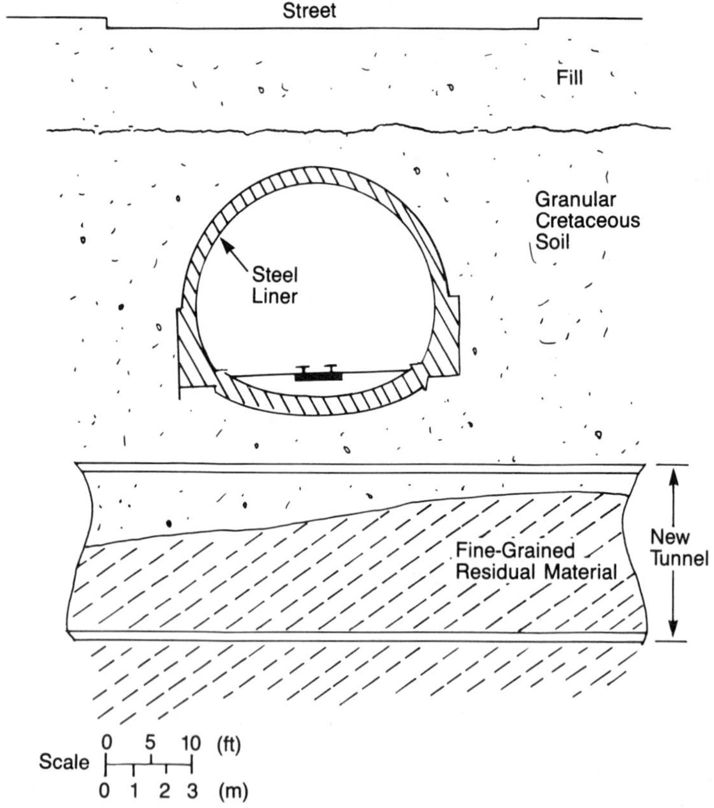

Fig. 2.   Subsurface conditions

typically 13% passing the #200 sieve (0.07 mm) (Fig. 3). The estimated permeability of that layer was of the order of 3×10^{-4} cm/sec.

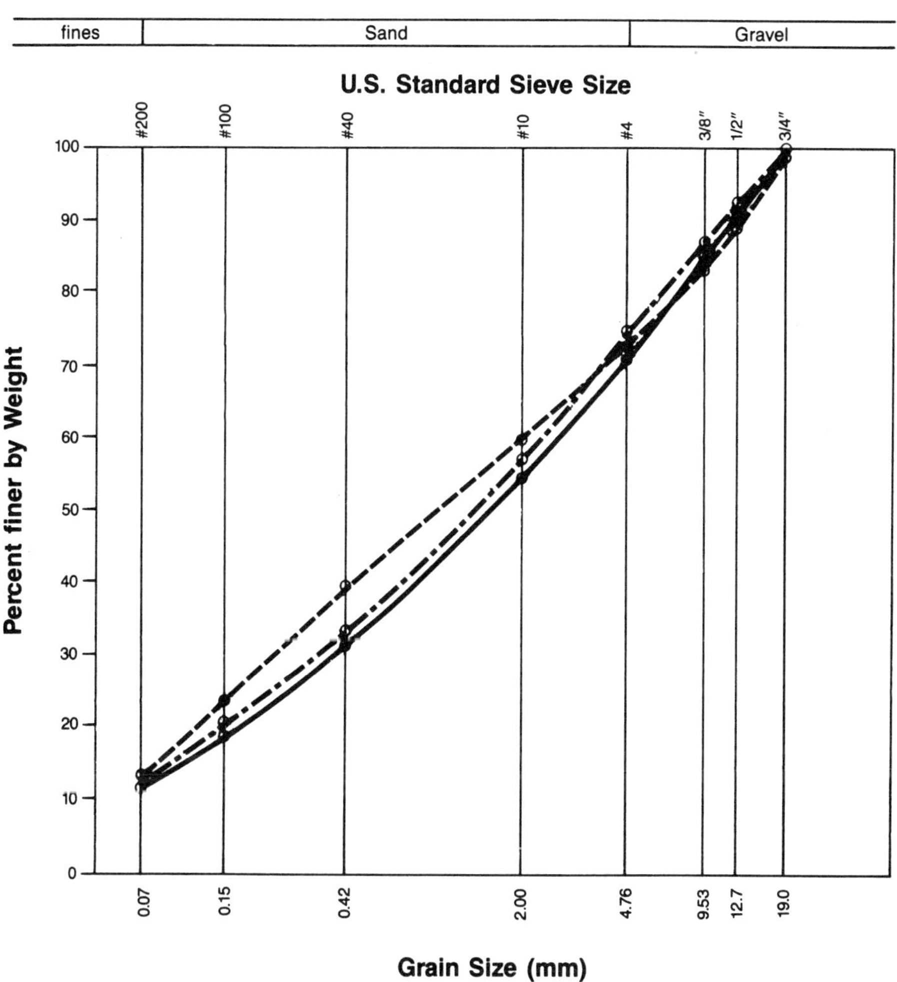

Fig. 3.   Grain size analysis data

GROUND SETTLEMENT

Settlement of a structure caused by tunneling underneath it is a function of the loss of ground during tunneling. The ground loss is a function of soil and groundwater characteristics, geometrical parameters and construction procedures. The two most important soil characteristics affecting ground loss are permeability and coherence (or cohesion). Because tunneling was done under compressed air, the impact of seepage on the ground loss was minimized. Coherence, on the other hand, was an important factor affecting ground loss, particularly in the Cretaceous granular soil layers existing directly beneath the B&O Railroad tunnel. The potential ground loss into the tunnel excavation was estimated according to Schmidt [1].

The distributions of ground settlements at the invert of the B&O Railroad tunnel and at the ground surface were estimated by the method described by Schmidt which equates the volume of the settlement trough to the volume of the ground loss minus any heave or bulking of the soil. The geometry of the settlement trough which has the same shape as a probability curve was defined by the depth of the tunnels, their diameters, and the distance between them. Based on that analysis, the free-field settlement of the B&O Railroad tunnel due to passage of the two transit tunnels beneath it was estimated to be 3.5 inches (88 mm) with the maximum settlement expected to occur directly above the crown of each of the new tunnels (fig. 4). Midway between the tunnels, the anticipated settlement was about 0.5 inch (13 mm). The maximum slope of the settlement trough was 2.3 percent. At the ground surface, the maximum calculated settlement was 1.5 inches (38 mm).

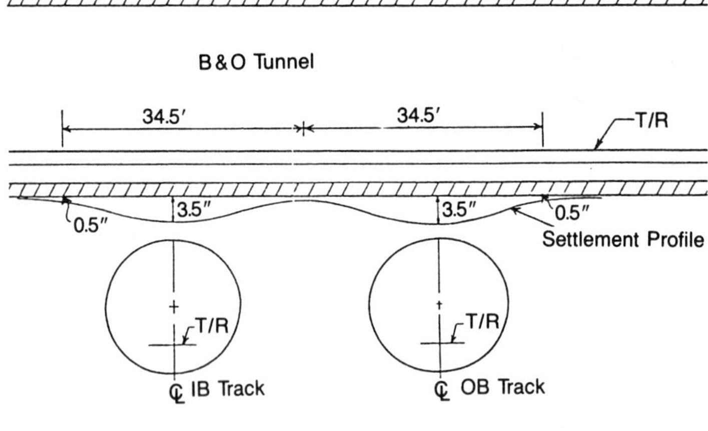

Fig. 4.   Settlement profile

The impact of the potential ground settlement on the structural integrity of the B&O Railroad tunnel was evaluated. Since the tunnel lining was of brick, it was feared that the articulated fashion of its deformation might cause loosening and fall out of the inner course of bricks. Therefore, a tunnel protection scheme was required.

TUNNEL PROTECTION ALTERNATIVES

The tunnel protection alternatives considered included: (1) structural underpinning, (2) maintenance and repair, (3) micro piles (as direct support or in-situ reinforcement), (4) ground freezing and (5) chemical grouting.

Underpinning involved bridging the tunnel structure over any created void. Heavy beams resting on drilled caissons would be placed inside the tunnel on each side of the track, with cross-members transferring the loads from the walls and the rail to these beams. A variation would be to support the tunnel directly on six caissons without longitudinal or cross members. The tracks would then be unsupported requiring ballast leveling maintenance. The direct underpinning schemes were disruptive to the tunnel operation and relatively expensive.

The maintenance schemes involved installation of a protection lining, consisting of liner plates or shotcrete, over the brick arch and walls, and providing maintenance, releveling of tracks and repair when necessary. Although relatively inexpensive, this scheme was rejected as a sole source of protection due to the negative impact of the required maintenance on the operation of the railroad tunnel.

The use of micro piles--as a direct support or in-situ reinforcement--was not feasible due to the geometry and spacing of the three tunnels (more than 200 micro piles were required in a very tight space). Ground freezing required a long lead time, significant surface disruption, and difficult tunnel excavation in solid frozen soil. It was also the most expensive scheme analyzed. Chemical grouting, on the other hand, was feasible and reasonably economical.

All alternatives were judged on their technical, operational and economical merits, and their presumed acceptability by the owners of both the transit and the railroad tunnels. Of all the accepted alternatives, chemical grouting was the most cost-effective. The cost of chemical grouting was estimated to be 63% that of structural underpinning and 32 percent that of ground freezing.

PROTECTION SCHEME

Fig. 5 illustrates the B&O Railroad tunnel

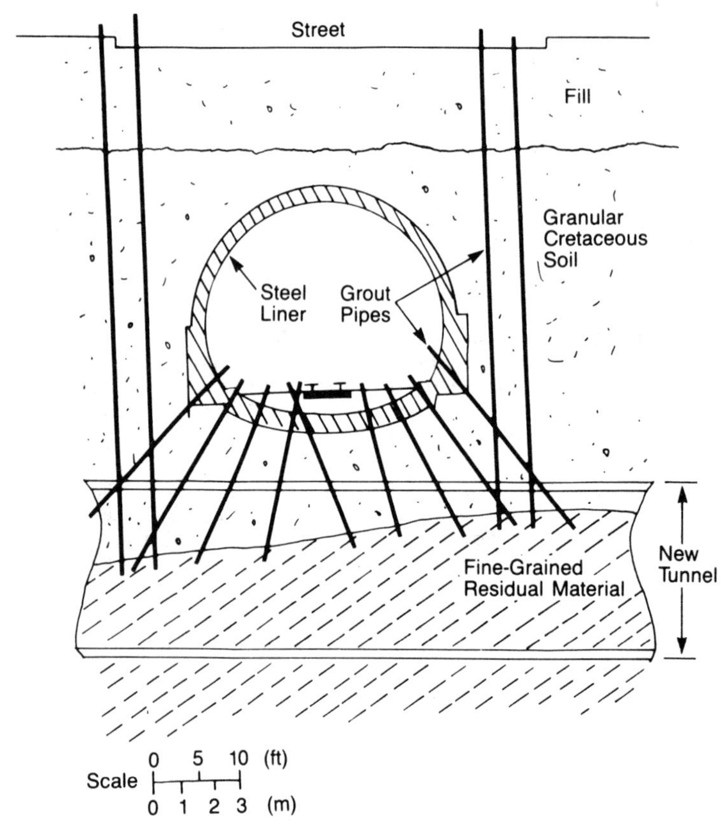

Fig. 5.   Tunnel protection scheme

protection scheme. It involved grouting the soil beneath and around the tunnel, and the use of a grouted metal lining inside the tunnel. The purpose of grouting was to increase the coherence of the soil layer between the old and the new tunnels so that the new construction can proceed with the least possible loss of ground. The grouted zone also served to support the B&O Railroad tunnel by strengthening the ground around it.

Grouting the soil between the new and old tunnels had an added advantage. Since the clearance between the two tunnels was 7 ft (2.1 m), there was a danger that the high air pressure used in the new tunnel might displace the water in the pores of the granular soil between the two tunnels and connect with the upper old tunnel. By filling the pores with grout, the danger of a blow in this area was minimized.

In addition to grouting, the protection scheme included structural reinforcement of the railroad tunnel using an interior lining of fabricated steel liner plates erected inside the tunnel. The liner plate arch was made an integral part of the tunnel lining by using a compound system of resin-anchored tie bolts and cement grout filler. The purpose of this added protection was to maintain the serviceability of the railroad tunnel in the event of deformation.

LABORATORY GROUTING TESTS

Laboratory and field grouting tests were performed to investigate the groutability of the in-situ soils, and determine the appropriate grout and grouting procedures. The laboratory tests included sieve analysis, X-ray diffraction, water and grout injection, and permeability and compressive strength determination of grouted samples.

Wet soil samples 4-inch(100 mm) long and 1-inch (25 mm) in diameter were packed tightly into coated glass tubes using rams and vibration (Fig. 6). Injection rates through the soil columns were measured using tap water and grout. The injection pressure and viscosity of the chemical grout were recorded for each injection test.

Compressive strength test specimens were cut from the injection test samples into lengths of one and one half times the I.D. of the glass tube. The compressive strengths of the grouted soil samples were measured using the Tinius Olsen compressive machine after the samples had cured at room temperature for 18 to 24 hours.

The permeabilities of the reconstituted soil samples as measured by the water injection tests were of the order of 10^{-5} cm/sec (lower than the average 3×10^{-4} cm/sec estimated previously for the Cretaceous soils at the site). Because of the relatively low permeability of the soil and the 13 percent fines measured by the sieve analyses, cement grouting was ruled out and four types of chemical grout--Siroc (40% silicate), Siroc (60% silicate), Herculox and PWG--were selected for grout injection testing.

Table 1 presents a summary of the laboratory grout test data. The test results showed rather low unconfined compressive strengths, probably in part due to the coarse-grained nature of the soils -- the coarser the soil, the more dominant are the characteristics of the pure gel. The unconfined compressive strength was lowest for the PWG grout and highest for the Herculox. For the samples injected with silicate-based grout (Siroc), more than fifty percent strength increase was experienced with increasing the silicate concentration from 40 to 60 percent. The silicate-based grout was selected for field testing.

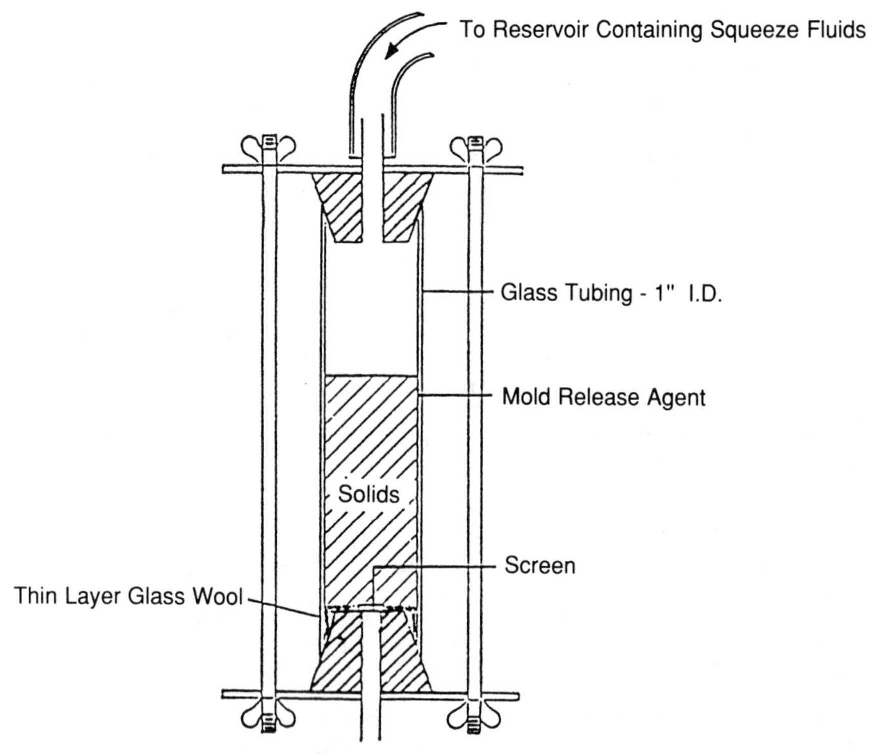

Fig. 6.   Grouting test aparatus

Table 1 -- Laboratory grout test data

Boring No. NWA-	6	7	7	7	11	11
Sample No.	21	21	21	21	11	11
Depth (ft)	49	48	48	48	43	43
Water Injection						
Pressure (psig)	19	19	19	19	15	15
Perm. (10^{-5} cm/sec.)	3.4	1.5	1.9	.53	3.5	2.1
Grout Injection						
Type	40%S	60%S	Herc	PWG	40%S	60%S
Pressure (psig)	19	19	17	19	19	19
Rate (ml/min)	0.1	0.12	0.13	0.07	0.1	0.03
Strength (psi)	15.5	22.4	586	12.0	9.0	20.7

Notes: S = Siroc   1 psi = 7 kPa
       Herc = Herculox   1 ft = 30 cm

FIELD TESTING

Two field injection tests were performed at the site. In the first test, a combined exploratory and grout hole was drilled using hollow stem augers with continuous split spoon sampling along the depth to be grouted. The hollow stem method of grout pipe placement was selected since it allowed soil samples to be taken in the same hole. A 1 inch (25 mm) diameter slotted plastic grout pipe was installed through the hollow stem of the augers and the annular space around the pipe was packed with sand over the bottom 10 feet (3 m), then sealed with cement mortar for several feet above the slotted section.

Following water testing of the grout pipe, a grout injection test was attempted with 40 percent sodium silicate grout. The grout test was abandoned after 10 minutes due to a low grout take of 3 gpm (11.5 lpm) even under a high grout pressure of 120 psi (840 kPa) which is about 3 times the vertical in-situ stress at the injection level. The results of this grout injection test were considered inconclusive due to the possibility of incorrect grout pipe placement or excessive smear effect from the auger method of drilling.

A second grouting test was performed at the site using a pre-slotted plastic grout pipe installed in a hole drilled using a self casing, water washed, rotary percussion method of drilling. The void between the pipe and the soil was filled with sand as the casing was withdrawn, and mortar seal was placed above the slotted section of the pipe.

After water testing the hole, 2833 gallons (10.7 kl) of sodium silicate chemical grout (GELOC-3) were injected. This grout had the same chemical characteristics as the grout Siroc used in the laboratory tests. At 10 to 40 psi (70 to 280 kPa) pressure, the grout take was 18 to 25 gpm (68.4 to 95 lpm), with 32 and 40 percent silicate solution and 25 to 12 minutes gel time. This grout take was considered satisfactory for commercial use.

The water takes prior to injection were 7.5 gpm (28.5 lpm) at 5 psi (35 kPa), 15 gpm (57 lpm) at 25 psi (175 kPa), and 20 gpm (76 lpm) at 60 psi (420 kPa). Very roughly, these numbers correspond to a permeability to water injection of the order of 4 to 6×10^{-4} cm/sec, or of the same order of magnitude as expected for the soil at the site.

Following the grout injection test, a boring was drilled to test the effectiveness of the grouting operation. The evidence of cemented soils was detected indirectly by the increased SPT values (greater than 100) and by odor. An attempt to obtain core samples of cemented soils was not successful, possibly due to the fragile nature of the silicate-grouted granular soil. Failure to obtain core samples has been the typical experience with soils injected with sodium silicate grouts, even though the grouted soils were proven by inspection or direct observation to be of a competent and coherent nature [2]. Where gravels are present, the task becomes virtually impossible with conventional coring equipment. The conventional core barrel is believed to have too large a side friction for the grouted sample to withstand the shearing and raveling forces that occur during sampling. The adjacent basements and the B&O Railroad tunnel were inspected and no evidence of grout invasion was observed.

GROUTING SPECIFICATIONS

To achieve the grouting program's objectives, it was very important to have a properly uniform distribution of grout --just a small coherence would greatly increase the soil's stability and stand-up time, both at the tunnel face and the tail void, resulting in reduced and uniform settlements. For this reason, the construction specifications were written to allow variable concentrations (and set-up times), using the maximum concentration compatible with the grout takes experienced. In areas of low grout take, the contractor was allowed to reduce the silicate concentration, or even switch to a lower viscosity grout.

CONSTRUCTION GROUTING

The chemical stabilization program included grouting from the street level and from within the railroad tunnel (Fig. 5). Approximately, 170 3-inch (76 mm) diameter grout holes were drilled from the street to 2 ft (0.6 m) into the fine grained residual soil beneath the tunnel. The surface holes were arranged in two staggered rows along each side of the tunnel with 5-ft (1.5 m) longitudinal and transverse spacings. The grout holes drilled from inside the tunnel were 2 inches (50 mm) in diameter, arranged in a "fan" fashion with 5-ft (1.5 m) longitudinal spacing along the railroad tunnel.

Approximately 354,000 gallons (1340 kl) of sodium silicate grout (GELOC-3) were pumped into the Cretaceous sands and gravels at pressures as high as 100 psi (690 kPa). During the initial stages of grouting, however, the high grouting pressures produced a 2-inch (50 mm) heave of the rail inside the tunnel. By reducing the pressure to 80 psi (550 kPa), no further heave was experienced. Grouting through a hole was terminated when the grout flow was reduced to less than 0.5 gallon (1.9 l) per minute [3].

Both the chemical stabilization measures and the

installation of the steel liner plates were completed
prior to passage of either of the transit tunnels. Ground
settlements and deformations were monitored along the
transit tunnels long before and at the intersection of
the B&O Railroad tunnel. Unfortunately, the geotechnical
instrumentation program did not monitor settlement of the
tunnel itself during that crossing. Settlement of the
street surface over the tunnel, however, was only 1/8
inch (3 mm), about 8% of that predicted. The lower-than-
anticipated settlement can be attributed to higher soil
heave during tunneling, and lower ground loss due to
increased soil coherence through chemical injection. No
structural distress in the tunnel liner or settlement of
the rails were visually observed in the B&O Railroad
tunnel during and after excavation of the transit
tunnels. Both the old and the new tunnels are functioning
properly.

CONCLUSIONS

1.   Chemical soil stabilization is a viable and cost-
      effective alternative to conventional underpinning
      for support of structures during tunneling. The
      ground loss at the tunnel face is reduced and the
      foundation   of   the   overlying   structure   is
      strengthened as a result of chemical injection.

2.   Cement grouting is not suitable for the Cretaceous
      granular soils at the site. A low-viscosity sodium
      silicate chemical grout is suitable and cost-
      effective.

3.   Chemical grouting of granular soils does not
      necessarily produce high strength due to the
      relatively weak nature of the grout gel--the coarser
      the soil, the more dominant are the characteristics
      of the pure gel. A 50 to 100 percent strength
      increase, however, was experienced with increasing
      the silicate concentration from 40 to 60 percent.

4.   An increase in soil coherence due to grouting
      greatly reduces the ground loss during tunneling and
      its associated settlement. The long-term settlement
      of the soil is also influenced by the initial ground
      heave during tunneling.

5.   To minimize ground loss and differential settlement
      along the tunnel alignment, it is important to have
      a properly uniform distribution of grout. The
      contractor should be allowed to adjust the silicate
      concentration, grout pressure, set-up time and
      viscosity, as needed, to achieve that goal.

6.   The hollow stem method of grout pipe placement is
      not recommended due to possible excessive smear
      effect from the auger method of drilling.

7. It is extremely difficult to obtain core samples from granular soils injected with sodium silicate grout, using conventional coring equipment, due to the fragile nature of the grouted soils. The effectiveness of the grouting operation can be tested by gradual reduction in grout take during injection and increased SPT values afterwards.

8. A pre-construction field injection test is a useful tool for determining the effectiveness of the chemical grout and establishing appropriate grouting specifications.

ACKNOWLEDGEMENT

Parsons Brinckerhoff Quade & Douglas, Inc. designed the Lexington Market Tunnels and the B&O Railroad protection scheme for the Maryland Department of Transportation. The grouting program was implemented by the Hayward Baker Company as subcontractor to the joint venture of Traylor Brothers, Morrison Knudson and Grow Tunneling. The laboratory grouting tests were performed by the Halliburton Services.

REFERENCES

[1] Schmidt, B., "Prediction of Settlements Due to Tunneling in Soil: Three Case Histories," Rapid Excavation and Tunneling Conference, Vol. 2., 1974.
[2] Baker, W.H., "Report of Field Grout Injection Tests for the MTA Lexington Market Subway Tunnels at B&O Railway Tunnel Undercrossing," Submitted to Parsons Brinckerhoff Quade & Douglas, Inc., 1976
[3] Ziegler, E.J., and Wirth, J.L., "Soil Stabilization by Grouting on Baltimore Subway," ASCE Specialty Conference on Grouting in Geotechnical Engineering, New Orleans, 1982

Raymond J. Castelli

VIBRATORY DEEP COMPACTION OF UNDERWATER FILL

REFERENCE: Castelli, R. J., "Vibratory Deep
Compaction of Underwater Fill," _Deep Foundation
Improvements: Design, Construction, and Testing_,
ASTM STP 1089, Melvin I. Esrig and Robert C. Bachus,
Eds., American Society for Testing and Materials,
Philadelphia, 1991.

ABSTRACT: A case history is presented describing
the use of vibratory deep compaction of
underwater fill placed behind an anchored steel
sheet pile bulkhead. The project included
extensive testing to evaluate a variety of
compaction equipment, procedures and backfill
materials. Quality control tests, including
static cone penetration tests, standard
penetration tests, and settlement measurements
were performed to verify compliance with
specified compaction criteria and to assess the
impact of deep compaction on the bulkhead
structure. The successful completion of this
project illustrates the effectiveness of
vibratory deep compaction for marine
applications.

KEY WORDS: bulkheads, deep compaction, field
tests, underwater fill, vibroflotation

INTRODUCTION

Vibratory deep compaction is a method now frequently
used for densification of loose granular soils to
effectively increase foundation bearing capacity, reduce
ground settlement, and improve soils susceptible to
liquefaction. One application of deep compaction is in
marine construction where frequently it is necessary to
place a significant depth of loose, underwater fill. For
such projects, gradual settlement of the fill can lead to
increased maintenance cost for repair of pavement and

Mr. Castelli is a Supervising Geotechnical Engineer at
Parsons Brinckerhoff Quade & Douglas, Inc., One Penn
Plaza, New York, N.Y. 10119

surface structures. As illustrated in the following case history, vibratory deep compaction is an ideal method which can be used to avoid these problems.

PROJECT DESCRIPTION

The Port of Kismayo is located 45 km (28 miles) below the Equator on the coast of the East African country of Somalia. The port facilities at Kismayo included a four berth wharf with a total length of 631 m (2070 ft). Originally constructed in the mid-1960's, the wharf consisted of a 18.3 m (60 ft) wide precast concrete platform supported by precast prestressed concrete piles.

Soon after completion of the structure, serious deterioration was observed. As a result, a complete rehabilitation of the port was implemented from 1986 to 1988. The deterioration of the original wharf structure, and the design of the replacement structure were described by Castelli and Secker [1].

As shown in Figure 1, the replacement structure included a 13.8 m (45.3 ft) high anchored sheet pile bulkhead located outboard of the existing wharf. To build this scheme the existing wharf platform was demolished and the tops of the existing concrete piles were cut off. A steel sheet pile bulkhead was then installed 10.7 m (35 ft) outboard of the existing platform, and supported by a continuous steel sheet pile deadman located near the back of the platform. Behind the bulkhead a granular backfill was placed underwater, then compacted from the surface using deep vibratory compaction. Deep compaction was used to minimize settlement of the backfill, particularly differential settlement in the vicinity of the cut-off piles where wide variability in the density of the loosely placed backfill was anticipated. The use of deep compaction would, therefore, minimize the need of future maintenance of the wharf's rigid concrete pavement, and avoid possible disruption to the surface drainage system. Deep compaction was also used to increase the passive soil resistance for support of the deadman anchorage.

FILL MATERIAL

Two types of underwater fill material were specified, including sand fill behind the bulkhead, and a select fill at the deadman for increased passive soil resistance. The sand fill used in construction consisted of a uniformly graded, medium to fine sized beach sand. The minimum and maximum dry densities of the sand fill were 1.62 and 1.73 g/cm^3 (101 and 108 pcf), respectively, when tested in accordance with ASTM D 2049-69. The select fill was a sand-gravel mix with about 30 percent gravel size. The minimum and maximum dry densities of the select fill were

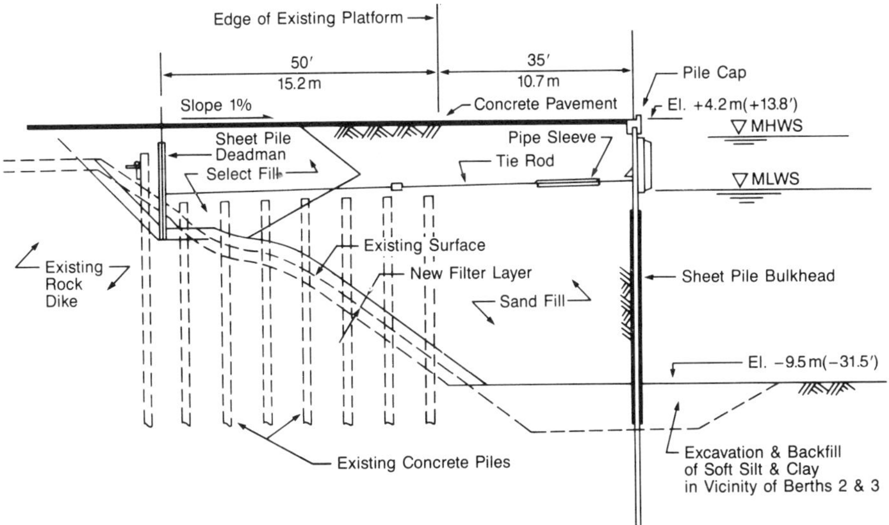

Fig. 1—Typical section of anchored sheet pile bulkhead

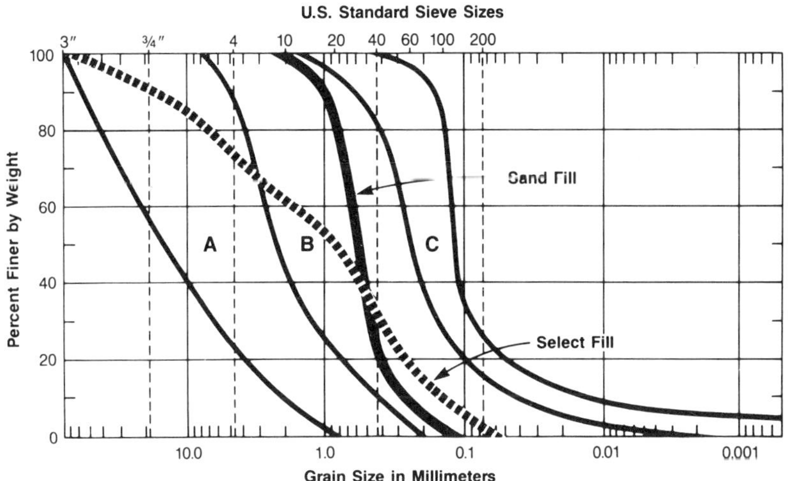

Zone Suitability (Brown, 1977)

A   Penetration very slow; vibratory compaction may be uneconomical
B   Soils best suited for vibratory compaction
C   Soils very difficult to compact

Fig. 2—Gradation curves for fill materials

1.66 and 1.94 g/cm^3 (104 and 121 pcf). Grain size distributions for both the sand fill and select fill are presented in Figure 2. Also shown, for comparison, are the ranges suggested by Brown [2] identifying the suitability of in situ soils for compaction by vibroflotation. The maximum height of the sand fill above the sea bottom was about 14 m (46 ft). The select fill had a depth of about 5.5 m (18 ft).

A variety of materials were evaluated for backfilling the probe hole during withdrawal of the vibratory probe, including sand fill and select fill, as described above, and a 25 mm (1.0 inch) size gravel. The suitability of these materials for backfilling the probe holes was initially assessed using criteria suggested by Brown [2] in which the suitability of a particular backfill material is related to the rate at which the material settles downward through the probe hole. A "suitability number" was defined by Brown based on the grain size distribution of the material. Using Brown's definition, a suitability number less than 10 indicates an "excellent" backfill material, while a value between 10 and 20 indicates a "good" backfill material. The gravel, sand fill and select fill had suitability numbers of 0.3, 8, and 15, respectively.

COMPACTION CRITERIA

The construction specifications required that the backfill material placed below water be compacted using deep vibratory compaction methods to achieve a minimum relative density of 80 percent as determined by in situ testing and the specified correlations described below. The contractor was given the freedom to select the appropriate compaction equipment and procedures, subject to the approval of the engineer and verification by an initial field testing program. In situ quality control tests performed during the initial field testing program (and periodically during production compaction operations) included both static cone penetration tests (CPT) and standard penetration tests (SPT). The CPT soundings were performed using an electric cone penetrometer with a diameter of 35.5 mm (1.40 inches), a friction sleeve length of 133 mm (5.24 inches), and a 60° cone tip. The relative density of the fill was determined from CPT cone resistance values using correlations suggested by Schmertmann [3]. (It should be noted, however, that newer correlations [4,5] between relative density and CPT cone resistance are now available which may provide improved results.) Relative densities were determined from SPT N-values using correlations suggested by Gibbs and Holtz [6]. The specifications required all SPT and CPT tests be performed at the center of four compaction probe holes.

The above correlations were well suited for the sand fill, but were not considered reliable for the select fill due to the presence of a significant amount of gravel size particles. In the select fill the above correlations were used only as an index of the densification achieved, and as a contractual target for acceptance of the work.

To prevent permanent sag of the bulkhead tie-rods, the deep compaction operations in the sand fill area were required to be performed in two stages. The first stage compaction, performed when the fill reached the tie-rod level, extended to the bottom of the fill. After completion of Stage I compaction and correcting any resulting sag of the tie-rods, the remaining fill was placed and Stage II compaction performed only to the tie-rod level.

The spacing of the compaction probes was determined from the initial field tests. However, to avoid any interference with the tie-rods, the probe spacing parallel to the face of the bulkhead was set equal to the tie-rod spacing of 2.0 m (6.6 ft).

FIELD TESTING OF SELECT FILL

Test Procedures

A total of eleven test trials were conducted to evaluate various types of compaction equipment, compaction procedures, and materials for backfilling the probe holes. The conditions for each of these test trials are summarized in Table 1.

The various types of compaction probes included:

o   260 x 290 mm (10.2 x 11.4 inch) steel I-beam, approximately 6 m (19.7 ft) long.

o   Winged probe consisting of a 170 mm (6.7 inch) diameter, 6 m (19.7 ft) long pipe with 310 mm (12.2 inch) long wings welded on in pairs at 500 mm (19.7 inch) intervals along the pipe.

o   750 mm (30 inch) diameter steel pipe with 15 mm (0.59 inch) wall thickness and a length of about 8 m (26 ft).

o   Ferro-Konstruckt RF 3000/5G hydraulically operated vibroflot.

o   GKN (Keller) A-Vibrator electrically operated vibroflot.

TABLE 1 -- Summary of initial test trials in Select Fill area

Trial No.	Probing Pattern (m)	Number of Probes	Compaction Probe	Jetting Agent	Backfill Material	Average Probe Penetration (m)	Remarks
1	1.5 x 2.0	6	I-Beam	None	None	3.0	(a)
2	2.0 x 2.0	6	Winged Probe	None	None	4.0	(a)
3	2.0 x 2.0	10	Winged Probe	Air	Select Fill	4.0	(a)
4	2.0 x 2.0	9	750 mm diameter Pipe	None	None	4.0	(a)
5	2.0 x 2.0	6	Single Ferro-Konstruckt	Air	Select Fill	4.0	(a)
6A	2.0 x 2.0	6	Single Ferro-Konstruckt	Water	Gravel	4.5	(b)
6B	2.0 x 2.0	4	Single Ferro-Konstruckt	Water	Sand	4.5	(a)
7	2.0 x 2.0	6	Single Ferro-Konstruckt	Air	None	4.5	(a)
8A	2.0 x 2.0	6	Single Ferro-Konstruckt	Water	Sand	4.5	(a)
8B	2.0 x 2.0	9	Single Ferro-Konstruckt	Water	Select Fill	4.5	(c)
9	2.0 x 2.1	(d)	Twin GKN	Water	Select Fill	4.2	(b)

Remarks:

(a) The test trial failed to obtain the specified compaction for the full depth of fill.

(b) Required compaction was obtained for the full depth of fill.

(c) Insufficient compaction at bottom of fill, but test accepted pending modification to procedures.

(d) Test Trial No. 9 was performed within area of production probings.

The I-beam, winged probe and pipe probe were advanced into the ground and vibrated on withdrawal using a 150 KW (200 HP) ICE 416 vibratory hammer which has an eccentric moment of 203 kN-mm (1800 inch-lb) and a maximum frequency of 1500 vpm. The 66 KW (90 HP) Ferro-Konstruckt vibroflot has a maximum operating speed of 3000 vpm with a corresponding maximum centrifugal force of 400 kN (88,180 lbs). The 50 KW (67 HP) GKN vibroflot has an operating speed of 2000 rpm with a corresponding centrifugal force of 160 kN (36,000 lbs). During compaction operations, the equipment operator monitored either the hydraulic pressure of the Ferro-Konstruckt vibroflot or the electrical resistance of the vibratory hammer and the GKN vibroflot to assess the degree of compaction obtained.

Deep compaction in the select fill area was performed using a mobile crane operating on the land side of the wharf. The compaction probes were advanced to the required depth by vibrating the probes with or without the aid of jetting. After reaching the required depth, the vibrator speed was increased, any jetting used was reduced, and the probe was then withdrawn in short lifts. Each lift involved raising the probe 1.0 m (3.3 ft) then lowering the probe 0.5 m (1.6 ft) and holding it in place for a duration of about half a minute.

In seven of the test trials in select fill, backfill material was added to the probe hole to compensate for the decrease in volume resulting from densification. As shown in Table 1, the various backfill materials included gravel, select fill and sand fill. For the remaining four tests, however, the probes were withdrawn without adding backfill material.

The compaction probe holes for the test trials in select fill were generally in a 2.0 x 2.0 m (6.6 x 6.6 ft) square pattern. However, in Test No. 1, with the I-beam compaction probe, the probe spacing was 1.5 x 2.0 m (4.9 x 6.6 ft). Generally, at least six compaction probings were performed for each test trial. A typical arrangement of compaction probe holes is shown in Figure 3.

Test Results

The I-beam, winged probe and pipe probe were unsuccessful in compacting the full depth of the select fill to the required density. Figure 4 shows CPT and SPT test results obtained from Test Trial No. 2, illustrating the deficiency of the winged probe in meeting the compaction criteria.

The poor performance of the I-beam, winged probe and pipe probe is attributed to (a) the lower compacting efficiency provided by their vertical vibrating motion, and (b) the absence of any significant water jetting

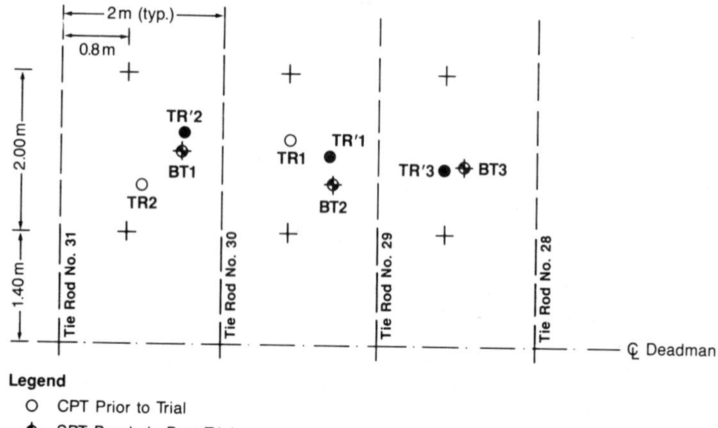

Fig. 3—Plan of test trial no. 2 in Select Fill area

capability. These equipment characteristics made it difficult to compact the top 2 to 4 m (6.6 to 13 ft) of fill which was above water level. For these conditions, the horizontal oscillations and water jetting capability of the GKN and Ferro-Konstruckt vibroflots proved to be more successful.

The most successful results were obtained from (a) Test Trial No. 9 using the twin GKN vibroflots with water jetting and select backfill material, and (b) Test Trial No. 6A using the Ferro-Konstruckt vibroflot with water jetting and gravel backfill material. As shown in Figure 5, both of these tests achieved the required densification for the full depth of fill. Based on these test results, the above compaction equipment and procedures were approved for production compaction operations in select fill areas.

In comparing the three different backfill materials added to the probe holes, the gravel and select fill provided better compaction than the sand fill. The observed results with gravel backfill are consistent with the "excellent" suitability rating defined for this coarse sized material. However, the select fill performed better than the sand fill, though it had a lower suitability rating. The better performance of the select fill is believed to be related to the stepped gradation of this material, and the presence of a large percentage of gravel

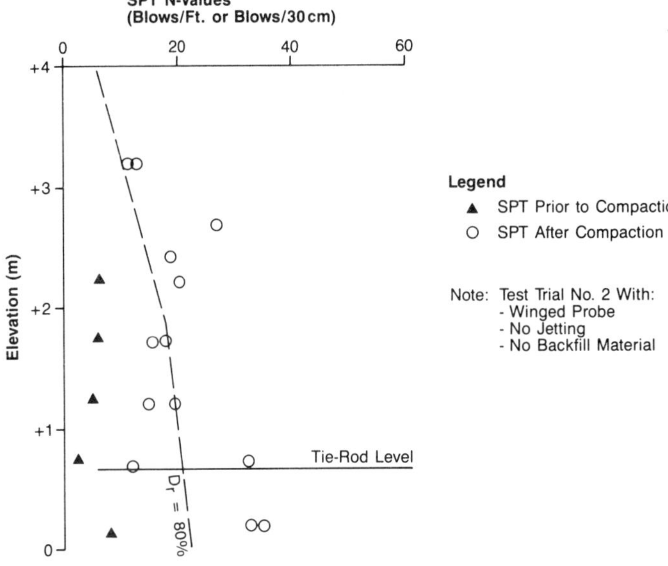

a) SPT Test Results

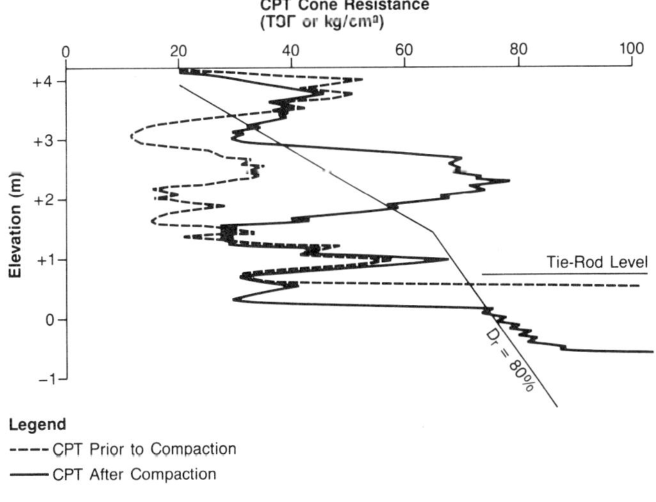

b) CPT Test Results

Fig. 4—SPT and CPT results for test trial no. 2 in Select Fill

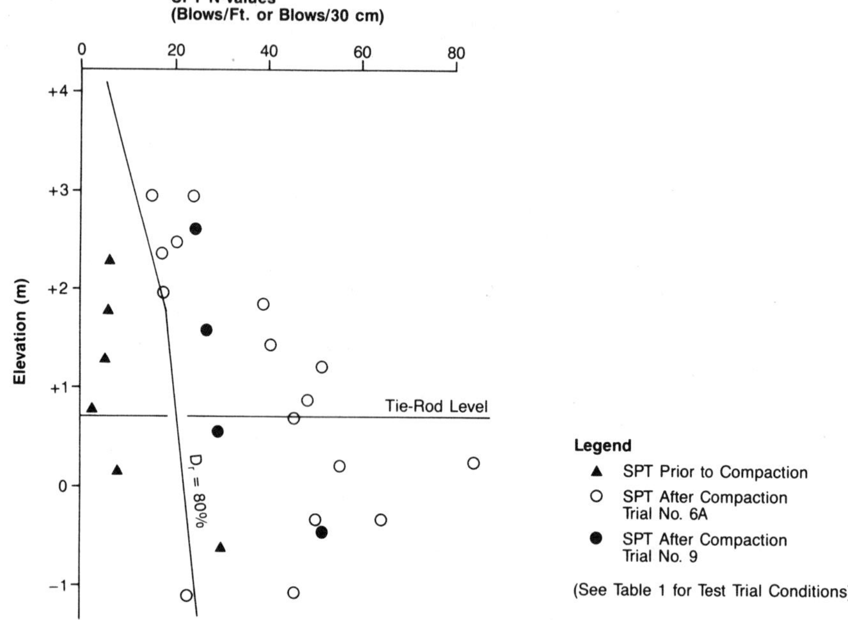

Fig. 5—SPT results for test trials no. 6A and 9 in Select Fill

size particles (Figure 2). Since the backfill suitability rating system suggested by Brown [2] accounts only for material at the D_{50} size and smaller, a large gravel fraction would not influence the calculated suitability number, but intuitively should improve the performance of the material as backfill. Based on the Kismayo Port test program, it is concluded that the backfill suitability rating system suggested by Brown is not applicable to step-graded materials. Further research is necessary to establish criteria for such materials.

FIELD TESTING OF SAND FILL

Five initial test trials were performed to assess Stage I compaction of sand fill near the wharf bulkhead. The compaction equipment and compaction procedures used for these test trials are summarized in Table 2. Only the Ferro-Konstruckt vibroflot and GKN vibroflot were used for these test trials. Deep compaction in the sand fill area was performed using a barge mounted crawler crane positioned outboard of the bulkhead.

Test Trial No. 1 was performed in three parts using the Ferro-Konstruckt vibroflot. The first part of the test, Trial No. 1A, was abandoned when CPT soundings

TABLE 2 -- Summary of initial test trials in Sand Fill area

Trial No	Probing Pattern	Number of Probes	Compaction Probe	Jetting Agent	Backfill Material	Average Probe Penetration (m)	Remarks
1A	2.0 x 2.0	10	Twin Ferro-Konstruckt	Water	None	--	(a)
1B	2.0 x 2.0	12	Twin Ferro-Konstruckt	Water	None	6.1	(b)
1C	2.0 x 2.0	9	Single Ferro-Konstruckt	Water	None	7.6	(b)
2	2.0 x 2.0	9	Single GKN	Water	None	11.7	(c)
3	2.0 x 2.1	12	Twin GKN	Water	None	12.0	(c)

Remarks: (a) Test abandoned due to presence of select fill.
(b) Probes unable to penetrate full depth of fill.
(c) Required compaction obtained for full depth of fill.

encountered select fill at the test area. Part B of the test was performed at a second location using two vibroflots simultaneously, with the vibroflots rigidly attached together and spaced 2.0 m (6.6 ft) apart. In Part C, a single vibroflot was used. A probe hole spacing of 2.0 x 2.0 m (6.6 x 6.6 ft) was used for all three parts of the test. No material was fed into the probe holes during vibrocompaction operations at these three test areas.

For both Test Trials 1B and 1C, the vibroflots failed to penetrate a dense layer within the sand fill. Although the fill extended to a depth of about 12.5 m (41 ft), the twin probes of Trial 1B penetrated only to depths of 5.0 to 7.0 m (16 to 23 ft). The single probe of Trial 1C succeeded in penetrating only to depths of 4.8 to 9.0 m (16 to 30 ft). CPT tests performed after compaction indicated relative densities exceeding the specified minimum value for the depth penetrated by the vibroflot. However, the bottom untreated portion of the fill had a relative density of only 50 to 60 percent as determined from the Schmertmann correlation [3]. The inability of the Ferro-Konstruckt probe to penetrate the required depth was believed to be caused by an inadequate water jet at the tip of the probe. Due to the limited success of the Ferro-Konstruckt vibroflot, it was not permitted for compaction in sand fill areas.

Test Trial No. 2 was performed using a single GKN vibroflot with compaction probe holes spaced 2.0 x 2.0 m (6.6 x 6.6 ft) in a square pattern. Test Trial No. 3 was performed using two GKN vibroflots rigidly attached together to provide a probe hole spacing of 2.1 x 2.0 m (6.9 x 6.6 ft). No material was fed into the probe holes for either of these test trials.

Both the single and twin GKN vibroflots successfully penetrated the full depth of sand fill, and both successfully compacted the fill to the specified minimum relative density of 80 percent as determined from the Schmertmann correlation [3]. CPT data from each of these test trials are presented in Figure 6, along with a typical CPT plot obtained prior to vibrocompaction. The post-compaction CPT data shown in Figure 6 are for tests performed at the centroid of four probe locations. Due to the high densification achieved, these CPT soundings were unable to penetrate more than about 6.5 m (21 ft) of the approximately 12 m (39 ft) depth of treatment. To facilitate quality control testing, the engineer permitted the contractor to perform CPT testing 1.1 m (3.6 ft) beyond the outer row of compaction probe holes. Tests performed at these locations, however, were still required to meet the specified minimum relative density criteria. As shown in Figure 7, this revised test procedure provided acceptable results while eliminating the problem of penetrating the densely compacted fill. Based on these

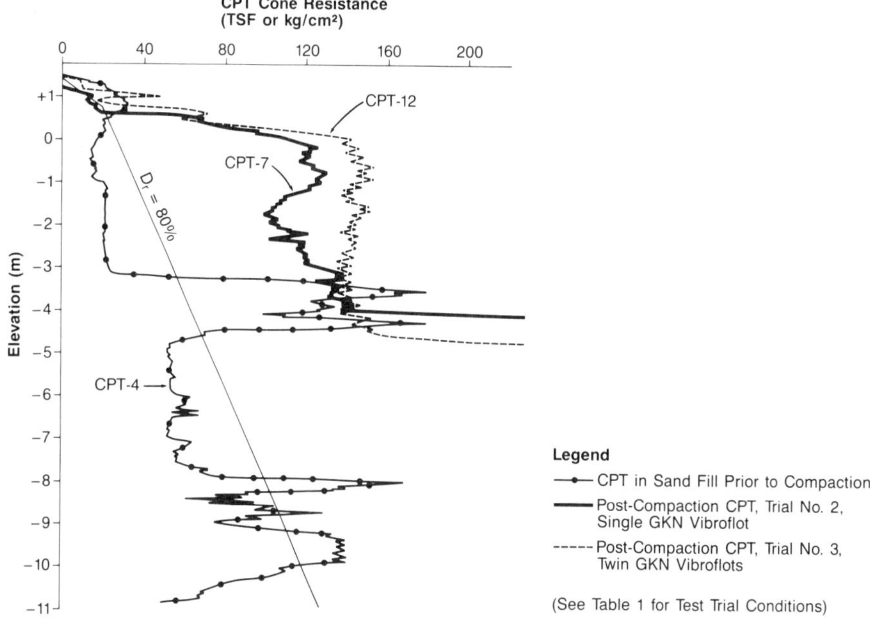

Fig. 6—CPT results for test trial no. 2 and 3 in Sand Fill

results, both the single and twin GKN vibroflots were approved for use in sand fill areas.

Volume Reduction of Sand Fill

In addition to CPT testing, the performance of Stage I deep compaction in the sand fill area was also assessed by determining the volume reduction of the fill. Since no backfill material was added to the probe holes during compaction operations, it was possible to estimate the volume reduction from ground surface settlement measurements. Settlements were determined by optical survey of the ground surface for a grid of points within the test area.

Table 3 presents a summary of measured ground settlements and estimated volume reductions (in percent) at the sand fill test areas. The twin GKN vibroflots, with twice the energy of the single GKN vibroflot, resulted in approximately 44 percent greater volume reduction of the fill. These results, as well as the CPT plots of Figure 6, illustrate the greater densification that can be achieved using twin vibroflots. In addition, the use of twin vibroflots can cut the time for compaction work almost in half.

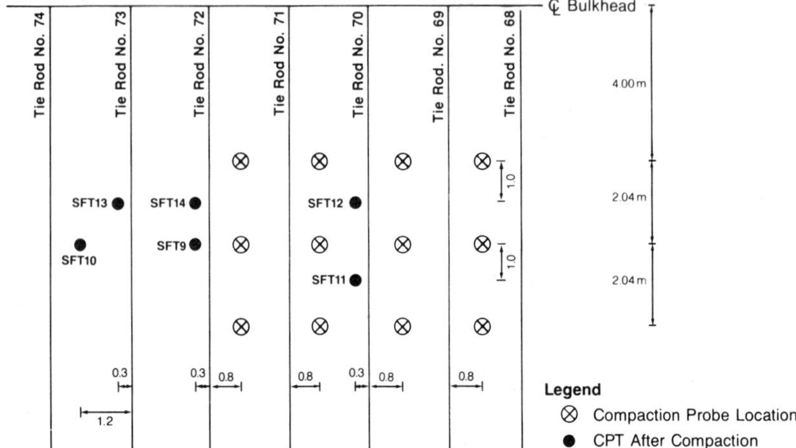

a) Plan of Test Area

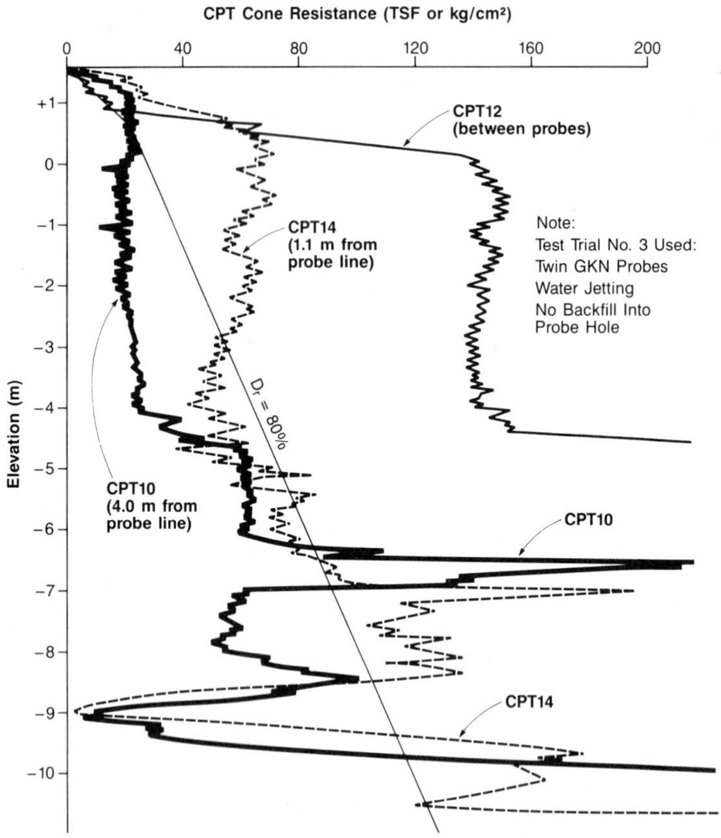

b) CPT Test Results

Fig. 7—Test trial no. 3 in Sand Fill

TABLE 3 — Percent volume reduction of Sand Fill

Trial No.	Equipment	Average Probe Penetration (m)	Average Penetration Time (min./m)	Average Compaction Time (min./m)	Average Settlement (m)	Volume Reduction (Percent)
1B	Twin Ferro-Konstruckt	6.1	0.30	0.77	0.63	10.3
1C	Single Ferro-Konstruckt	7.6	1.12	0.80	0.86	11.3
2	Single GKN	11.7	0.29	1.38	0.88	7.5
3	Twin GKN	12.0	0.23	0.76	1.30	10.8

Table 3 shows that the single and twin Ferro-Konstruckt vibroflots both obtained a percent volume reduction approximately the same as the twin GKN vibroflots. However, the high volume reduction obtained by the single Ferro-Konstruckt vibroflot in Test No. 1C is attributed to the significantly longer penetration time used for this test in an unsuccessful attempt to achieve a greater depth of penetration. Although the twin Ferro-Konstruckt vibroflots in Test No. 1B obtained a percent volume reduction comparable to the GKN vibroflots, this test failed because of the inability of the vibroflots to fully penetrate the sand fill.

Influence of Compaction on Bulkhead

Concern was raised during design that deep compaction of the backfill may result in a significant increase in the lateral earth pressures acting on the bulkhead. Rather than increase the strength of the bulkhead and its anchorage to resist the additional pressures, the construction specifications stipulated that no compaction be done immediately behind the bulkhead for a distance to be determined from field testing. The determination of this distance was made using CPT tests.

Figure 7 (a) presents a plan showing the location of compaction probes and CPT soundings for Test Trial No. 3 in the sand fill area. For this test, the CPT soundings were performed within the area of treatment, and at distances of 1.1, 3.1 and 4.0 m (3.6, 10.2, and 13 ft) beyond the outer row of compaction probes. A summary plot of CPT cone resistance versus depth and distance from the treated area is presented in Figure 7 (b).

Except for a dense zone from El. -3.0 to -4.5 m (-10 to -15 ft) in the pre-compaction CPT, there is little difference between the cone resistance values obtained prior to compaction (Figure 6) and the post-compaction values shown in Figure 7 (b) for CPT-10 located 4.0 m (13 ft) beyond the last row of compaction probes. Using CPT cone resistance values as an index of in situ strain and stress conditions, it was concluded from these results that beyond a distance of about 4.0 m (13 ft) there was little or no increase in lateral ground stresses developed as a result of vibratory deep compaction operations. These results, and similar findings from other test trials, were the basis for locating the first row of compaction probes at a distance of 4 m (13 ft) from the centerline of the bulkhead.

For future projects utilizing vibratory deep compaction adjacent to steel sheet pile bulkheads consideration should be given to the use of strain gages and inclinometer casing attached to the sheet piling to

determine stresses and deflections of the bulkhead with
depth, and from these data to estimate the corresponding
lateral earth pressures.

CONCLUSIONS

The Port of Kismayo project illustrates the successful
use of vibratory deep compaction to densify loose,
underwater fill behind a marine bulkhead. This technique
resulted in significant compaction of the fill without
impacting the bulkhead or its anchorage system. As a
result of this work, it is anticipated that post-
construction settlement and related maintenance costs will
be substantially reduced.

The successful use of vibratory deep compaction at the
Port of Kismayo suggests that it is a practical
construction technique which may be appropriate for a
broad range of marine applications. Accordingly, the use
of vibratory deep compaction should be routinely
considered in the design of all marine projects involving
placement of underwater fill.

A quality control program is an essential component of
the vibratory deep compaction method. As part of this
program, in situ testing must be conducted at the start of
compaction operations to verify that the selected
compaction equipment, procedures and backfill materials
will achieve the required densification. The Port of
Kismayo project presents an example of the standard types
of tests generally used for such a quality control
program.

Specific findings of the Port of Kismayo project are:

o   A wide variety of compaction equipment, compaction
     procedures and backfill material are available for
     vibratory deep compaction.

o   Specialized equipment, particularly the high
     horsepower, horizontally oscillating vibroflots now
     available, provide greater densification than probes
     powered by vibratory hammers.

o   Gravel and other coarse sized material performed well
     as backfill for compaction probe holes. However, new
     criteria need to be established to rate the
     suitability of step-graded material for backfill.

o   Compaction of clean, uniform sized underwater fill can
     be successfully performed without adding fill material
     to the probe hole.

o   Twin vibroflots operated simultaneously provide greater soil densification and significantly shorten the duration of deep compaction work.

Acknowledgement

Funding for the Port of Kismayo Rehabilitation was provided by the U.S. Agency for International Development. Parsons Brinckerhoff was the design engineer and provided field services during construction. The George A. Fuller Company was the contractor for the project and performed the vibratory deep compaction work reported herein.

This paper is dedicated to the memory of Warren M. Buser, Parsons Brinckerhoff's project manager for the Kismayo Port Project, who was a victim of the crash of Pan Am Flight 103 on December 21, 1988.

References

[1]   Castelli, R.J. and Secker, N., "Port of Kismayo (Somalia) Rehabilitation," American Society of Civil Engineers, Specialty Conference, Ports '86, Oakland, California, 1986.

[2]   Brown, R.E., "Vibroflotation Compaction of Cohesionless Soils," _Journal of the Geotechnical Engineering Division_, American Society of Civil Engineers, Vol. 103, No. GT12, 1977, pp 1437-1451.

[3]   Schmertmann, J.H., "Measurement of In-Situ Shear Strength," State-of-the-Art Paper, American Society of Civil Engineers, Specialty Conference on In-situ Measurement of Soil Properties, Vol II, Raleigh, N.C., 1975.

[4]   Robertson, P.K. and Campanella, R.G., "Interpretation of Cone Penetration Tests. Part I: Sand," _Canadian Geotechnical Journal_, Vol. 20, No. 4, Nov. 1983, pp. 718-733.

[5]   Villet, W.C. and Mitchell, J.K., "Cone Resistance, Relative Density and Friction Angle," American Society of Civil Engineers, Symposium on Cone Penetration Testing and Experience, St. Louis, 1981, pp. 178-208.

[6]   Gibbs, H.J. and Holtz, W.G., "Research on Determining the Density of Sands by Spoon Penetration Testing," Fourth International Conference on Soil Mechanics and Foundation Engineering, Vol. 1, London, England, 1957.

K. Rainer Massarsch

DEEP SOIL COMPACTION USING VIBRATORY PROBES

REFERENCE: Massarsch, K. R., "Deep Soil Compaction
Using Vibratory Probes," Deep Foundation Improvements: Design, Construction, and Testing, ASTM STP
1089, Melvin I. Esrig and Robert C. Bachus, Eds.,
American Society for Testing and Materials, Philadelphia,
1991.

ABSTRACT: A new approach to deep compaction of
granular soils is presented, which makes it possible
to take into account the site-specific conditions in
the design of the compaction process. Special vibratory probes have been developed, the shape and dynamic properties of which are chosen to achieve optimal transfer of the compaction energy to the soil.
The operating frequency of the vibrator, which is
attached to the top of the probe, can be varied to
achieve optimal soil densification. Results from
extensive field measurements from compaction projects are presented. Settlements after compaction
range between 5 and 10 % of layer thickness. An
increase of penetration resistance of between 50 and
300 % has been observed. A rational design concept
for the resonant compaction technique is presented.

KEYWORDS: compaction, sands, vibrations, resonance,
wave velocity, penetration tests, compressibility,
settlements, earthquakes, liquefaction, permeability

Various reasons can exist for deep compaction of
cohesionless soils. i.e. to increase the shear strength,
to reduce the compressibility, to decrease the permeability, or to modify the dynamic properties of the soil
deposit. The optimal solution for a specific project will
be influenced by a number of factors, such as logistic
aspects, the size and location of the project, the geotechnical and geohydrological conditions, the required
improvement of the soil deposit and the time available
for execution of the project.

Dr. K. Rainer Massarsch, GEO ENGINEERING SA, 26,
avenue des Petits Champs, B-1410 WATERLOO, Belgium.

Soil compaction can be achieved by different methods, such as impact loading (blasting, falling weight, pile driving), vibratory action (vibroflotation, vibratory probes, vibratory rollers), soil reinforcement by stone columns, piles, or by static preloading, grouting and injection (infiltration and/or soil displacement using hardening liquids), etc. Each of these methods has its particular advantages and limitations.

The present paper describes a new concept for deep compaction of granular soils using vibratory probes, which takes advantage of the amplified ground response, which occurs when a soil layer is excited at its resonant frequency. By this approach, it is possible to adopt the compaction process to the site-specific geotechnical and geodynamic conditions. The concept has several advantage over conventional vibratory compaction, both with respect to compaction effect and efficiency of project execution.

VIBRATORY COMPACTION

Compaction Equipment

Vibratory compaction uses a specially designed steel probe, to the top of which is clamped a heavy vibrator, which can generate either vertical or torsional oscillations. The soil is compacted as a result of repeated insertion and withdrawal of the probe. The Terraprobe was developed in North America and employs a vibro-piledriver, attached to the top of a 76 cm diameter open tube (Fig. 1a). The Japanese Vibro-rod system, which is similar, uses a steel rod, which is provided with short ribs (Fig. 1b). The vibratory compaction concept was further developed in Belgium and in Sweden, respectively [1].

The star-shaped Franki Y-Probe (also referred to as Tri Star), consists of three 0.5 meter wide steel blades, which are welded together at an angle of 120 degrees (Fig. 1c). Small horizontal ribs are attached to the blades in order to increase the friction between the probe and the soil. The probe can be up to 25 m long.

The Swedish Vibro Wing system utilizes an about 15 m long steel rod, which has approximately 0.8 meter long wings, spaced about 0.5 m apart (Fig. 2). When probe penetration is slow, it is possible to reduce the friction resistance by jetting. The probe can also be provided with drainage tubes to facilitate dissipation of the excess pore water pressure which is generated during the compaction process. This effect can be of special significance in soil deposits with restricted drainage, e.g. if a sand deposit contains horizontal layers of silt or clay, or when compaction is to be carried out inside watertight caissons [2]. The compaction probe can either be suspended from a crane (Y-probe) or guided in the mast of a piling rig (Vibro Wing). While the crane has advantages from an operational

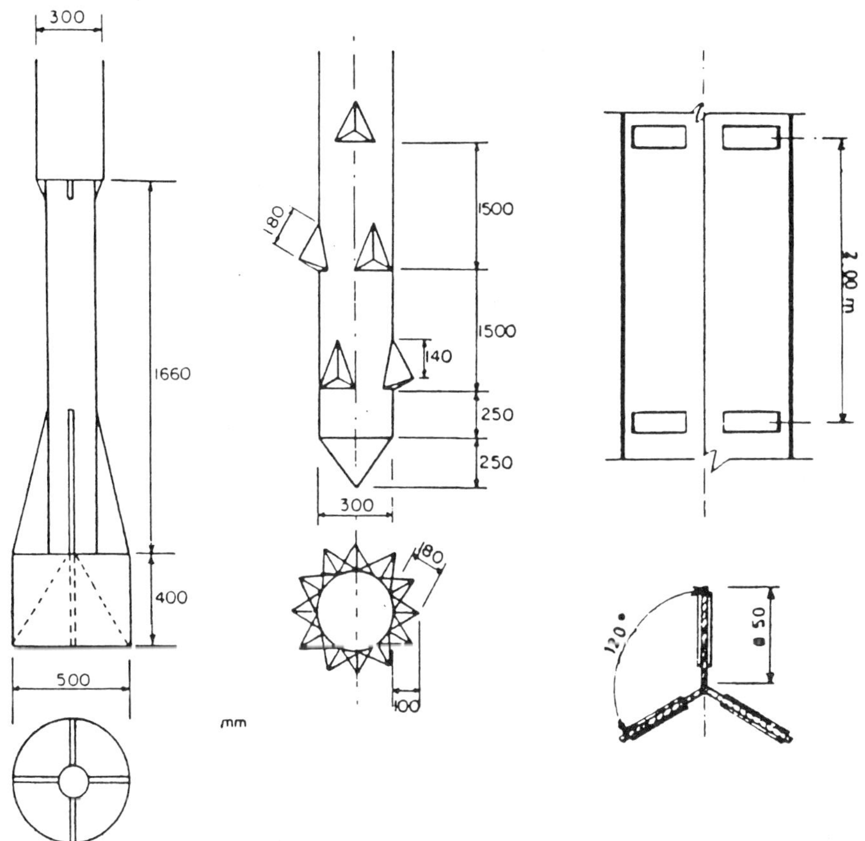

FIG. 1 -- FOSTER probe (a), Japanese Vibro-rod (b) and
          Belgian FRANKI Y-Probe (c)

viewpoint, the piling rig offers better control of verticality and minimizes eccentric forces on the probe.

Compaction Process

Three parameters are of importance for the design of
the compaction project: the spacing between the compaction
points, the duration of compaction in each point, and the
mode of probe operation (insertion, suspension and withdrawal). A more rational design concept for vibratory
compaction is needed, as the empirically chosen parameters
have great economic impact on the project. In addition,
factors such as initial density of the soil, depth of the

FIG. 2 -- Swedish VIBRO WING system

soil layer to be compacted or degree of soil compaction to be achieved, must be taken into consideration. The grid spacing ranges typically between 1.5 and 4.5 meters, and the duration of vibration in each compaction point varies generally between 5 and 35 minutes.

In loose, saturated sands, initial densification occurs mainly as a result of the sudden increase of pore water pressure ("liquefaction") in a zone adjacent to the vibrating probe. The compaction effect is largest when the overburden pressure is high, and the efficiency of compaction increases thus with depth. This fact distinguishes vibratory probes from other compaction methods. However, only little compaction is normally achieved in the zone close to the ground surface, especially in the case of partially saturated soils above the ground water table.

The soil displacement effect during probe insertion contributes also significantly to soil compaction, which results in an increase of vertical and lateral stresses. Also the number of penetration cycles at different depths, and the probe movement are important. The probe is normally inserted to full depth and thereafter withdrawn in steps. This procedure is repeated until the required compaction effect has been achieved.

The optimal spacing between compaction points depends on the shape and size of the compaction probe. It is preferable to use a narrow grid spacing, and to reduce instead the duration of compaction, which will result in more homogeneous soil densification.

Another important factor is the sequence in which compaction is performed. It is advantageous to execute compaction in two passes, working at first at a coarser grid spacing. The vertical insertion of the compaction probe will be facilitated when the grid spacing is larger. After the first compaction pass, the soil deposit should be given sufficient time for reconsolidation, before the second pass is started. Experience has shown that during the second pass, compaction time is often significantly shorter and the densified soil columns will guide the probe into the loose zone, yet to be compacted.

The geotechnical and geohydrological (ground water) conditions, soil deposition and stratification are of importance. Vibratory compaction should be limited to granular soils, which are free-draining. It is normally safe to follow the recommendation by Mitchell [3], who has proposed a boundary for grain size distributions curves, which identify soils, suitable for vibrocompaction, (Fig. 3).

Even relatively thin layers of silt and clay in a sand deposit can negatively affect the densification process. It is therefore recommended to base the compaction design on detailed geotechnical investigations, including electric cone penetration tests with friction sleeve measurements and/or pore pressure soundings ("piezocone"), which can better detect soil stratification.

Good compaction results can generally be expected, when the friction ratio from electric cone penetration tests (local sleeve friction as percentage of point resistance) is below 0.8 %. When the friction ratio exceeds 1.5 %, then vibratory compaction is usually not efficient.

vibratory probes from other compaction methods. However, only little compaction is normally achieved in the zone close to the ground surface, especially in the case of partially saturated soils above the ground water table.

The soil displacement effect during probe insertion contributes also significantly to soil compaction, which results in an increase of vertical and lateral stresses. Also the number of penetration cycles at different depths, and the probe movement are important. The probe is normally inserted to full depth and thereafter withdrawn in steps. This procedure is repeated until the required compaction effect has been achieved.

The optimal spacing between compaction points depends on the shape and size of the compaction probe. It is preferable to use a narrow grid spacing, and to reduce instead the duration of compaction, which will result in more homogeneous soil densification.

Another important factor is the sequence in which compaction is performed. It is advantageous to execute compaction in two passes, working at first at a coarser grid spacing. The vertical insertion of the compaction probe will be facilitated when the grid spacing is larger. After the first compaction pass, the soil deposit should be given sufficient time for reconsolidation, before the second pass is started. Experience has shown that during the second pass, compaction time is often significantly shorter and the densified soil columns will guide the probe into the loose zone, yet to be compacted.

The geotechnical and geohydrological (ground water) conditions, soil deposition and stratification are of importance. Vibratory compaction should be limited to granular soils, which are free-draining. It is normally safe to follow the recommendation by Mitchell [3], who has proposed a boundary for grain size distributions curves, which identify soils, suitable for vibrocompaction, (Fig. 3).

Even relatively thin layers of silt and clay in a sand deposit can negatively affect the densification process. It is therefore recommended to base the compaction design on detailed geotechnical investigations, including electric cone penetration tests with friction sleeve measurements and/or pore pressure soundings ("piezocone"), which can better detect soil stratification.

Good compaction results can generally be expected, when the friction ratio from electric cone penetration tests (local sleeve friction as percentage of point resistance) is below 0.8 %. When the friction ratio exceeds 1.5 %, then vibratory compaction is usually not efficient.

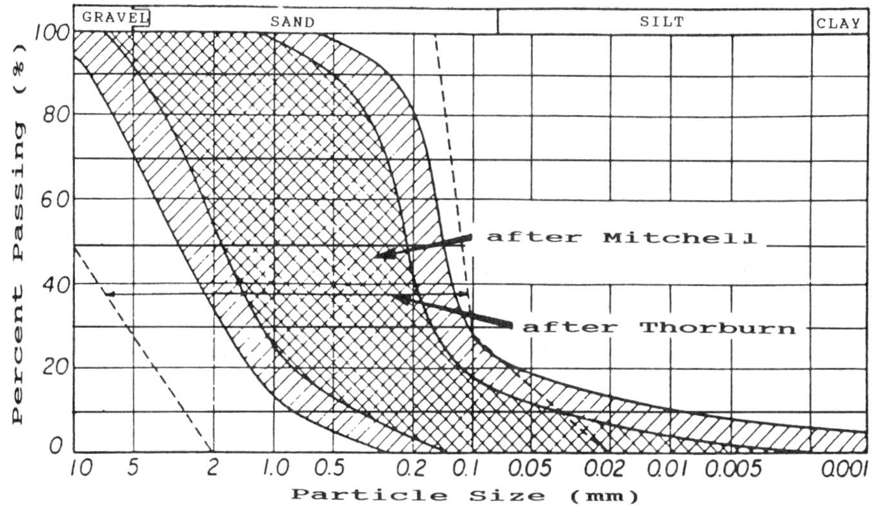

FIG. 3 -- Range of particle size distribution for
        soils suitable for densification by
        vibrocompaction [3]

CASE HISTORIES

The present paper summarizes the experience from more
than 20 compaction projects in different parts of the
world, using either the Vibro Wing or the Y-Probe (Tri
Star). Vibratory compaction has been applied to a wide
variety of problems. In addition to conventional soil
densification projects for industrial or residential
buildings, bridges etc, also hydraulic fills have been
densified, e.g. for land reclamation, harbours and quay
structures [1,2]. Such projects included densification of
hydraulic sand fill on and off shore, and inside sheet pile
walls or concrete caissons [2,4,5]. Other applications
concerned the compaction of sand fills under water, working
from a barge [5,6].

In a number of cases, soil deposits were densified in
order to reduce the length of bored and driven piles. This
was achieved by creating a firm soil stratum at shallow
depth, on which shorter precast or cast in situ piles could
be founded. In this way, the pile length could in some
cases be reduced by more than 10 meters [7]. Experience has
shown that deep compaction, in combination with pile
foundations can be economical, if the pile length can be
reduced by about 5 meters. Another important application of
vibratory compaction concerned the stabilization of loose,
saturated sands, which are susceptible to liquefaction
during earthquakes [8].

Table 1 presents such projects, where the compaction process was monitored by various types of field measurements. Table 2 provides information concerning the vibrator and the compaction procedure employed, and summarizes the results from various field measurements. It can be seen that the duration of compaction and the grid spacing varied within rather wide limits. The reason can be found in the diverse soil conditions and different compaction requirements. At several large projects, it was decided to optimize the compaction procedure by field test during the initial phase of the project.

Based on the above information, it can be concluded that depending on soil type, grid spacing and duration of compaction, the penetration resistance could be increased typically by a factor of 1.5 to 5. As can be expected, a soil deposit, which is initially in a very loose state, can be densified to a higher relative compaction value, than an already dense soil. As mentioned above, only a low compaction effect can be expected in the soil layer close to the ground surface [1].

RESONANT COMPACTION

The results from extensive field tests have been used to develop a new approach to soil compaction by vibratory probes, which will be described in the following. Resonant compaction offers a rational design concept, taking advantage of the amplified ground response, which occurs when a soil layer is excited at a resonant frequency. This can be achieved by adjusting the vibrator frequency to one of the resonant frequencies of the soil-probe system. At resonance, the probe achieves an optimal transfer of vibration energy to the surrounding soil. As will be shown below, this results in an improved compaction effect. It should be noted, however, that the resonance concept is not applicable without modification to other soil compaction techniques such as surface rollers, plate vibrators or vibroflotation.

Resonant frequency concept

The objective of resonant compaction is to excite the soil layer at its resonant frequency. It is possible to calculate theoretically the resonant frequency of an elastic horizontal soil layer, resting on an infinitely rigid base,

$$f = C / 4 H \qquad (1)$$

where C is the wave velocity of the soil and H is the layer thickness. However, in practice, it is more convenient to measure the resonant frequency directly on site during various phases of compaction, and to adjust the vibrator

TABLE 1 -- Deep Compaction Projects

No	Site location	Project Type	Date	Probe Type	Size, m^2	References
1	Broechem, Belgium	Reservoir embankment	1977	Y-Probe	–	[6]
2	Lilla Edet, Sweden	Sheet pile caissons	1977	Vibro Wing	1 700	[5]
3	Boom, Belgium	Coffer dam	1978	Y-Probe	–	[6]
4	Zeebrugge, Belgium	Breakwater for harbour	1980	Y-Probe	–	[6]
5	Perth, Australia	Compaction test	1981	Y-Probe	–	[6]
6	Fredricia, Denmark	Quay strucure	1981	Vibro Wing	20 000	[–]
7	Rostock, GDR	Harbour quay structure	1982	Vibro Wing	120 000	[3]
8	Verrebroek, Belgium	Compaction test	1983	Y-Probe	–	[10]
9	Strandberg, Sweden	Compaction test	1983	Vibro Wing	–	[9]
10	Hamburg, Germany	Bridge abutment	1983	Vibro Wing	1 000	[–]
11	Zeit Bay, Egypt	Quay structure	1984	Vibro Wing	10 000	[–]
12	Aker Verdal, Norway	Offshore structure	1984	Vibro Wing	25 000	[–]
13	Hamburg, Germany	Pile foundation	1986	Tri Star	12 400	[–]
14	Sydney, Australia	Factory foundation	1988	Tri Star	–	[–]
15	Annacis Island, Canada	Liquefaction	1988	Tri Star	800	[8]
16	Sundsvall, Sweden	Bridge abutment	1988	Vibro Wing	8 000	[–]
17	Aden, South Yemen	Quay structure	1988	Vibro Wing	2 000	[–]
18	Düsseldorf, Germany	Quay structure	1988	Vibro Wing	2 000	[–]
19	Storebaelt, Denmark	Artificial island	1988	Vibro Wing	70 000	[–]
20	Hong Kong	Container terminal	1989	Tri Star	75 000	[–]
21	Dordrecht, Holland	Pile foundation	1989	Tri Star	700	[–]

TABLE 2 -- Compaction Parameters

No.	Soil conditions	Depth (m)	Spacing (m)	Vibrator	Vibr. time (min)	Cone penetr. resist. bef. (MPa)	Improvem. ratio[b]	Settlement (m)
1	Hydraulic sand fill	10	3.5	PTC 40A2	–	3.4	2.4	–
2	Gravel fill	11	2	?	20	–	–	0.30
3	Hydraulic sand fill	10	3	PTC 40A2	–	2	4.8	–
4	Hydraulic sand fill	7	3.2	PTC 40A2	–	4.8-6.1	2 – 3	–
5	Unsaturated sand	8	1.5	TOMEN 5000	–	5.1	3.25	–
6	Hydraulic sand fill	8-10	2	?	10	3-5	2	0.4
7	Hydr. fill/natural sand	12	2.5	TOMEN 4000	5	2-5	3-5	0.55
8	Hydr. fill/fine silty sand	8	1.75	PTC 40A2	15	2	5	0.3-0.65
9	Loose sand and silt	13	2	TOMEN 4000	13	1-3	2.5	0.8-0.9
10	Natural sand deposit	8	2	?	10	10 (N)[a]	2	0.3
11	Hydraulic sand fill	10-12	1.5	TOMEN 4000	12	2	3.5	0.5-0.7
12	Hydr. sand fill/silty sand	12	2	TOMEN 4000	20	3-4	1.5	0.3
13	Fine Sand	13	2.75	PTC 40A2	8	2.2-7.5	1.2-2.6	0.4
14	Sand	6	2-4.5	TOMEN 5000	15-35	2.5-6	1-1.3	–
15	Loose sand	12	1.75	ICE 812	13	4-8	1.8-2	0.25
16	Sand with silt layers	12	1.5	TOMEN 5000	10	1-3	2	–
17	Hydraulic sand fill	8	2.5	TOMEN 5000	10	2-4	1.5-2	0.35
18	Hydraulic sand fill	9	1.5	?	10	15 (N)[a]	1.5-2	0.4
19	Dredged sand/nat. sand	6-12	3	ICE 416,815	10	5	2-4	–
20	Hydraulic fill on sand	16	3	ICE 1412	8	2.5-5.5	2.2-4.7	1.1
21	Dry Sand	23	2.2	ICE 812	10	6-14	1.7-2.3	0.5

[a] Dynamic penetrometer test
[b] Ratio of penetration resistance (after/before compaction)

frequency accordingly. Field measurements are also more reliable than theoretical estimates as, during the compaction process, the wave velocity changes, resulting usually in an increase of the resonant frequency.

Figure 4 shows the ground response, measured with a geophone, located at 2.5 m distance from the probe during the switch-off phase.

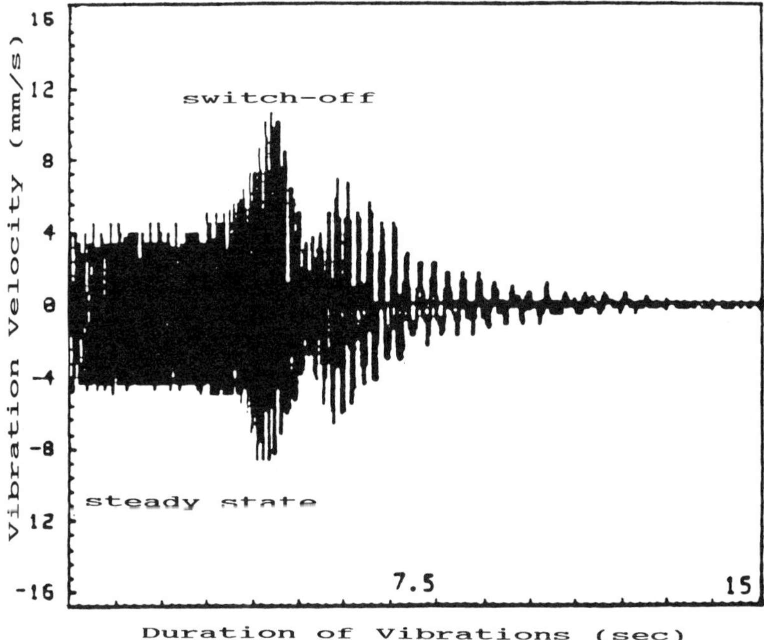

FIG. 4 -- Variation of ground vibration amplitude (RMS) during switch-off of vibrator from 22 Hz

It is apparent that significant vibration amplification occurs at two distinct frequencies, which are lower than the maximum operating frequency (22 Hz). In order to achieve a maximum transfer of vibration energy to the soil, the operating frequency should be kept within a range which corresponds to one of the resonant peaks of the probe-soil system.

A frequency analysis of the ground response, performed during various phases of a compaction project shows several resonant peaks, suggesting that soil layers of varying stiffness exist (Fig. 4). With progressing compaction, however, the higher vibration modes tend to disappear, indicating that more homogeneous soil conditions have been achieved.

Extensive field test were performed to check the effect of vibration frequency on soil densification [3,7,8,9,10]. In one case, compaction tests were carried out in an 8 m deep deposit of saturated fine sand and silt [10]. The compaction points were arranged in a triangular grid and the densification effect was measured by cone penetration tests in the centre points of the grid, before and after compaction. Average values of cone resistance were determined for different depth intervals. Figure 5 shows result from compaction tests at 14 and 17 Hz, respectively. In spite of the scatter of data points, it is apparent that a higher cone resistance was achieved at the lower vibration frequency (14 Hz).

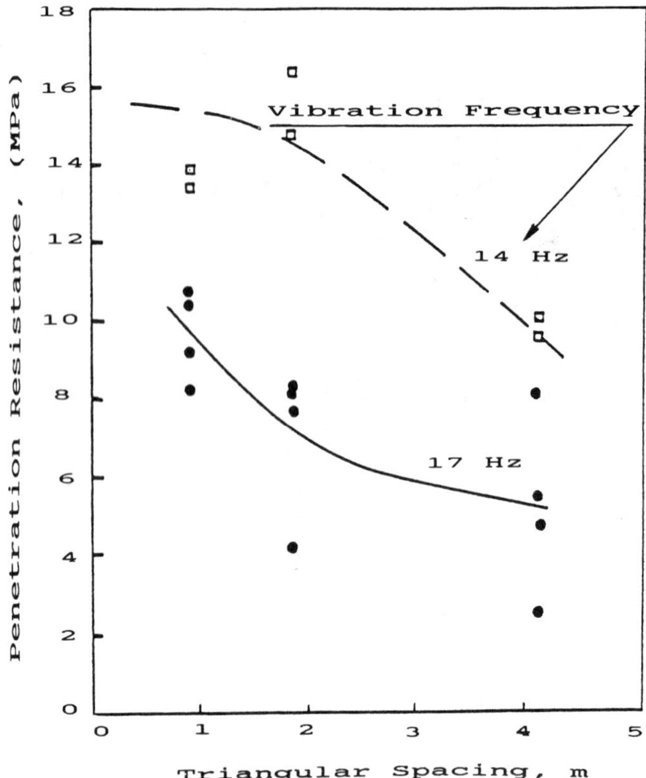

FIG. 5 -- Densification effect measured by cone
         penetration tests at two vibration
         frequencies [10]

These and several similar tests have confirmed that better soil densification can be achieved when the excitation frequency of the vibrator is reduced, approaching the resonant frequency of the soil layer. The concept of using a lower vibration frequency is in contrast to the general opinion, that high centrifugal forces and thus maximum vibrator power is needed to achieve the best compaction effect. However, it should be pointed out that for resonant compaction, the capacity of the vibrator must be sufficient to excite the soil layer, to be compacted at the lower vibration frequency.

Vibration at resonance increases also the soil volume, affected by the compaction process, resulting in more homogeneous soil densification. It was thus found that as a result of resonant compaction, soil densification could be extended to a zone several meters below the maximum depth of probe penetration [8].

The concept of dynamic shear stress ratio, developed in earthquake engineering for the assessment of soil liquefaction, has been used to establish empirically the ground acceleration, required to induce liquefaction during compaction [11]. Field vibration measurements from several compaction projects in sand are the basis for the empirical relationship shown in figure 6. The ground acceleration, required to cause liquefaction, can be estimated, if the initial cone penetration resistance and the soil layer thickness are known. It can be seen that for an initially medium dense sand, a horizontal ground acceleration of about 0.10 g is required.

PROPOSED COMPACTION PROCEDURE

The compaction procedure consists of two phases, probe insertion/extraction and actual soil compaction. Penetration and extraction is carried out most efficiently when slippage occurs between the probe and the surrounding soil particles. Therefore, during this phase, a high vibration frequency should be used in order to generate maximum centrifugal force.

<u>Selection of compaction frequency</u>

Once the probe has penetrated into the soil layer to be compacted, the actual densification phase starts. During this phase, the vibration energy should be transferred as efficiently as possible from the probe to the surrounding soil. This can be achieved by adjusting the operating speed of the vibrator to one of the resonant frequencies of the probe-soil system, i. e. the fundamental vibration mode or an overtone thereof. At resonance, probe penetration is markedly reduced, and ground vibration response increases.

The development of simple and accurate seismic measuring techniques has made it practical to determine on site the optimal vibration frequency and thus to control the compaction process accurately in the field. The monitoring equipment required for resonant compaction consists of a vibration sensor, an amplitude recording device and preferably also a frequency analyzer. In field practice, geophones (velocity transducers) are often used, as they are rugged, cheap and simple to handle.

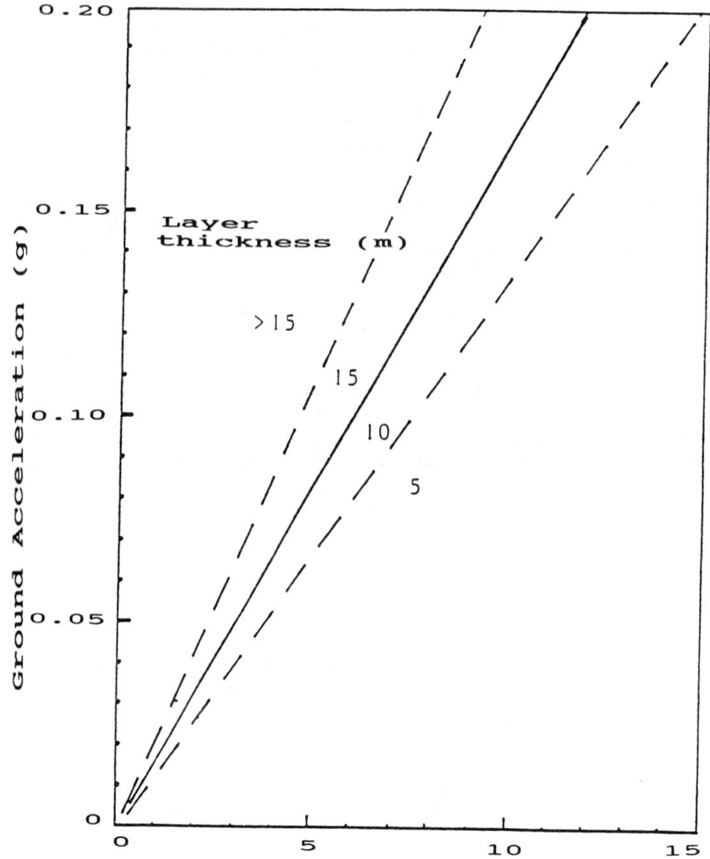

FIG. 6 -- Ground acceleration, required to induce liquefaction in sand during vibratory compaction, as a function of initial cone penetration resistance and compaction depth

The optimal compaction frequency can be readily determined on site by varying the vibrator speed and measuring its effect on ground response. This process is relatively simple and does not require elaborate analyses techniques. The optimal compaction frequency can change during the compaction process, as it is a function of the wave velocity of the soil deposit (cf. equation 1), but depends also on the variation of soil layer thickness within the compacted area. Thus, progressive vibration monitoring may be required at larger compaction projects.

Monitoring of probe movement

A more difficult parameter to be established on site, however, is the optimal sequence of probe movement. The number of insertion cycles, their duration and respective penetration depth are influenced by various factors such as the geotechnical conditions, soil layering and the compaction effect to be achieved in the respective layer. Ground vibration measurements provide again valuable information concerning the most effective compaction procedure. Vibration monitoring makes it is possible to follow qualitatively the progress of soil compaction during different phases of probe insertion and extraction. Figure 7 shows the variation of vertical ground vibration velocity (RMS-values), during penetration, suspension and extraction of the probe, respectively.

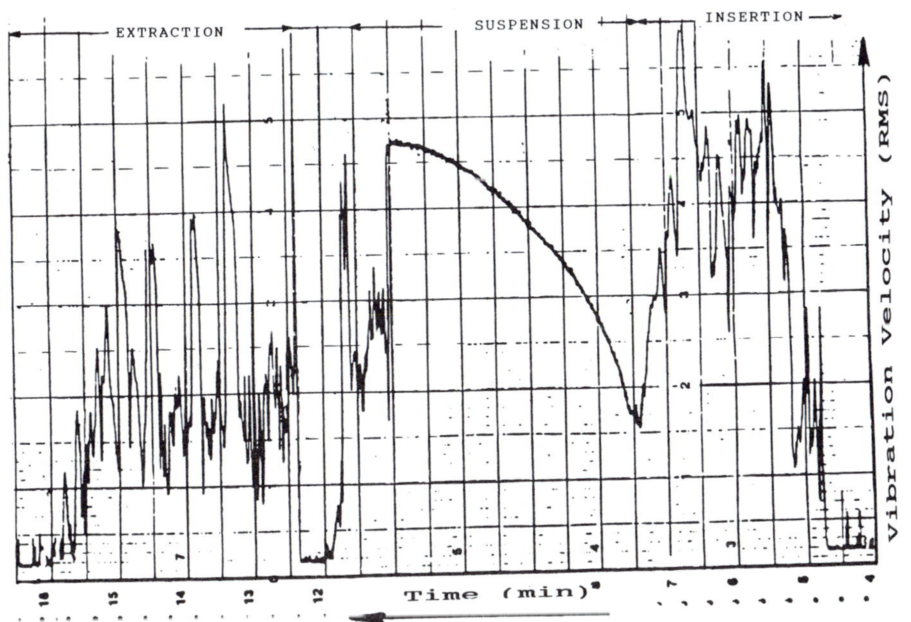

FIG. 7 -- Variation of vibration velocity during penetration, suspension and extraction

It can be seen that in this particular case, the highest vibration amplitude is obtained when the probe penetrates into the soil. During the initial phase of compaction, the sand liquefies and the vibration amplitude drops sharply. Within a few minutes, however, the soil reconsolidates and the vibration amplitude increase gradually. During the step-wise extraction of the probe, the vibration amplitude shows peaks, but is generally lower than during insertion.

COMPACTION MONITORING

Settlement measurements

A simple but very useful compaction control method is the measurement of ground settlements. These are usually largest close to the probe insertion point, and decrease with increasing distance. As a rule of thumb, the largest settlements can be expected in a zone with a radius corresponding to about twice the probe diameter [3].

Probe penetration

Also the rate of penetration of the vibrating probe can be used as a measure of the compaction effect. Unfortunately, this simple but very useful compaction monitoring information is rarely recorded on site. The compaction probe can also be equipped with a load cell, mounted on top of the compaction probe. The measured penetration and pull-out resistance can be used as an indicator of soil densification. This information is of practical importance in order to avoid overcompaction, which especially in coars-grained soils can cause difficulties during extraction of the probe.

Penetration tests

The most common method to check compaction in the field is by penetration tests. Usually, compaction specifications are given in terms of minimum penetration resistance, as this value can be related to relative density. Penetration tests should be carried out in the centre between the points of a triangular grid, and not adjacent to single compaction points, or at the perimeter of a compacted area, as in this case, penetration tests usually give significantly lower values than in test points located inside the compacted area. The tests can also be influenced by the lateral deflection of the penetrometer away from the compacted zone. At one occasion, test with a penetrometer, equipped with an inclinometer in the tip, showed at 15 m depth a lateral deflection of the cone in excess of 2.5 m.

Pore pressure measurements

Valuable information can also be obtained from pore water pressure measurements, during and after soil compaction. Pore pressure measurements from several projects suggest that normally, the highest pore water pressure is generated during the initial insertion of the probe and dissipates gradually, when the probe is kept in suspension or is withdrawn. As a result of compaction, the excess pore water pressure, generated by probe penetration and extraction, can even become negative (dilating soil). Pore water pressure measurements can also be a valuable indicator of the rate of reconsolidation after compaction. This information is of particular importance when compaction is carried out in two phases, and in the case of soil deposits with impermeable layers.

Wave velocity measurements

The compaction effect can also be monitored by cross hole tests before and after compaction [2]. The shear wave velocity, or the surface (Rayleigh) wave velocity, which for most practical purposes is equal to the shear wave velocity, can be used to estimate the increase in soil stiffness. If the shear wave velocity C_S is known, the shear modulus G_S at small strain can be calculated from

$$G_S = C_S^2 \cdot \rho \tag{2}$$

where ρ is the bulk density of the soil. Figure 8 shows result from cross hole measurements, with shear wave velocities before and after compaction [4].

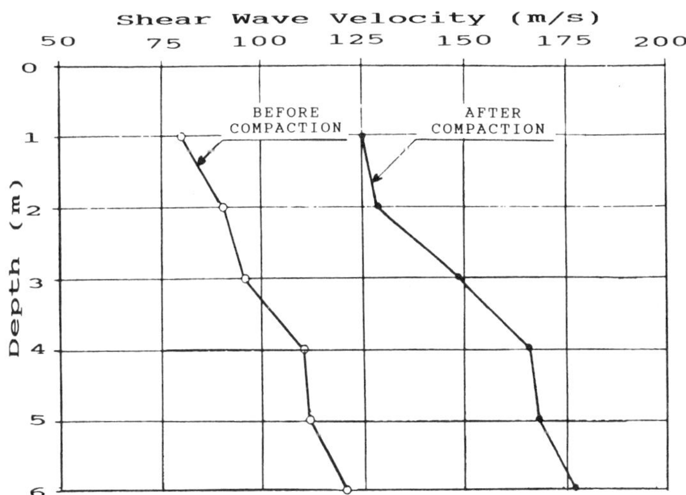

FIG. 8 -- Variation of shear wave velocity with depth, determined from cross-hole tests [4]

A doubling of the shear wave velocity implies that the shear modulus increases by a factor of four, equation (2). The variation of wave velocity during and after compaction

can also be determined from ground response measurements.
According to equation (1), a change in resonant frequency
is directly related to the wave velocity, which offers an
additional way of checking the compaction effect. Another
advantage of wave velocity measurements is, that the
compression properties of a relatively large soil volume
are obtained, compared to penetration tests in individual
locations, which may not be representative for the whole
area.

Vibration amplitude

It is also possible to measure the variation of vibration velocity on the ground surface and at different depth
intervals, as a function of distance from the compaction
probe.

Figure 9 shows the variation of vertical and horizontal vibration velocity at increasing distance from the
compaction probe [2]. It is interesting to note that in
spite of the fact, that the probe was excited only
vertically, the horizontal vibration components were larger
than the vertical ones. As in the actual case the vibration
frequency was kept constant (20 Hz), the equivalent ground
acceleration can be readily determined.

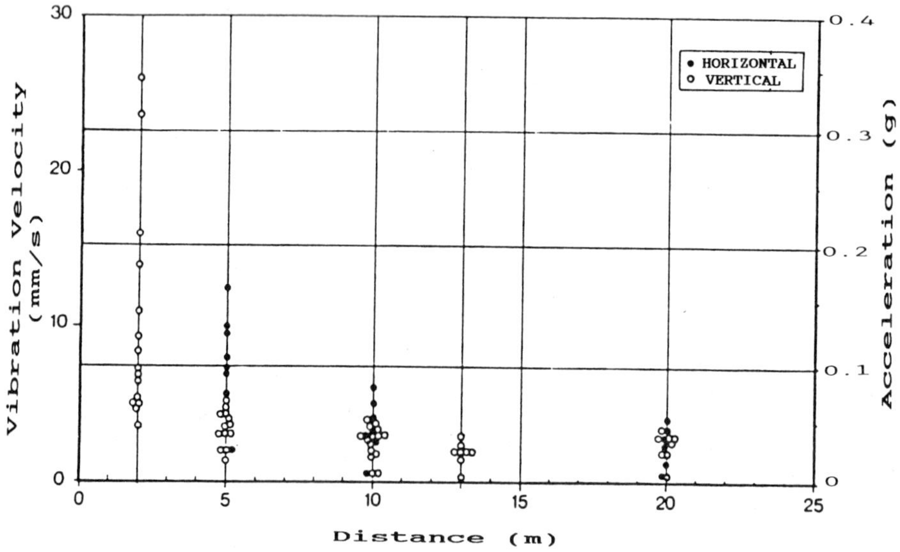

FIG. 9 -- Variation of horizontal and vertical vibration
         amplitude (single amplitude) as a function of
         distance from the compaction point [2]

Based on settlement measurement in the respective
points, a direct correlation can be established between
observed settlements and maximum ground acceleration.
Settlements after compaction vary typically between 4 to
10 % of layer thickness (excluding the surface layer).

Figure 10 presents an empirical relationship between
relative settlements in the compressible layer, initial
cone penetration resistance and required ground
acceleration.

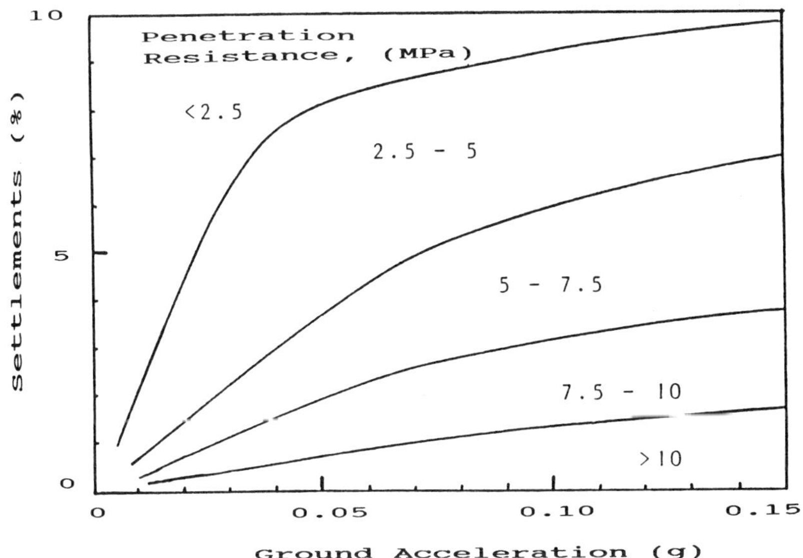

FIG. 10 -- Empirical relationship between average relative
           settlements of the compressible soil layer,
           ground acceleration and initial cone penetration
           resistance

Another important factor for the evaluation of the
efficiency of vibratory compaction is the variation of vibration amplitude with depth [7]. In order to investigate
this aspect, geophones were placed at various depth intervals, 2.5 m from the penetrating probe, and the vibration
response was measured when the probe penetrated the respective depth level. Figure 11 shows that the vibration
amplitude does increase slightly with depth.

DESIGN CONSIDERATIONS AND CONCLUSIONS

Time effect

Although most compaction projects listed above were carried out in clean sands, a marked time effect could be observed . In some cases, the penetration resistance increased within 2 to 8 days by more than 50 % [7]. This time effect occurred without any measurable ground settlements and after dissipation of pore water pressure following compaction. However, the increase of penetration resistance with time was always most pronounced in soils with layers of silt and clay. It is thus recommended that

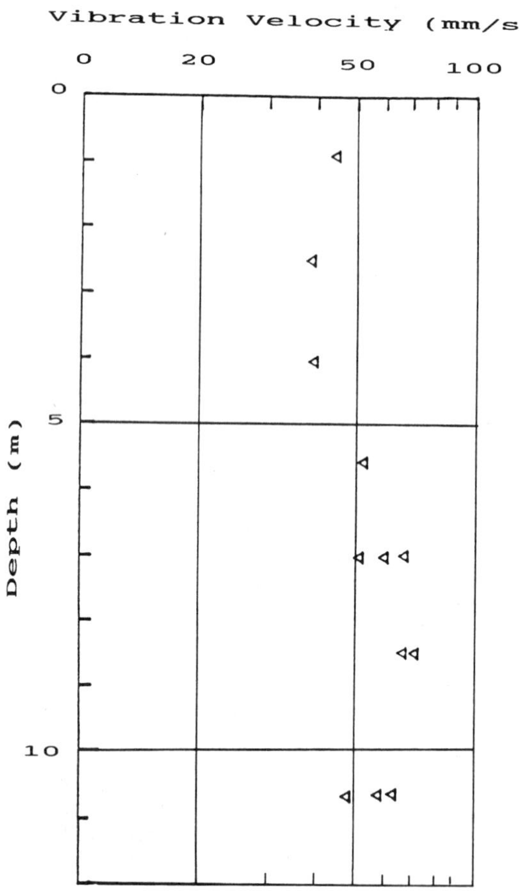

FIG. 11 -- Variation of ground vibration amplitude (single amplitude) as a function of depth (note logarithmic scale) [7]

control tests are carried out at least 2 days after completion of compaction.

Sequence of compaction

Experience from several compaction projects suggests that the sequence in which compaction is carried out in an area, is of considerable importance for the compaction effect [8]. The best results are obtained when the project is carried out in two phases. At first, compaction is performed at a coarser grid spacing. After initial compaction, the soil deposit should be allowed to rest for a few days, before the intermediate points are compacted.

This method has also several other practical advantages. By monitoring ground settlements during the first compaction pass, it is possible to adjust compaction time during the second pass, based on a comparison of ground settlements during the first and second pass.

Also the time for probe penetration can be used as an indication for optimizing the compaction procedure, as discussed above. As discussed above, the compaction probe itself can also be used as a penetrometer, by measuring the rate of penetration during subsequent penetration cycles. With increasing densification, the time for probe penetration at each penetration cycle increases. This simple information can be used to monitor the duration of compaction during the various densification passes.

Based on the above information, the following design recommendations concerning resonant compaction can be made:

1. The vibrator should be powerful enough to permit efficient probe penetration and extraction.

2. For resonant compaction, the vibrator should have variable frequency regulation.

3. The shape and impedance of the compaction probe should be such to facilitate transfer of compaction energy to the soil.

4. It is preferable to work at closer grid spacing in order to obtain more homogeneous compaction.

5. Compaction should be carried out in two passes, with a rest period in between.

6. Compaction tests with penetrometers should not be performed earlier than 3 days after completion of compaction.

7. Compaction problems can be expected, if the soil has a friction ratio (electric cone) above 1.5 %, or if the soil deposit contains impermeable layers.

8. By monitoring the ground vibration response, valuable information concerning the optimal compaction process, and densification results can be obtained.

9. Little or no compaction can be expected in the layer close to the ground surface.

ACKNOWLEDGEMENTS

The author wishes to acknowledge the permission from Beazer Asia, Franki International, Frankipfahl Baugesellschaft and Hercules Grundläggning, and their clients, to include results from their compaction projects in this report. The opinions expressed, however, do not necessarily represent those of any organisation, referred to in this report.

Part of the research program was supported by the Swedish Building Research Council and the Royal Institute of Technology (KTH), Stockholm, and by internal research projects of Franki International and Hercules Grundläggning.

REFERENCES

[1]   Massarsch, K. R., "Deep Compaction of Sand using Vibratory probes," <u>Third International Geotechnical Seminar: Soil Improvement Methods</u>, Nanyang Technological Institute, 27 - 29 November, 1985, Proceedings, 9 p.

[2]   Massarsch, K. R. and Lindberg, B., "Deep Compaction by Vibro Wing Method," <u>8. World Conference on Earthquake Engineering</u>, San Fransisco, July 21 - 28, 1984, Proceedings, 8 p.

[3]   Mitchell, J. K., "Soil Improvement-State-of-the-Art Report," <u>X. International Conference of Soil Mechanics and Foundation Engineering</u>, Stockholm, 15 - 19 June, 1981, Volume 4, pp. 509 - 565.

[4]   Massarsch, K. R. and Broms, B. B., "Soil Compaction by Vibro Wing Method," <u>8. Europ. Conf. Soil Mech. and Found. Eng. Helsinki, 1983</u>, Proceedings, Volume 1, pp. 275 - 278.

[5]   Näsman et al., "Fångdamm med cirkulära spontceller," <u>Byggforskningen, Rapport R148:1979</u>, pp. 51 - 79.

[6] Wallays, M., "Deep Compaction by Vertical and
     Horizontal Vibration," _Symposium on Soil and Rock
     Improvement Techniques_, Bangkok, 29th Nov. - 3rd
     Dec. 1982, 17 p.

[7] Franki Research Report, "Report on Vibrocompaction
     Tests Carried out at Hamburg," _Franki International_,
     Internal Project Report, Germany, 1986, 42 p.

[8] Massarsch, K. R. and Vanneste, G., "Tri Star Vibro
     Compaction, Annacis Island", _Franki International
     Technology_, Internal FIT Report, 1988, 23 p.

[9] Bjerin, L., "Djuppackning med Vibro-Wingmetoden,"
     Swedish Geotechnical Institute, Internal Report
     2-206/83, 1984, 11 p.

[10] Wiesner, T., "Report on Vibrocompaction Tests carried
     out at Verrebroek," _Franki International R & D
     Section_, 1983, Report TW/MH/805, 56 p.

[11] Seed, H. B., "Evaluation of Soil Liquefaction Effects
     on Level Ground during Earthquakes," _ASCE National
     Convention, 1976_, Liquefaction Problems in
     Geotechnical Engineering, pp. 1 - 104.

William J. Neely and David A. Leroy

DENSIFICATION OF SAND USING A VARIABLE FREQUENCY VIBRATORY PROBE

REFERENCE: Neely, W. J. and Leroy, D. A., "Densification of Sand Using a Variable Frequency Vibratory Probe," <u>Deep Foundation Improvements: Design, Construction, and Testing</u>, <u>ASTM STP 1089</u>, Melvin I. Esrig and Robert C. Bachus, Eds., American Society for Testing and Materials, Philadelphia, 1991.

ABSTRACT: The first application of a new vibratory compaction method in North America is described. A variable frequency piling vibrator is used to insert a probe of Y-shaped cross section into the soil to be densified. A unique feature of the system is that on-site monitoring can be used to 'tune' the probe so that compaction is carried out at optimum frequency, vibration time and spacing of treatment point locations. The method is simple, fast and economical and, in contrast to other conventional methods, no material is added during the compaction process.

KEYWORDS: sand, densification, probe, vibratory, testing, settlement, liquefaction.

INTRODUCTION

Deep compaction of loose cohesionless soils is usually required to prevent excessive settlements or to minimize the potential for liquefaction during earthquake loading. A variety of methods has been used for deep compaction, e.g. vibroflotation, dynamic compaction, compaction piles and blasting. This paper describes the first application in North America of a relatively new vibratory compaction technique which involves inserting a low-displacement probe of Y-shaped cross section into the soil using a powerful piling vibrator. The energy of the vertically excited probe is transferred to the soil through a series of small ribs which function as individual pounders. The shape of the probe optimizes transfer of vibration energy to the soil and minimizes undesirable decompression during extraction. The process is simple, fast and economical and, in contrast to other in situ compaction methods such as vibroflotation, no backfill material is added during the compaction process, resulting in lower costs.

Dr. W.J. Neely, formerly with Franki Northwest Company, is now Chief Engineer of DBM Contractors, Inc., 1220 S. 356th, Federal Way, WA 98003 and Mr. D.A. Leroy is Manager of Franki Canada Limited, 8268 River Way, R.R. #7, Delta, B.C. Canada

The paper documents the influence of a number of factors, such as spacing of compaction point locations, method of inserting and extracting the probe, and frequency of the vibrator on the performance of the Y-probe (also referred to as the TriStar probe) in densifying loose sands. A case history from the Vancouver, B.C. area is used to illustrate use of the TriStar probe to densify loose alluvial sands susceptible to liquefaction.

SUITABILITY OF SOILS FOR VIBROCOMPACTION

Massarsch [1] pointed out that in situ densification by any vibratory compaction method depends on soil type, degree of saturation, initial relative density, initial in situ stresses, and soil structure.

Soil Type

Vibratory compaction is best suited to granular soils relatively free of fines (material passing the No. 200 sieve), since the smaller the fines content, the easier it is to densify the soil. As a general guide, the range of suitable grain sizes is the same as that given by Mitchell and Katti [2] for vibroflotation. Densification is possible in soils containing up to about 15% fines, although a very much smaller amount of plastic, clay-size particles can significantly reduce the effectiveness of any vibratory compaction method. Experience with compaction piles indicates that the reduced compaction effect caused by 1% clay-size particles is equivalent to that of 10% silt-size particles [3].

Initial Relative Density

The looser the soil, the easier it is to produce a given increase in relative density. The ratio of penetration resistances after and before treatment is greatest for initially loose soils but decreases with increasing fines content.

Degree of Saturation

For best results, the soil to be compacted should be fully saturated. In partially saturated soils, capillary cohesion forces act to prevent rearrangement of soil particles making compaction more difficult. The TriStar probe was used on a site in Australia in partially saturated, medium to coarse sand with about 5% fines, but the spacing between treatment point locations had to be reduced to about 40% of that for a similar sand located below the water table.

In Situ Stresses

This factor seldom receives much attention in the design and specification of densification work, but can have a marked influence on the effectiveness of the compaction method. Since vibratory compaction creates temporary liquefaction, the compaction effect will increase with increasing overburden pressure during reconsolidation from the liquefied state. Although not investigated, it seems reasonable that temporary surcharging of the ground surface could be used to improve the effectiveness of the compaction, particularly at shallow depths.

Soil Structure

Vibratory compaction may produce undesirable effects, e.g. in cemented soils whose structure may be destroyed during dynamic loading. The elimination of secondary forces at contact points between individual grains or loss of stiffness due to aging, etc., may make it difficult to determine the effectiveness of the compaction process [4]. Field trials indicate that there is little, if any, loosening due to overvibration, other than close to ground level where overburden pressures are small.

THE PROBE, ITS OPERATION AND TYPICAL RESULTS

Details of the Probe

The TriStar probe (Fig. 1), which was developed and patented in the late 1970's, consists of three steel plates, 500mm wide and 20mm thick, welded together at 120° angles. The overall length of the probe can be up to 20m. Each plate is fitted with 300mm x 50mm x 20mm ribs at intervals of 2m. The probe is attached to a variable frequency piling vibrator and vibrated to the required depth.

FIG. 1   Y-PROBE AND VIBRATOR

Operation

The probe is penetrated as rapidly as possible to minimize compaction and to avoid getting stuck before reaching the appropriate depth. Most of the compaction occurs as the probe is extracted from the ground. Several extraction methods are available; one of the most effective is referred to as surging in which the probe is held at the maximum depth for a certain time, then raised in a similar time. This process is repeated several times or for a certain total treatment time.

The unique feature of the probe is that it is possible, based on ground vibration measurements, to determine the optimum vibration frequency, vibration time and spacing of compaction point locations which are site-specific parameters. The vibrations caused by the probe can be measured using one and three-dimensional geophones. The A.C. voltage generated is transformed into a root mean square (RMS) velocity; the higher the RMS velocity, the greater the compaction.

Performance Evaluation

In order to evaluate performance of the probe, a series of tests was carried out on a site underlain by hydraulically-placed fine sand (containing up to 30% silt) with a gradation on the fine side of the limits given in [2]. The test program comprised 18 identical patterns, each consisting of 10 compaction point locations and 6 cone penetrometer (CPT) soundings after compaction, across the site. Each pattern consisted of 3 pairs of equilateral triangles with spacings between compaction point locations of 1.75m, 3.75m and 2.5m (Fig. 2). Compaction at each location generally involved lowering the vibrating probe to a depth of 8m in about one minute, compacting at this depth for one minute, withdrawing the probe and repeating the operation. The test program was designed to investigate the separate influence of the method of insertion and withdrawal, time of compaction and vibrator frequency. Only one factor was varied separately for each full pattern.

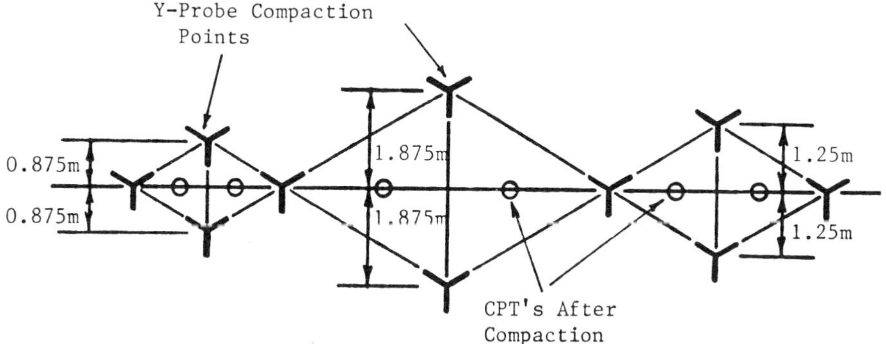

FIG. 2   TEST PATTERN SHOWING COMPACTION POINT LOCATIONS

The influence of spacing of compaction point locations is shown in Fig. 3 where the average cone resistance after compaction is plotted against the tributary area, defined as the area of the two equilateral triangles for each spacing, for two identical test patterns. Even in the very silty sand, cone resistance was improved by a factor of as much as 4. However, individual values of the average cone resistance after compaction vary over a fairly wide range, particularly as the tributary area increases. This probably reflects local variations in the soil, which might be expected to become more pronounced as spacing is increased.

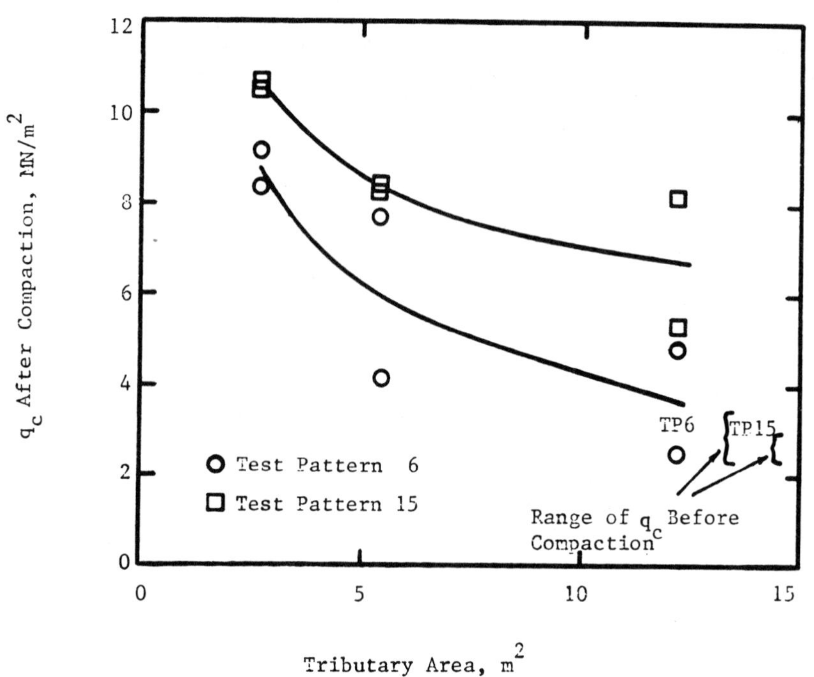

FIG. 3   EFFECT OF SPACING OF COMPACTION
POINT LOCATIONS ON CONE RESISTANCE

Spacing of compaction point locations and vibration time are the most important factors in determining the total cost of a compaction project. The surging method, which involved inserting the probe to 8m at a rate of 2m in 15 seconds, compacting at 8m for one minute and then withdrawing the probe in 2m steps every 15 seconds, was used at 6 test locations for total vibration times ranging from 7 1/2 to 30 minutes. For these tests, the vibrator frequency was 17 Hertz and the probe was fitted with 20mm thick ribs at 2m intervals. The results given in Fig. 4 show that maximum improvement in cone resistance occurred after a total treatment time of 15 minutes for all spacings.

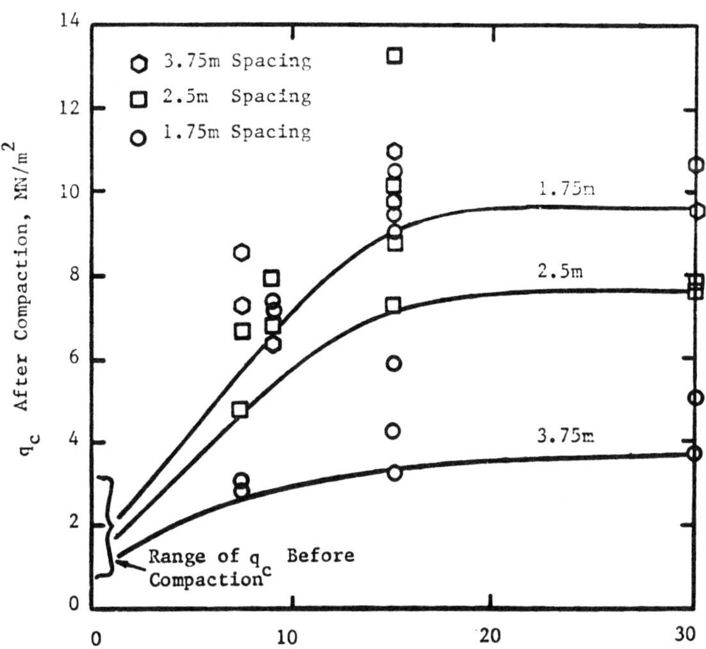

Total Vibration Time per Compaction Point Location, mins.

FIG. 4   EFFECT OF VIBRATION TIME AND SPACING OF COMPACTION
POINT LOCATIONS ON CONE RESISTANCE

The results of tests designed to investigate the effect of vibrator
frequency showed that the greatest improvement occurred at the lowest
test frequency. At a spacing of compaction point locations of 1.75m,
the average cone resistance was about 20% higher at a frequency of 14
Hertz than for the same treatment at the maximum frequency of 26 Hertz.
Different vibrators operating at the same frequency had little effect
on the measured cone resistance even though the centrifugal force of
the vibrators varied from 20 to 31 tonnes at 14 Hz.

Results from other test patterns indicated that the surging method
described previously, and a continuous surging method, in which the
probe is repeatedly inserted and withdrawn every 1 1/2 minutes produced
the greatest improvement, probably because the up-down motion produced
more vertical displacement of the soil. It was also established that
increasing the thickness of the ribs on the probe was slightly more
effective than increasing the number of ribs, presumably because
thicker ribs promote greater vertical displacement of the soil.

Wallays [3] introduced the concept of an improvement factor f,
defined as the average cone resistance after treatment, q_{cm}^b, divided by
the average cone resistance before treatment, q_{cm}^b, as a basis for
interpreting field data; typical results from several TriStar
compaction projects are summarized in Fig. 5. Tests on two sites in
Belgium show that compaction with the Y-probe is more efficient than
the conventional vibroflotation method, Fig. 6.

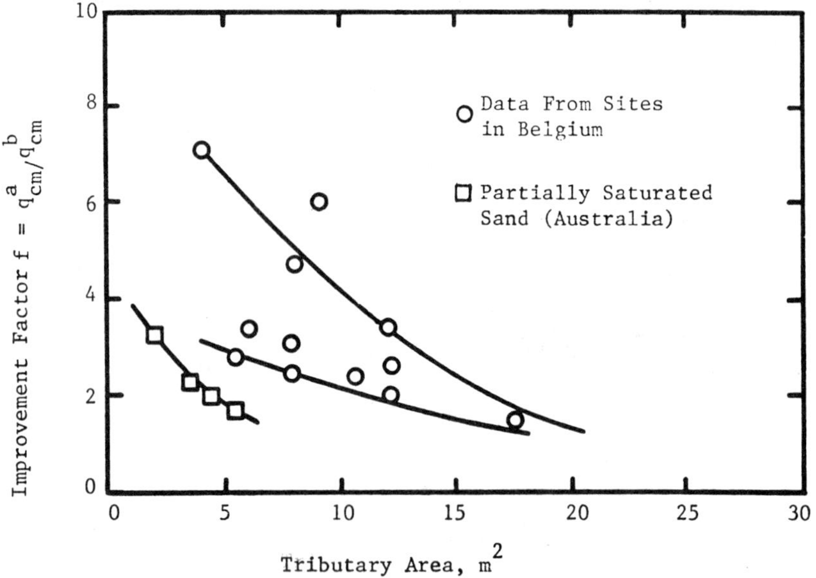

FIG. 5   IMPROVEMENT FACTOR VS TRIBUTARY AREA

CASE HISTORY

General

Development of a site on Annacis Island, B.C. involved the construction of buildings adjacent to the Annacis Channel. Because of the liquefaction potential of the subsoils, it was decided to undertake a densification program to reduce the possibility for large lateral movements towards the open water. This was accomplished using the TriStar probe to form a dyke of densified soil, 230m long, 3 - 4.5m wide and 10 - 11m deep, between the buildings and Annacis Channel.

Soil Conditions

Exploratory borings showed subsurface conditions to consist of 1.8 - 2.4m of sand fill, 2.4 - 3.9m of clayey silt underlain by some 30m of loose alluvial sand. Standard penetration N-values in the alluvial sand averaged 12 in the range of 5 to 24 blows/0.3m. To prevent liquefaction, the sand in the dyke area was to be densified to produce minimum N-values varying from 14 at 4.5m to 17 at 9m below ground level. The grain size characteristics of the alluvial sand are compared with Mitchell and Katti's limits in Fig. 7. The mean grain size D_{50} is about 0.3mm and the fines content is around 10%. For D_{50} = 0.3mm, the SPT-CPT correlation gives $q_c/N=5$, where q_c is the CPT point resistance in bars [5].

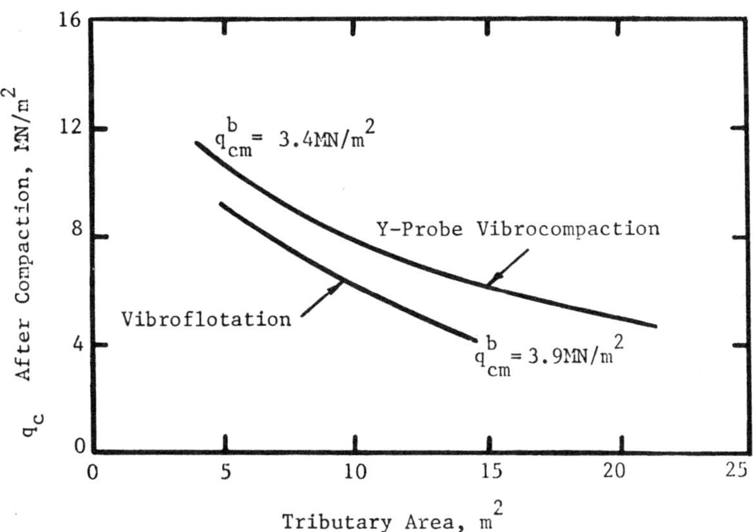

FIG. 6   COMPARISON OF Y-PROBE AND VIBROFLOTATION
COMPACTION

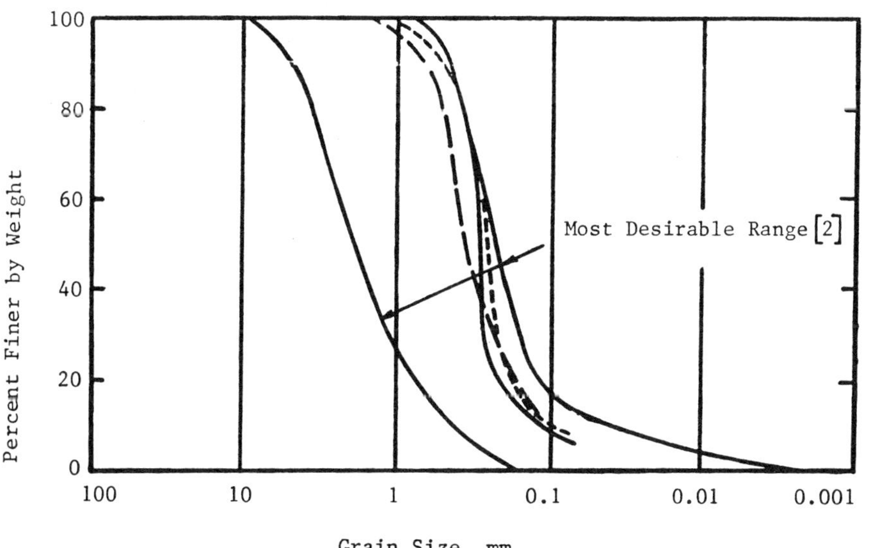

FIG. 7   GRAIN SIZE CURVES SHOWING MITCHELL AND KATTI'S
RANGE FOR VIBROCOMPACTION

Field Trials

Because the soils were considered by others to be marginally compactable by vibratory methods, densification trials were carried out to determine optimum vibrator frequency, vibration time and spacing of compaction point locations. The probe was 12m long and was attached to an ICE 812 variable frequency vibrator (range: 6.6–26.6Hz). The compaction process consisted of lowering the probe to 10m in 2 – 3 minutes, followed by a steady state phase in which the probe was vibrated at a constant, but lower frequency, at 10m. The probe was then withdrawn in a series of steps (referred to as step-surging) allowing time for steady state vibration before withdrawing to the next step.

In order to determine the optimum vibrator frequency and time of steady state vibration, ground surface vibrations were measured by one and three-directional geophones. The AC-voltage generated by the vibrations was transformed into an RMS (root mean square) velocity which can be used as an indicator of the effectiveness of the process; the higher the RMS velocity, the greater the compactive effort. A typical vertical vibration record is presented in Fig. 8. It can be seen that the maximum vibration level occurred during penetration, dropped dramatically at the beginning of the steady state phase at 10m and then increased during steady state vibration at 10m. In some cases, the RMS velocity-time signal during steady state vibration increased at first and then reached a constant level but, in general, it was found that 5 minutes of steady state vibration at 10m was sufficient to satisfy the densification requirements. At some points, the vibrator speed was varied during the steady state phase to determine the optimum frequency. Typical results are presented in Fig. 9 indicating an optimum frequency of about 13 Hz (speed of rotation in revolutions per minute divided by 60) for this site which is much less than the maximum speed 1600 rpm (or 26.6 Hz) of the vibrator.

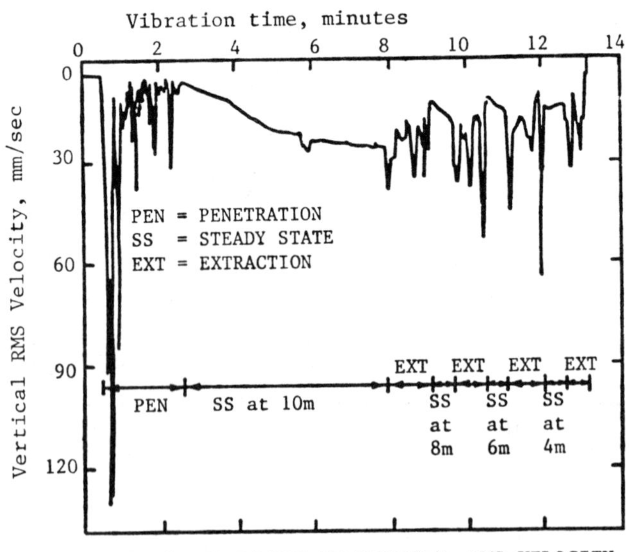

FIG. 8   VARIATION IN VERTICAL RMS VELOCITY DURING Y-PROBE OPERATION

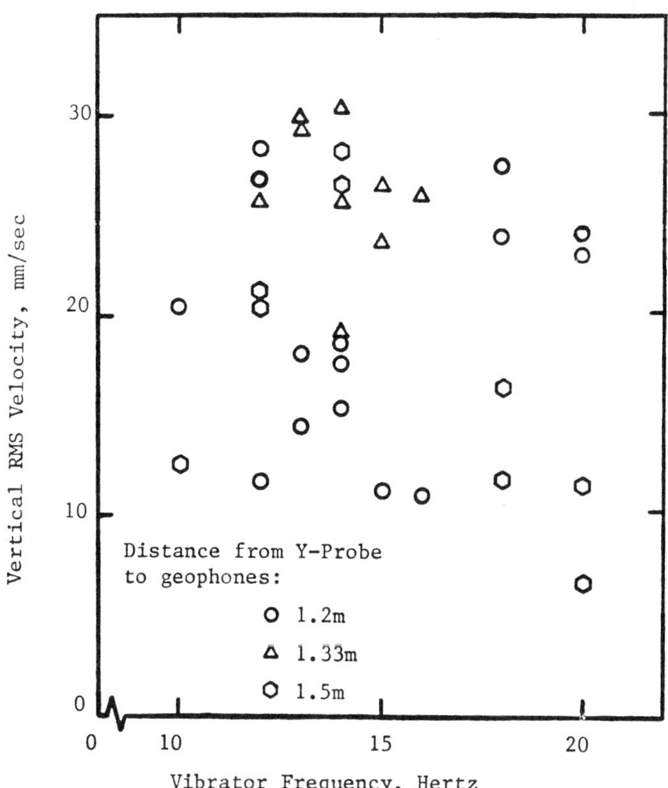

FIG. 9   VARIATION IN VERTICAL RMS VELOCITY
WITH VIBRATOR FREQUENCY

One test area comprised 7 compaction point locations at 2.3m spacing; the CPT tests before and after compaction are compared in Fig. 10 where it can be seen that a small decrease in q_c was measured shortly after compaction. However, a large increase in cone resistance – in excess of the 7 – 8.5 MN/m² required – was observed 3 1/2 days later. Low q_c values were measured in the silt layer above the alluvial sand but, significantly, the essentially zero excess porewater pressures before compaction were found to be strongly negative after compaction indicating that some densification of the silt did occur. Due to liquefaction, positive pore pressures are developed which dissipate with time, resulting in higher cone resistance. Increases in penetration resistance continue to occur even after dissipation of pore pressures as aging produces secondary forces at contact points between individual grains and the stiffness increases.

Production Densification

Production compaction extended to depths of 10 - 11m on a triangular pattern at 2m spacing. During penetration of the probe, which took 2 1/2 minutes, the operating frequency was 20 Hz which was reduced to 13 Hz during the 3 - 4 minutes of steady state vibration at full depth. Extraction was by the step-surging method with intermediate steady state vibration at about 9m and 6m below ground level. Total vibration time at each compaction point location ranged from 13 to 15 minutes. Two and three rows of compaction point locations were used in the 3m and 4.5m wide dyke zones, respectively.

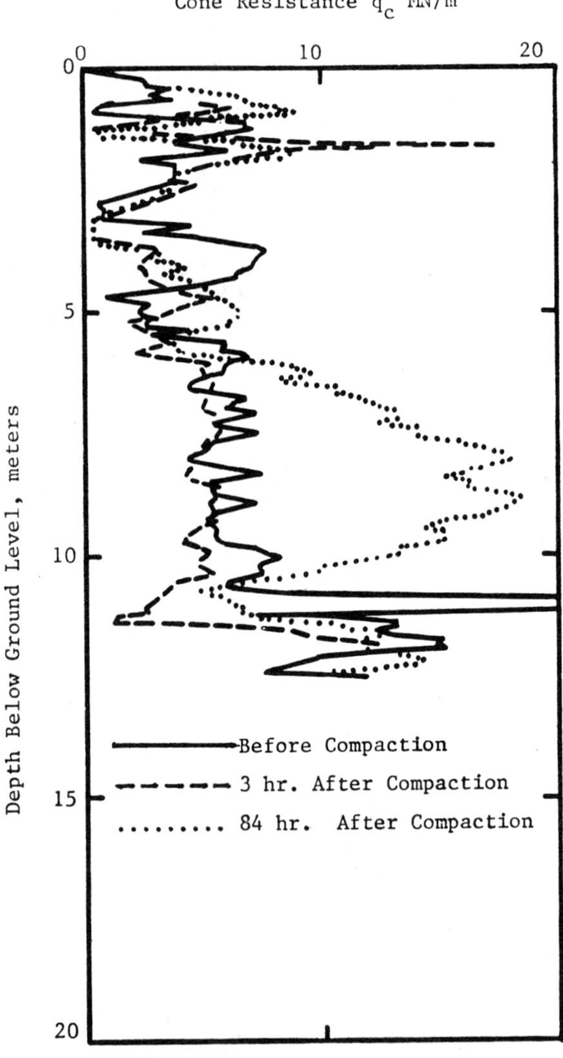

FIG. 10 CONE RESISTANCE PROFILES
         IN TEST AREA

Ground surface settlements were measured at all compaction point locations and along the centerline for 2 rows and midway between the outer rows in the 4.5m wide zone. Average settlements at compaction point locations ranged from 0.34m to 0.49m, depending on location, while corresponding average settlements between rows of compaction point locations ranged from 0.31m to 0.45m.

Standard penetration N-values before and after compaction, together with the required minimum blowcounts, are shown in Fig. 11. It is evident that vibrocompaction using the TriStar was very effective at this site as the average N-value of 12 before compaction was increased to about 52, i.e. an improvement factor of f = 4.3, which is in line with that anticipated from Fig. 5. Additional work indicates that the relative density of the sand between 5.5m and 10m was increased from 55% to 85 – 90% [6]. Tests with the flat dilatometer suggest that the increase in relative density implied by the increased penetration resistance, is due more or less equally to an increase in lateral stress and relative density.

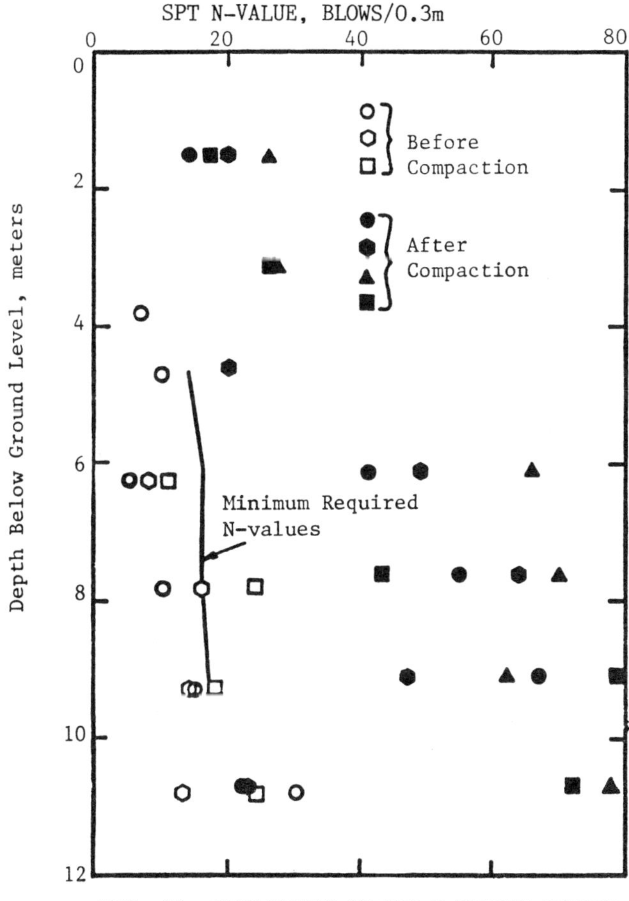

FIG. 11   COMPARISON OF SPT N-VALUES BEFORE
AND AFTER COMPACTION

CONCLUSIONS

The TriStar probe, attached to a variable frequency piling vibrator, was used to densify initially loose alluvial sand to a depth of 11m to create a 230m long by 3 - 4.5m wide dyke adjacent to open water. The purpose of the dyke was to prevent large lateral movement of the retained soil towards the open water in the event of a major earthquake.

The improvement in penetration resistance that was actually achieved was considerably greater than required. Part of the reason for this was inherent conservatism because the soils were considered marginal for densification with conventional vibrocompaction methods. Furthermore, the relatively narrow zone to be densified resulted in somewhat closer compaction point locations than would be necessary in an area of greater lateral extent.

REFERENCES

[1] Massarsch, K.R., (1985) "Deep Compaction of Sand Using Vibratory Probes," Third International Geotechnical Seminar: Soil Improvement Methods, Singapore.
[2] Mitchell, J.K. and Katti, R.K., (1981) "Soil Improvement: State-of-the-Art Report," Proc. Tenth International Conference on Soil Mechanics and Foundation Engineering, Stockholm, 261-317.
[3] Wallays, M., (1982) "Deep Compaction by Casing Driving," Symposium on Soil and Rock Improvement Techniques, Bangkok, A6.1-A6.20.
[4] Mitchell, J.K. and Solymar, Z.V., (1984) "Time-Dependent Strength Gain in Freshly Deposited or Densified Sand," _Journal of Geotechnical Engineering_, American Society of Civil Engineers, 110(11), 1559-1576.
[5] Robertson, P.K., Campanella, R.G. and Wightman, A., (1983) "SPT-CPT Correlations," _Journal of Geotechnical Engineering_, American Society of Civil Engineers, 109(11), 1449-1459.
[6] Brown, D.F., (1989) "Evaluation of the TriStar Vibrocompaction Probe," M.A. Sc. Thesis, Department of Civil Engineering, University of British Columbia, Vancouver, Canada, 162p.

Author Index

Subject Index